Style and Tradition
Studies in Honor of Wendell Clausen

Beiträge zur Altertumskunde

Herausgegeben von
Michael Erler, Ernst Heitsch, Ludwig Koenen,
Reinhold Merkelbach, Clemens Zintzen

Band 92

B. G. Teubner Stuttgart und Leipzig

Style and Tradition

Studies in Honor of Wendell Clausen

Edited by
Peter Knox and Clive Foss

B. G. Teubner Stuttgart und Leipzig 1998

Die Deutsche Bibliothek – CIP-Einheitsaufnahme

Style and tradition: studies in honor of Wendell Clausen /
ed. by Peter Knox and Clive Foss. – Stuttgart; Leipzig:
Teubner, 1998
(Beiträge zur Altertumskunde; Bd. 92)
ISBN 3-519-07641-1

Printed in Germany
Druck und Bindung: Röck, Weinsberg

Contents

Preface

The essays contributed to this volume by students, friends, and colleagues of Wendell Clausen reflect a broad range of approaches to the literature of Greek and Roman antiquity. Each is distinguished by the concern with precise verbal scholarship that readers will recognize as the hallmark of his style as a scholar. Latin poetry has been the chief focus of his career and the ranks of our profession include many whose first impressions of its qualities were formed in classroom of Wendell Clausen, *qui solus legit poetas*.

The first group of essays explores the poetry of the late Republic and early Empire, while the second deals particularly with Virgil. The twin themes of style and tradition manifest themselves in the observation of philological detail which then leads to larger consequences. This manner is applicable not only to Latin poetry, but to ancient literature broadly conceived, as is reflected in the essays on Latin prose in the third section of this volume. Several contributors take us beyond Roman poetry to the tradition of Greek literature that informed it, a central concern in the scholarship of Wendell Clausen. Others explore topics in the reception of Latin literature from late antiquity to the twentieth century. Above all, the technique of composition and the art of reading are the focus of these sudies.

"Hath literature been thy choice?" Coleridge wrote, in a passage once memorably recalled by our honorand. "And hast thou food and raiment? Be thankful, be amazed at thy good fortune." This volume represents a token of gratitude to Wendell Clausen for sharing his good fortune with us and others.

PEK
CF

Publications of Wendell Clausen

"The Scorched Earth Policy, Ancient and Modern," *CJ* 40 (1945) 298-99.

"Bede and the British Pearl," *CJ* 42 (1947) 277-80.

"Notes on Sallust's *Historiae*," *AJP* 68 (1947) 293-301.

Erchanberti Frisingensis 'Tractatus Super Donatum' (Diss. University of Chicago 1949).

"Three Notes," *AJP* 70 (1949) 309-15.

"Codex Vat. Reginensis 1560 of Persius," *TAPA* 80 (1949) 238-44.

"Two Notes on Juvenal," *CR* 1 (1951) 73-74.

Review of M. Schuster, ed., *Catulli Veronensis liber* (Leipzig 1949), in *CP* 47 (1952) 57.

"Silva coniecturarum," *AJP* 86 (1955) 47-62.

"Schraderiana," *Mnem.* 8 (1955) 49-52.

A. Persi Flacci Saturarum liber (Oxford 1956).

Review of H. Heusch, *Das Archaische in der Sprache Catulls* (Bonn 1954), in *CP* 51 (1956) 204-5.

Review of N. Scivoletto, *Auli Persi Flacci Saturae* (Florence 1956), in *CP* 53 (1958) 141-42.

A. Persi Flacci et D. Iuni Iuuenalis Saturae (Oxford 1959).

"An Interpolated Verse in Horace," *Philol.* 106 (1962) 205-6.

"Two Conjectures," *AJP* 84 (1963) 415-17.

"Sabinus' MS of Persius," *Hermes* 91 (1963) 252-56.

"Crater, Cratera, Creterra," *CQ* 13 (1963) 85-87.

"An Interpretation of the *Aeneid*," *HSCP* 68 (1964) 139-47.

"The Textual Tradition of the *Culex*," *HSCP* 68 (1964) 119-38.

"Callimachus and Latin Poetry," *GRBS* 5 (1964) 181-96.

"Concaua uerba," *CP* 59 (1964) 38.

"On Editing the *Ciris*," *CP* 59 (1964) 90-101.

"Lucretius," Introduction to *On the Nature of Things*, trans. H.A.J. Munro (New York 1965).

Review of P.J. Enk, *Sex. Propertii Elegiarum Liber Secundus* (Leiden 1962), in *AJP* 86 (1965) 95-102.

"Cato, *De agricult.* 14.5," *Philol.* 110 (1966) 306.

Appendix Vergiliana, with R.F.D. Goodyear, E.J. Kenney, and J.A. Richmond (Oxford 1966).

"Adnotatiunculae in Seruium," *HSCP* 71 (1966) 57-58.

"Statius, *Thebaid* 10.299," *Philol.* 111 (1967) 146.

"Catullus and Callimachus," *HSCP* 74 (1968) 85-94.

"Duellum," *HSCP* 75 (1971) 69-72.

"On the Date of the FIrst Eclogue," *HSCP* 76 (1972) 201-5.

"Propertius 4.11.53," *AJP* 96 (1975) 271.

Review of D. Bo, *A. Persi Flacci Saturarum liber* (Turin 1969), in *Gnom.* 47 (1975) 142-45.

"Catulli Veronensis Liber," *CP* 71 (1976) 37-43.

"On Seneca the Elder, *Suas.* 5.2," *AJP* 97 (1976) 1-2.

"Cynthius," *AJP* 97 (1976) 245-47.

"Juvenal and Virgil," *HSCP* 80 (1976) 181-86.

"Virgil and Parthenius," *HSCP* 80 (1976) 179.

Review of J.C. Bramble, *Persius and the Programmatic Satire: A Study in Form and Imagery* (Cambridge 1974), in *Gnom.* 49 (1977) 311-13.

"Ariadne's Leave-Taking. Catullus 64.116-20," *ICS* 2 (1977) 219-23.

"Cynthius. An Addendum," *AJP* 98 (1977) 362.

Review of R. Goujard, *Caton, De l'Agriculture* (Paris 1975), in *CP* 73 (1978) 164-66.

"Ovid, *Met.* 15.90," *AJP* 100 (1979) 247-49.

The Cambridge History of Classical Literature II. Latin Literature, ed. with E.J. Kenney (Cambridge 1982).

"The New Direction in Poetry," in *The Cambridge History of Classical Literature II. Latin Literature* (Cambridge 1982) 172-206.

"Theocritus and Virgil," in *The Cambridge History of Classical Literature II. Latin Literature* (Cambridge 1982) 301-19.

"Cicero and the New Poetry," *HSCP* 90 (1986) 159-70.

Virgil's Aeneid and the Tradition of Hellenistic Poetry (Berkeley 1987).

"Catulliana," *BICS,* Suppl. 51 (1988) 13-17.

"Sir Roger Aubrey Baskerville Mynors," *Vergilius* 35 (1989) 3-7.

"Philology," *Comparative Literature Studies* 27 (1990) 13-15.

"Virgil's Messianic Eclogue," in *Poetry and Prophecy: The Beginnings of a Literary Tradition*, ed. J.L. Kugel (Ithaca 1990) 65-74.

"Three Notes on Lucretius," *CQ* 41 (1991) 544-45.

A. Persi Flacci et D. Iuni Iuuenalis Saturae, rev. ed. (Oxford 1992).

Virgil: Eclogues. Edited with an Introduction and Commentary (Oxford 1994).

"Horace, *Sermones* 2.3.208," *Mnem.* 48 (1995) 202-3.

I

Latin Poetry

VENUS AND MARS (*DE RERUM NATURA* 1.31-40)

THOMAS COLE

Lucretius is obviously composing some sort of allegory when he asks "Venus" to make use of her influence with "Mars," lovesick and longingly drinking in her gaze while reclining on her lap, to bring an end to the *fera moenera militiai* that menace the peace of Rome. But what are the "tenors" that correspond to his two "vehicles"? Love and Strife is the answer most often given—on the assumption that this allegory, like those of the Magna Mater and Phaethon's ride found later in the poem (2.600-43 and 5.396-406), is either "physiological" or moral-philosophical in character, and that here, as in the celebration of Sicily's most distinguished son found at 1.716-33, Lucretius is revealing his indebtedness to Empedocles.[1]

The possibility that Venus and Mars—like the *iuuenis* of Virgil's First *Eclogue* or the deified Daphnis of the Fifth—are figures with at least partial counterparts in the realm of Roman politics has not, so far as I know, been suggested. Yet both the context into which the two divinities are introduced and the ostensible reason for the poet's interest in it are clearly political: unless Venus intervenes to end or avoid a *patriai tempus iniquum* at Rome, the poet's patron Memmius will have to devote to the service of his country time which might otherwise be spent studying philosophy (1.41-43). And if one looks more closely at that *tempus* and asks whether it contained any erotic relationship on whose continued success the peace of Rome depended, a very crucial one comes immediately to mind. Julius Caesar's daughter Julia was betrothed to Pompey in 59 and died in 54, after five years of apparently blissful marriage. During that five-year period (a period in which most scholars would be inclined to place the composition of the bulk of

---

[1] Cf., for recent restatements of the view, David Furley, "Variations on Themes from Empedocles in Lucretius' Proem," *BICS* 17 (1970) 55-64, and David Sedley, who goes so far as to suggest ("The Proems of Lucretius and Empedocles," *GRBS* 30 [1989] 290-91) that Empedocles' own proem contained a similar prayer for the victory of Love/Aphrodite over Strife/Ares.

Lucretius' poem[2]) one could fairly say to Julia what Lucretius says to Venus: ... *tu sola potes tranquilla pace iuuare* | *mortales*—or what Lucan does in fact say to her a century later, perhaps going back by way of his principal source, Livy, to contemporary assessments recorded in Asinius Pollio's history (1.114-18):[3]

> ... *tu sola* furentem
> inde uirum *poteras* atque hinc retinere parentem
> armatasque manus excusso iungere ferro
> ut generos soceris mediae iunxere Sabinae.

The historical parallel Lucan adduces may have occurred to Lucretius as well (see below, p. 11), but another precedent from the annals of early Rome would have been even more important. It involved a *gener–socer* relationship between one of the two figures that appear in Lucretius' allegory and a son of the other:[4]

Venus
|
Aeneas
|
Ilia = Mars

---

[2] A natural assumption, given the usually accepted date (55 or 53 B.C.) for the poet's death and the fact that the poem is unfinished. Luciano Canfora's recent attempt to redate Lucretius' death to the early forties seems to me much the weakest part of his valuable reexamination and reevaluation of ancient and modern speculation relating to the character of Cicero's familiarity with the poet and his work (*Vita di Lucrezio* [Palermo 1993]).

[3] Lucan's general dependence on Livy is clear from the close parallels between the choice and arrangement of subject matter throughout his poem and that found in the *periochae* of *Ab urbe condita* 109-12. Livy's use of Asinius Pollio as a principal source for his treatment of the *motum ex Metello consule ciuicum* that began in 60 B.C. is generallly assumed on the basis of parallels between his and Pollio's accounts of Cicero's death, both quoted by Seneca the Elder (*Suas.* 6.22 ff.). For Livian echoes in this particular section of Lucan—his exposition of the *causas ... tantarum rerum* (1.67 ff.)—cf. the statement of subject matter in the *perioche* to book 109: *causae ciuilium armorum ... referuntur*, and, from Livy's general preface, *res est ... immensi operis* (= Lucan 1.68: *immensumque aperitur opus*), *iam magnitudine laboret sua* (= 1.71: *graues sub pondere lapsus*), and *nec uitia nec remedia nostra pati possumus* (= 1.72: *nec se Roma ferens*).

[4] The genealogy is, of course, that found in Ennius (Servius *ad Aen.* 6.777), and perhaps the one with which Lucretius and his readers would have been most familiar.

If Lucretius is in fact composing an allegory of peace maintained or restored through a union of the Roman Mars (Caesar's Gallic campaigns not having lasted long enough at any point in the early fifties to create a challenge to Pompey on this score) with the most important *Veneris neptis* among his contemporaries, Mars' earlier involvement with the Aeneadae is unlikely to be far from his mind. Peace is to come through recreating the union from which the Roman race had sprung in the first place, and perhaps not only peace but the future guarantor of peace as well. Had Caesar's daughter borne Pompey a son, he would surely have been hailed by partisans of both *socer* and *gener* as a new Romulus; and had that son lived, monarchy might well have come to Rome rather earlier than it did, thereby avoiding the Civil War. Lucretius' prayer for peace remained, as it turned out, unanswered; but his choice of addressee would have been a perfectly reasonable one at the time. Fifteen years later another marriage between a descendant of Venus (Octavian's sister) and the leading general of the day (Marc Antony) was to call forth similar hopes for the future and another allegorical figure, the *puer* of Virgil's Fourth *Eclogue*, to symbolize them.[5]

That such women could be expected to use their influence with their husbands to further not only the general good, but that of their own family as well, went without saying. Thus Horace, writing in the 20's, felt no need to explain that, when the uxorious river Tiber overflowed his banks, the complaints of his wife Ilia to which he was responding (*Carm.* 1.2.13-20) stemmed from her desire to see a murdered relative, Julius Caesar, avenged.[6] How far such expectations were shared by Lucretius is uncertain—but probably not far enough to make his proem a piece of narrow political partisanship. The invocation to *Aeneadum genetrix* at the outset is often taken as a reference to the *Venus Genetrix*, whose "cult ... was traditional in the family of the Iulii".[7] If so, the nar-

[5] Virgil's allegory was, and continues to be, variously explained; but it is hard not to agree with Wendell Clausen when he insists (*A Commentary on Virgil's Eclogues* [Oxford 1994] 121-22) that, for readers in 40 B.C. at any rate, the *puer* could only be the expected offspring of Octavia's marriage to Antony.

[6] The explanation is provided by the commentator Porphyrio. For a defense of his view (against the doubts expressed by Nisbet-Hubbard *ad loc.*) see Gordon Williams, *Figures of Thought in Roman Poetry* (New Haven 1980) 9-11.

[7] Bailey *ad loc.* The earliest clear evidence comes from the last years of Caesar's life: a denarius from 46 B. C. which shows the scene of Aeneas fleeing Troy

rower alternative meaning thereby imparted to *Aeneadum* looks forward to the role one particular member of the family is to play in the allegory and may give a certain Caesarian slant to the triumviral propaganda it contains. This might be explained as a tribute to Caesar's Epicurean sympathies and/or as a reflection of the fact that Memmius, though a Pompeian in the 60's and a vocal critic of the triumvirate and Caesar in 59-58,[8] was already at the time of writing contemplating the shift of allegiance which would make him Caesar's candidate for the consulship of 53.[9] But the allegory would have been intended in the first instance to promote concord rather than the interests of one triumvir over the other. What appealed to Lucretius about the triumvirate was presumably the very thing that repelled Cicero—the seductive prospect of *otium sine dignitate* (or the need to worry about it) which it held out to Roman citizens of all classes.[10] He would have been inconsistent had he shown

(reproduced as Fig. 27a in Paul Zanker's *The Power of Images in the Age of Augustus* [Engl. trans. Ann Arbor 1988] 35) and the Lex Coloniae Genetricis Iuliae Ursonensis (=*ILS* 6087; cited by Zvi Yavetz, *Julius Caesar in his Public Image* [Engl. trans. Ithaca 1983] 145, in his discussion of the Caesarean colonies dating to the period of the dictatorship). But descent from Venus as a means of establishing Caesar's genealogical credentials appears as early as 69 B.C., in the *laudatio funebris* for his aunt Julia, widow of Marius (fr. 29 Malcovati)

[8] E. S. Gruen, "Pompey, the Roman Aristocracy, and the Conference of Luca," *Historia* 8 (1969) 106.

[9] Perhaps in response to his failure in the two years (56-55) following his return from Bithynia to make any further advances along the *cursus honorum* (cf. Canfora [above, n. 2] 48). This is certainly the date of composition which the character of Lucretius' references to Memmius' role in politics would suggest. The prayer to Venus to create the sort of situation in Rome which would make active participation in politics unnecessary might well have seemed tactless to a man either holding high office (as Memmius was until the end of 57) or in the process of seeking a higher one (as Memmius was in 54). On the other hand it could have been most welcome to someone involuntarily excluded from office-seeking—a way of suggesting desire for the *uita contemplatiua* as a reason for his retirement from public life, while at the same time providing an honorable explanation (public duty taking precedence over private inclination) for an eventual return to politics—if such a return should become possible. This date is also the one suggested by Lucretius' concern over the *fera moenera militiai* which Venus is to suppress or control—concern more natural during or immediately after the threatened rift between Caesar and Pompey which was patched up by the Pact of Luca in January of 56 than at an earlier period in their relationship. There is no reason to connect the anti-triumviral phase of Memmius' career with anything in the proem. *Communi deesse saluti* (1.43) need not carry the aristocratic, senatorial overtones which Brown (*ad loc.*) finds in the phrase.

[10] To *certare ingenio, contendere nobilitate, | noctes atque dies niti praestante labore | ad summas emergere opes rerumque potiri*, as the *pectora caeca* apostrophized at

himself overly concerned with the relative standings, measured in *dignitas*, of its two principal members.

The modern—or post-Virgilian—reader may well find the allegory posited by this interpretation heavy-handed and intrusive.[11] But intrusive is exactly what the Venus and Mars passage is, however interpreted. It replaces the highly impersonal goddess of the poem's opening lines with a disturbingly anthropomorphic successor. Her concern is no longer to bring swift, non-problematic sexual satisfaction to the animal world but to withhold it from her divine suitor until he agrees to help her human descendants. The interpretation suggested here has the advantage of supplying a motive—political opportunism, political conviction, or some mixture of the two—for what would otherwise be a gratuitous dissonance. The demands of the traditional poet-patron relationship are often invoked in similar fashion to explain and justify the equally intrusive introduction of Memmius into the proem, and the fulsome terms in which *Memmi clara propago* is addressed.

Once considered apart from its immediate Lucretian context the allegory becomes even less problematic. There are certainly contemporary parallels from the visual arts to the way it invites comparison of a Roman statesman and commander to a Greek god.[12] The fashion obviously derived from the Greek East; and Pompey, who prided himself on his resemblance, both in appearance and achievements, to Alexander the Great,[13] was in the vanguard of fashion. The Hellenistic lavishness with which he chose to memorialize these achievements is well attested—most fully in Appian's description (*Bell. Mithrad.* 17.116-17) of the triumph he celebrated in 61. On that occasion pictures or perhaps, as has recently

2.11-15 are taken to task for doing, is to do exactly what Cicero insists seekers after true *gloria must* do: *aliis debent otium et uoluptates quaerere, non sibi. sudandum est eis pro communibus commodis, adeundae inimicitiae, subeundae saepe pro re publica tempestates, cum multis audacibus, improbis, nonnumquam etiam potentibus dimicandum* (*Pro Sestio* 139).

[11] It has, for example, none of the "conciseness and indirection" (Clausen [above, n. 5] 123) of the allegory in the Fourth *Eclogue* and none of the "allusiveness and complexity" evident in the way one finds "Octavian implied in the First *Eclogue* and Caesar adumbrated in the Fifth" (*ibid.*, xxv and 152, n. 5)

[12] See the discussion in Zanker (above, n. 7) 5-18.

[13] Plutarch, *Life of Pompey* 2.1

been suggested,[14] sculptural tableaux depicted the more spectacular moments in the downfall of the vanquished Mithradates—his final defeat in battle, his nighttime flight from his besieged capital, the fate of the sons and daughters who predeceased him, the suicide of virgins that accompanied his own death. If, as many have assumed, the immediate inspiration for Lucretius' picture of Venus and Mars was a particular work of art, it may have been a similar tableau—allegorical rather than historical—displayed at some point in the festivities that accompanied Pompey's wedding with Julia two years later.[15] The tableau would doubtless have struck some observers as curious or even absurd—as the mingling of divine and human traits in Pompey's Copenhagen portrait, going back to an original from the mid 50's, still does.[16] But Cicero's allusion (*Ad Att.* 2.17.1) to the *repentina adfinitatis coniunctio* with Caesar as yet another instance of the way "His oriental highness is running amuck" (*turbat Sampsiceramus*) is at least compatible with the possibility that the affair was a highly un-Roman extravaganza, in questionable taste.

As it is obviously impossible, by any stretch of the imagination, to link Julia and Pompey to Love and Strife, the identification of Venus and Mars proposed here makes the "Empedoclean" reading of the myth no more than, at most, marginally relevant to the understanding of Lucretius' proem. Yet even without this identification, the Empedoclean interpretation faces formidable difficulties. Contrary to what is implied by some commentators, Ares is never set in one-to-one opposition to

---

[14] J.J. Pollitt, *Art in the Hellenistic Age* (Cambridge 1993) 284.

[15] The reference in line 35 to Mars' "smooth neck pillowed" on Venus' lap (*tereti ceruice reposta*) is sometimes cited (cf. Monica Gale, *Myth and Poetry in Lucretius* [Cambridge 1994] 83) to support the view that the poet is thinking of human skin as represented in bronze or marble. It is equally possible, unless *teres* here means "round and shapely" (so Bailey, following Munro *ad loc.*), that he is insisting on Mars' beardlessness—not necessary for a god, but *de rigueur* for a Roman aristocrat of the 1st century B.C., particularly one who wanted himself thought of as the Roman Alexander.

[16] "Curiously" is the adverb used in Zanker's description ([above, n. 7] 10) of the way "Alexander's flowing hair, like a lion's mane, descends over Pompey's simple peasant face." Cf. Pollitt (above, n. 14) 34 on the "element of absurdity", imparted to "Pompey's all too human face—mundane, middle-aged, perhaps a trifle overweight," through the "affectation of the features of Alexander" evidenced in his portrait.

Aphrodite in any of the surviving fragments of Empedocles himself,[17] and the only ancient text to gloss the mythological tale in a way that might have been acceptable to Empedocles is the pseudo-Plutarchean *Vita Homeri* (101). Such an interpretation would have required that the goddess Harmony be identified with Love (as she seems to be at Empedocles B 27.4 and 122.2 [cf. Plutarch, *De Tranqu. animi* 15.474c and *De Is. et Osir.* 48.370e]), whereas the standard genealogy[18] made her Ares' and Aphrodite's daughter—a decidedly un-Empedoclean issue for any conjunction between Love and Strife. Eustathius (ad *Od.* 8.367) avoids this difficulty, but only by eliminating the equation of Love and Aphrodite altogether. Love is what brings about her union with Ares, and Strife is personified in Hephaestus. Equally far removed from Empedocles is Rutilius Namatianus' praise (1.67-71) of the rule of Rome as a combination of *vis* (Mars) and *clementia* (Venus).

Much the same may be said of Lucretius' own poem, with the possible exception of the present passage. However one reconstructs the Empedoclean cosmic cycles in which the workings of Love and Strife are displayed,[19] they do not allow the existence of the world as we know it except as a result of the combined action of both principles. The only alternatives would be the monistic sphere where Love holds full sway, and division and diversity do not exist, or a discordant collection of the element particles or element masses whose combination Strife makes impossible. For Lucretius, on the other hand, the unchecked operation of the forces working for combination and growth over which Venus is seen as presiding in the proem would not produce undifferentiated oneness but harmony: the ordering of diverse entities that exists in the world as we know it during periods when the forces of growth and combination prevail over those of disintegration and decay, but a process of ordering

---

[17] Fr. 128 B.1-3 in Diels-Kranz, sometimes cited in this context, refers to a time in human history when Aphrodite was worshipped and Ares was not—but Ares is merely one member of a group of absent divinities, the others being Kudoimos, Zeus, Cronos and Poseidon.

[18] Present, on occasion, even in texts (Heraclitus, *Alleg. Hom.* 69.8-9, Ps.-Plutarch, *Vita Homeri* 102) which claim to be offering an Empedoclean reading of the myth.

[19] See, for the various hypotheses advanced, the summary discussion in C.H. Kahn's review (*Gnomon* 41 [1969] 349-57) of the first volume of Jean Bollack's *Empédocle*.

made somehow permanent and irreversible—as it is, presumably, in the eternal world of the gods and, possibly, in the *aeternis uersibus* of certain poets.

Some, though not all, reconstructions of Empedocles' system assign to Love (in the initial phases of his rule) a role that is analogous to that which Lucretius assigns to Venus—to the extent that Love is both a kind of *summum bonum* and a personification of the cosmic forces working for combination, order and harmony. Other reconstructions reject this notion, positing instead a system of world cycles that bears a certain external resemblance to the alternation of periods of growth and decay which Lucretius on occasion sees as operative in the course of human and cosmic history.[20] No reconstruction, however, suggests that Lucretius' view of the *causes* of the alternation resembled that of Empedocles in any way whatsoever.[21]

It is just conceivable, all this notwithstanding, that Lucretius' starting point was, on the one hand, a kind of Manichean conception of the alternation of growth and decay—pleasure principle and death wish—in the life of man and the cosmos and, on the other, an eccentric interpretation of the Love-Strife dichotomy; and that he merged these two notions into a heretical piece of syncretism that was neither Empedoclean or Epicurean. But if this was what occurred, one would expect the vision of Mars and Venus to have been constructed along rather different lines. Assuming that Lucretius' prayer is equivalent to a wish that the Roman people will decide to make love, not war, it would have been sufficient to ask Venus to captivate Mars with her charms—as Hera captivates Zeus in *Iliad* 14 when she wants him to cease making war on the Greeks.[22] In-

[20] The extent to which Lucretius thought in terms of such cycles is a matter of debate. See, for an extreme statement of the "cyclicist" position, Richard Minadeo, *The Lyre of Science* (Detroit 1969), who divides the entire content of the poem into successive "movements" proceeding alternatively in the direction of generation or destruction.

[21] Cf. Furley (above, n. 1) 59-60, where the absence from Lucretius of any notion of "transcendent Duality" is noted, as is the poet's failure to mention Love or Strife in the passages where he speaks of the origin of animals or the eventual disintegration of the cosmos.

[22] Alternatively, he might have had Venus address, not Mars, but the *Mauortia gens* descended from him, as she does in Sedley's reconstruction (above [n. 1] 291) of the "literal" content of her prayer: "Romans, let your belief in a peaceful god over-

stead Venus appears both as a personification of the pleasures of love and as a suppliant urging on Mars the course of action which must be adopted if he is to enjoy them. She ceases, in other words, to be a philosophical or psychological abstraction and becomes simply a woman playing the sort of conciliatory role which Lucan attributes to Julia (above, p. 4) when he compares her to the Sabine wives intervening between *soceri* and *generi* for the sake of both. And Mars ceases to be the instinct to make war, which has only to be neutralized by an instinct of opposite character for there to be peace, and becomes instead the force that controls and, if need be, represses that instinct (1.32-33 *belli fera moenera* ... | *armipotens regit*) in response to outside persuasion. Whether the comparison with the intervention of the Sabine women was in Lucretius' mind or not,[23] it is difficult to believe that the introduction of the allegory signals an abandonment of the immediate personal and historical frame of reference that enters the proem with Memmius in favor of a brief return, via a glimpse of the cosmic Love-Strife dichotomy, to the universal one with which it began.

The universal perspective only returns, if at all, in the lines (44-49) which immediately follow the prayer for peace. And to find its presence there one must, of course, accept the view of those scholars—probably a majority among contemporary commentators on the passage—who believe that the six lines in question, though identical with 2.646-51 and obviously composed originally for that spot,[24] contain at least the substance—subject to minor verbal adjustments—of what Lucretius wanted to appear in his poem at this point as well.[25] If, as these scholars suggest, the intention of the passage is to justify invoking a divinity to bring peace on the grounds that peace is the thing that most characterizes the life of

---

come your belief in a warlike god, because peacefulness alone is the true nature of godlike happiness."

[23] The reverse comparison may have been already in Livy's mind when in the early 20's he composed his account of the Sabine episode. There the antagonists involved, as if in anticipation of the first-century replay, are referred to only as *patres* and *uiri* or *soceri generique* (*Ab urbe condita* 1.13.2), never as *uiri* and *fratres*.

[24] They contain, as David West has pointed out (*The Imagery and Poetry of Lucretius* [Edinburgh 1969] 111-12), a systematic refutation of the theological allegory described in the lines immediately preceding.

[25] Cf. for example, Bailey *ad loc.*, following Bignone, Friedländer, Regenbogen and Martin.

the gods, its adjustment to its present position could be made fairly simply, through the deletion of two lines:

> omnis enim per se divom natura necessest
> immortali aevo summa cum pace fruatur
> [semota ab nostris rebus seiunctaque longe];
> nam privata dolore omni, privata periclis,
> [ipsa suis pollens opibus, nil indiga nostri,]
> nec bene promeritis capitur neque tangitur ira.

So curtailed the passage can be read as no more than an indication of those aspects of the life of the gods (immortaliity, painlessness, security, freedom from *ira* and *studio*) which make their existence one of supreme peace. Their indifference to human concerns, hence human prayers, which is referred to in the two deleted lines, is a possible, but not necessary, corollary of this. We are once again within the universal frame of reference with which the poem began, and within this frame of reference to invoke a god to bring peace is as reasonable as to invoke Venus to supply *lepor* to Lucretius' verses. Without Venus, the poet explains (23), *neque fit laetum neque amabile quicquam*; and without the gods, he might have added, *neque tranquillum neque fiet | pacatum quicquam.*

What *is* unreasonable—without extensive textual changes—is the conjunction of this justification for asking Venus to bring peace (*omnis enim diuom natura ... necessest | ... aevo cum pace fruatur*) with the one immediately preceding (*nam tu sola potes tranquilla pace iuuare*). Peace is first something that Venus alone can bring about—through a personal appeal to her lover Mars; then it is something Venus can bring about through qualities she shares with *omnis diuom natura.* It is only at this point that the universal perspective with which the proem began is restored—via a transition so abrupt and unlikely that it requires either the deletion of lines 44-49 or, failing that, some special explanation.

If the deletion alternative is rejected, a simple, if mechanical, solution to the difficulty would be to regard lines 31-40 and 44-49 as a doublet:

A		B
29	effice ut interea fera moenera militiai	29
30	per maria ac terras omnis sopita quiescant	30
31	nam tu sola potes tranquilla pace iuuare	
	mortalis, quoniam belli fera moenera Mauors	
	armipotens regit, in gremium qui saepe tuum se	
	reicit …	
	hunc tu diua, tuo recubantem corpore sancto	
	circumfusa super, suauis ex ore loquellas	
40	funde petens placidam Romanis, incluta, pacem	
41	nam neque nos agere hoc patriai tempore iniquo	41
	possumus aequo animo nec Memmi clara propago	
43	talibus in rebus communi deesse saluti	43
	omnis enim per se diuom natura necessest	44
	immortali aeuo summa cum pace fruatur…	
	nam priuata dolore omni, priuata periclis…	
	nec bene promeritis capitur neque tangitur ira.	49

The two groups of lines belong to different and contradictory ways of moving beyond the original focus (1-28) on Venus alone so that the prayer to her may be one for peace as well as for poetic inspiration. In version B (*effice … nam neque nos … omnis enim…*) she becomes simply one instance of divine nature—because such a nature, if it is conceived of as an example, and therefore source, of the social blessings of peace and harmony, must be thought of as existing alongside other natures like itself. In version A (*effice … nam tu sola … nam neque nos…*) she takes her place alongside the contrasting figure of Mars—as peace becomes a matter, not of assimilating oneself to others of like mind, but of countering the efforts of differently-minded others to destroy it. The former sequence of passages is compatible with the universal focus of the first prayer. The latter one abandons it—perhaps out of a desire to express a judgment, or pay tribute to a patron's judgment, about the prospects for peace in their contemporary Roman setting, perhaps through a combination of such considerations with memories of the sort of Empedoclean gloss on the mythological tale of Venus and Mars that appears in Ps.-Plutarch's *Vita Homeri*. Either group of lines, 31-40 or

44-49 (suitably modified), makes sense with the rest of the opening invocation, but not both. Assuming that the poet had slated one or the other for omission, he may well have decided to let the course of events at Rome decide which one it would be. The relevance of 44-49 is timeless. The relevance of 31-40 could be easily terminated—by death, divorce or a shift of alliances.[26]

Such a solution is obviously possible; equally obviously, it runs the risk of oversimplifying complex and perhaps not completely resolvable problems created by a desire to reconcile the needs of poetry, philosophy and patronage within the confines of a single proem. But however these needs and the strategies to meet them were sorted out in the poet's own mind, it is easy to see why, to a perceptive audience in the early 50's, the figures of Mars and his divine partner might have suggested *praesentes deos* rather than philosophical abstractions. The explanation, and an excellent gloss on the whole passage, is inadvertently provided by Pompey's most recent biographer, when at the end of his first volume he comes to describing the situation in which the great general found himself at the end of 59. The new alliance with Caesar was already "paying handsome dividends," but "handsomest of all in Pompey's tender eye was his new wife, with whom he had fallen passionately in love. With his political struggles seemingly over, his popularity restored, and a delightful companion for his leisure, the same man who had so recently mocked Lucullus for retiring into a life of luxurious ease was himself now only too happy to desert the political scene for the peace of his villas and gardens..."[27] Such were the prospects enjoyed by Rome's most celebrated man of war at or fairly near the time when Lucretius was composing his proem. Such are the prospects which the new wife's divine ancestress offers the Roman god of war, provided he take up the cause of peace, in lines 31-40 of that proem. Could any reader fail to see the parallel, or fail to be impressed—before the advent of Augustan refinements in the art of

---

[26] Conceivably, the termination occurred during Lucretius' lifetime, in which case the decision to include 44-49 here as well as in the context in Book 2 for which they were originally composed will be the result of Julia's death in 54. The intention, never realized because of the poet's own death, would have been to use the lines, appropriately revised, as a replacement for 31-40.

[27] Peter Greenhalgh, *Pompey the Roman Alexander* (London 1980) 227.

assimilating human and divine figures within a literary context—by the author's skill in taking the fabrications of *ueteres Graium poetae* and employing them, *bene et eximie disposta*, in the service of a timely Roman message?

A PARACLAUSITHYRON FROM POMPEII

G.P. GOOLD

Though written a quarter of a century after the event Pliny's letters to Tacitus on the eruption of Vesuvius (6.16, 6.20) still read as vivid and exciting accounts of the most spectacular natural disaster to befall the Roman world. What a pity that we do not possess the relevant portion of the *Histories* to see what he made of them! A severe earthquake at Pompeii in February 62, described by Seneca in his *Natural Questions* (6.1.1-3), caused intense damage, some of it never repaired, and was a prologue to the catastrophe. But no one realized the danger. Vesuvius had been inactive during historical times, and the great eruption of 17 years later came out of a cloudless sky. In the forenoon of August 24, 79 the long-dormant volcano blew up and by the evening some six feet of lethal ash had already fallen. Twenty-four hours later over twice that amount had accumulated: the cities of Pompeii, Herculaneum, and Stabiae had been suffocated by the choking mantle of volcanic debris beneath which they were to lie entombed for more than sixteen centuries. More than two thousand people perished, most of whom will have underrated the danger or misjudged the best route to safety. Even so it is estimated that over 90 percent of the population managed to get away. For a while survivors returned to salvage what they could. But so much was found to be beyond recovery that the whole area was abandoned. Not until the eighteenth century, when it became a site for aristocratic treasure-hunting, was the path opened up for systematic excavation and in turn for the publication in *CIL* 4 and its supplements of every inscription found.

This paper, dedicated to Wendell Clausen, reviews the Pompeian graffito, *CIL* 4.5296, which Frank O. Copley, with an article in *AJP* 60 (1939) 333-349, made the starting point of his study of the paraclausithyron, the lament of the excluded lover. His resulting monograph, *Exclusus Amator* (*APA* 1956), still seems to me the best introduction to the subject, and I cannot better his summary of the basic scenario, which he gives on the opening page:

> The lover has been at a symposium where he has engaged in drinking; now, warmed with wine, he goes to seek out the girl by whose beauty and charm he has been attracted. Somewhat maudlin, a garland on his head, he takes up a torch, and either alone or in the company of a friend or two goes through the streets to the girl's house. He knocks at the door, begs for admission, and finds that he has been locked out. He pleads and threatens, but to no effect. Then he sings his song, in which he may combine a plea that the girl will yet relent, a warning of the lonely days to come, when she will be too old for love, a protestation against her cruelty, and a picture of his own sufferings. Sometimes he threatens suicide. In the end he may hang his garland on her door, or throw it at the threshold; sometimes he scribbles on the door some scraps of verse. Then he lies down in the doorway, there to remain until morning.

The paraclausithyron is not tied to a particular genre or meter. The earliest examples of it are found in Greek drama, the *Ecclesiazusae* of Aristophanes, for instance, and the *Cyclops* of Euripides. A remarkable variation occurs in a Hellenistic papyrus, commonly referred to as the *Alexandrian Erotic Fragment* (*Coll. Alex.*, p.177), a dochmiac dirge of some sixty lines: in it by a subtle touch the unrequited lover is female, and reminds us of the *Pharmaceutria* of Theocritus. The elements of the paraclausithyron recur like leitmotifs in the epigrams of Callimachus, Asclepiades, and Meleager. Closer to our concern are its appearances in Latin literature, not so much in comedy, where the first 164 lines of Plautus' *Curculio* furnish a much extended version, or even in Horace's *Odes*, which regale us with an astonishing amount of paraclausithyron imagery, but rather in Latin elegy.

The 67th poem of Catullus, while not a formal paraclausithyron, is an unmistakable variation of it, since the Door therein becomes a character with a speaking role. In his second poem Tibullus employs the theme to express his own feelings: he represents himself as outside Delia's door, which he addresses in alternately angry and servile tones; characteristically of this poet his initial *mise en scène* is soon forgotten, disconnected reflections are poured forth, and it is doubtful if the motif of the door continues to the end. No such uncertainty is raised by Propertius in 1.16: possibly inspired by Catullus, he has placed the whole poem in the mouth of the personified door; and by a brilliant conception the lover's own serenade becomes a poem within a poem, being narrated by the door itself as an example of what it has to endure. Propertius refrains from identifying himself as the lover and the girl inside as Cynthia, and thereby conceals the artificiality of the composition, for his scene is thoroughly convincing. In *Amores* 1.6 on the other hand we are entertained with suspirations directed at the janitor; but Ovid's elegance and facetiousness less fit a lover shivering on the doorstep than a balladeer seeking laughs, detached and composing in comfort. In short, his version is pure parody and suggests that by now the theme had been exploited to the full. At any rate no further examples of it occur in classical Latin literature.

Well, not in serious literature. Just over a century ago, however, in 1887, there was found scratched with a stylus on the right doorway of a house in Pompeii ("the doctor's house," Reg. IX Ins. 8) what purports to be an actual as opposed to a literary paraclausithyron. It is unfortunately no longer extant, for shortly after its discovery the writing was washed away and totally erased by a rainstorm, so that for the text we must rely entirely on the transcript printed in 1898 by Mattheo Della Corte in *CIL* 4 Suppl. (1898) 5296. The text, however, had already been published, notably by Buecheler in the previous year as poem 950 of his *Carmina Latina Epigraphica* (and it most recently appears in E. Courtney's *Musa Lapidaria* [1995], page 98). I append herewith a tracing of Della Corte's copy together with a transcript arranged in verses.

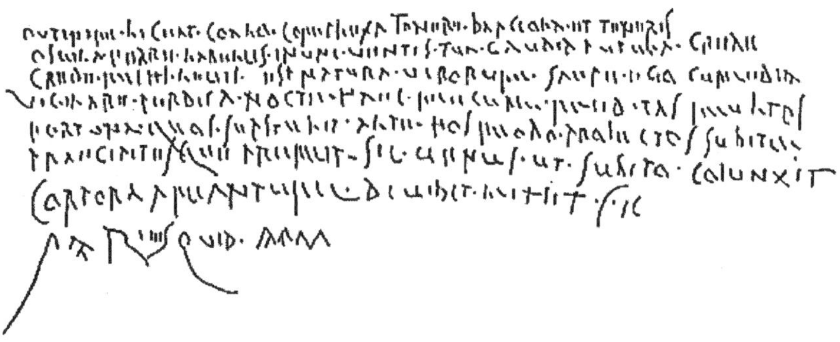

1 O utinam liceat collo complexa tenere
braciola, et teneris | oscula ferre labellis.
i nunc, uentis tua gaudia, pupula, crede; |
crede mihi, leuis est natura uirorum.
5 saepe ego cu(m) media | uigilare(m) perdita nocte,
haec mecum medita(n)s: multos | Fortuna quos supstulit alte,
hos modo proiectos subito | praecipitesque premit;
sic Venus ut subito co(n)iunxit | corpora amantum,
9 diuidit lux et se | paries. quid aam?

Of course, an inscription, being an autograph, does not allow of emendation, except insofar as it can be argued that a mistake has occurred in the act of writing, which will thus not reflect the original intent. In more than one place this seems to be the case here: like Buecheler and others I believe that in verse 3 the intended words were *i nunc <et> ventis* (cf. Juvenal 12.57); that by a kind of haplography *pupula* has been omitted at the beginning of verse 4; and that in verse 7 *subito* is a faulty anticipation of the same word in the next line. The final verse poses difficulties, which I defer for the moment, contenting myself here with a provisional translation (and interpreting the last word as *a'am* = *agam*):

1 O that I might entwine my arms about your neck
and with tender lips give you kisses!
Go now, little girl, and trust your hopes to the winds;
<little girl,> trust me, fickle are the ways of men.
5 Oft at the midnight hour, as I poor girl lay awake,
musing thus with myself: many whom Fortune has exalted,
these she now [suddenly] rejects and casts headlong down.
So after Love has suddenly joined the bodies of lovers,
9 dawn separates them and a wall. Oh what am I to do?

The speaker, who assumes the persona of an excluded lover, laments being parted from the beloved, but in lines 3 and 4 denounces the fickleness of men: the persona, therefore, is that of a woman, and this is confirmed by the adjective *perdita* in line 5. In his article Copley came to the conclusion (p.337) that "this graffito represents the sentiments of a girl shut out from the house of her lover. She writes on his doorway verses whose form is in outline that of the conventional paraclausithyron; the poem is intended, as the paraclausithyron almost invariably was, both to advertise her vigil and to rouse the pity of her obdurate lover." Her poem was, he thinks, modeled on a lost prototype, and its irregularities are to be attributed to the fact that in her emotion she misremembered or miscopied the hexameters she intended; the composition reflected the influence of Catullus, and a major source of its inspiration was the lament of the deserted Ariadne in the *Peleus and Thetis*.

Closer acquaintance with Pompeian graffiti, however, will guide our steps in a quite different direction. To begin with a fundamental question: does our graffito reflect a real-life situation? Now most of the inscriptions and graffiti found in Pompeii arose from actual experience (numbers refer to the inscriptions in *CIL* 4). Real voters are urged to elect real people:

7346 Popidio Rufo feliciter: dignus est
"Good luck to Popidius Rufus: he's worth your vote."

The gladiators are real and not imaginary:

2483 Mansuetus provocator victor Veneri parmam feret
"The challenger Mansuetus will dedicate his shield to Venus if he wins."

The advertisements are likewise serious:

807 Hospitium: hic locatur triclinium cum tribus lectis et comm(odis)

"Hotel: dining room with three couches and all amenities."

Even the scribblings spring from authentic situations:

4764 Perari, fur es (Perarius is a real person)
2409 Stronnius nil scit (Stronnius is not a pseudonym)
2224 Felix cum Fortunata (Actual live people)

What makes *CIL* 4.5296 impossible to take seriously as an authentic expression of emotion is its lack of particularity, its total lack of reference to detail in the real world. The poem is devoid of names; there is no mention of the author being shut out by a door; and the vague sequence of erotic platitudes feeds the suspicion that what we have here is a composition devoid of personal involvement. With the realization that the graffito does not reflect a real-life situation disappears all likelihood that it was composed or inscribed by a girl. We have no evidence that girls participated in formal education beyond a rudimentary level, whereas speeches on the lips of women abound in literature written by men from Homer to Heliodorus. True, we have in the Corpus Tibullianum a charming set of elegiacs written by a *culta puella*: but Sulpicia is unique, and the odds against female authorship of our poem must be deemed overwhelming.

The Pompeian graffiti that manifestly do not arise from real-life situations fall into two classes: (1) copies of the alphabet and literary quotations used as models for teaching the art of writing; and (2) compositions for amusement.

The copies of the alphabet are so numerous as to cry out for explanation. Just one or two examples:

9222 ABCDEFGHIKLMNOPQRSTUX (forwards);
5499 AXBUCTDSERFQGPHOINKML (both ways interlocked):
2541 the same as the above as far as SER, but upside down.

We also encounter alphabets with mistakes in the letter-order, and this provides the explanation we are seeking. Learning to read and write in school, boys were required to copy out the alphabet, again and again; and possibly with some pride in their newly acquired skill they scrawled it on walls outside the classroom. There is a passage in Manilius (*Astronomica* 2.755ff) where he refers to the stages of the curriculum: "Children who have not yet begun their lessons are first shown the shape and name of a letter, and then its value is explained: then a syllable is formed by the linking of letters; next comes the building up of a word by reading its component syllables; afterwards the meaning of expressions and the rules of grammar are taught, and then verses come into being and rise up on feet of their own." A quaint piece of abracadabra (its text is found more than once) deserves comment:

4235 Barbara barbaribus barbabant barbara barbis

This takes me back sixty-five years to a time when, with my Latin master Eric Knight thumping out the ictus on his desk, the whole class solemnly chanted after him *Down in a deep, dark dell sat an old cow munching a beanstalk*, for this was my introduction to the rhythm of the hexameter no less than *Barbara barbaribus* was to the boys from Pompeii.

A connection with the classroom is confirmed by over fifty tags and quotations from literature, Virgil with half of them easily leading the field. For example:

4832 Arma uirumque cano Troiae qui primus ab oris (*Aen.* 1.1)
3889 Conticuere omnes intentique (*Aen.* 2.1)

These two occur often enough to make us wonder why of the first lines of other books of the Aeneid we only find:

3796 Aeneiaque nutrix (*Aen.* 7.1)

Perhaps this was chosen as giving an example of the letter X, as the following (a rare example of X and Y) was brought in from the Eclogues:

1982 Carminibus Circe socios mutauit Olyxis (*Ecl.*8.70)

Figures of speech could have been illustrated from

4409 Quos ego sed (*Aen.*1.135)
2310 Tu, dea, tu pr(a)ese(n)s nostro succurre labori (*Aen.* 9.404)

All these point to the schoolroom. The schoolmaster, in teaching his charges how to write, has taken a memorable quotation as an example for practice, which some of the boys, exulting in their incipient literacy, have been unable to refrain from recording on some bare patch of wall. Occasionally the master has ventured to vary the author:

7353 Romulus in caelo (Ennius, *Annals* 110)
3072 Aeneadum genetrix (Lucretius 1.1)

The second class of graffiti unconnected with situations of the moment I have labeled compositions for amusement. Naturally, since they came into being at the whim of the writer, they evince a wide variety and do not permit of detailed classification. But a surprising number are related to the classroom and all in a general way to education. The most arresting is 2331, the Minotaur graffito: it needs little by way of commentary, for it is a young person's drawing of a maze labeled with the legend LABYRINTHVS: HIC HABITAT MINOTAVRVS and will have been the sequel to some lesson in mythology. Many are in verse, and the following, found with slight variations in three separate places (1904 the basilica, 2487 the amphitheater, and 2461 [first verse only] the theatrum maius) and here printed as it appears in *CLE* 957, seems to have been the prototype of an epigraphical commonplace:

admiror, paries, te non cecidisse ruina,
qui tot scriptorum taedia sustineas.

Elsewhere we meet with a couplet that is not so much a free composition as an adapted quotation:

1645 si quis forte meam cupiet uiolare puellam,
illum in desertis montibus urat amor.

This is hardly independent of what was written on the wall of a brothel on the Palatine (*CLE* 954):

Crescens, quisque meam futuet riualis amicam,
illum secretis montibus ursus edat.

Some wag in the fullers' district (Reg. IX Ins. 13) was moved to a skit on the first verse of the *Aeneid*:

9131 Fullones ululamque cano, non arma virumque

—not unlike one of my schoolfellows, who by adding but three strokes transformed the title page in his copy of a set-book to read not LIVY but SLIMY.

We also find quotations from, or verses modeled on, the elegiac poets, and here it is necessary to be more precise in our use of the term school. Alphabet graffiti and quotations for writing practice must illustrate education at an early stage, when school means elementary instruction in grammar and literature by the *litterator* or *grammaticus*; but erotic quotations and free compositions, neither a likely product of the lower school, point to education under the *rhetor*, training for the law and public life and the nearest equivalent of our colleges and universities. It was here that Ovid's verbal and rhetorical skills were honed to an unparalleled degree. To be sure, the elegiac poets were not studied in the rhetorical academies, but it was there that they will have been first met and where they found a ready circulation. Hence these couplets from Propertius:

1950 quisquis amator erit, Scythiae licet ambulet oris,
nemo adeo ut feriat barbarus esse uolet. (3.16.13f)
1894 ianitor ad dantis uigilet: si pulsat inanis,

surdus in obductam somniet usque seram. (4.5.47f)

Note well that apart from the trivializing *Scythiae* (instead of the more artistic *Scythicis*) the text of these graffiti is in every respect superior to the medieval manuscript tradition of Propertius. No doubt it was the erotic content that prompted the copying; and this was most likely perpetrated by young adults, not young enough to be just practising writing skills, not old enough to have left their schooling far behind them. Similarly with the following graffito, in which the first line derives from Propertius 1.1.5 and the second reproduces Ovid, *Amores* 3.11.35:

1520 candida me docuit nigras odisse puellas.
odero si potero; si non, inuitus amabo.

Other Ovidian quotations may well have been picked up in the years of rhetorical training:

1893 surda sit oranti tua ianua, laxa ferenti;
audiat exclusi uerba receptus amans. (*Am.*1.8.77f)

1895 quid pote tan durum saxso aut quid mollius unda?
dura tamen molli saxsa cauantur aqua. (*Ars* 1.475f)

Incidentally, there is in these erotic quotations one curious omission: Catullus. Upon the development of Augustan poetry no one exercised a greater influence. Can it really be that by the year 79 his poems had ceased to be read by the literate, whether at school or later? The answer is, unarguably: yes. To judge from the scholia on Virgil, Horace, and other writers, and from the grammarians and lexicographers—practically all their material depends upon compilation in the first century of the empire—Catullus was never studied at school at all. Not that this should surprise us. We can recover the reading list of the syllabus in the early empire from the collective index of passages in the seventh volume of Keil's *Grammatici Latini.* A little reflection makes it obvious that later writers like Aulus Gellius and Macrobius, who quote Catullus, did not possess texts of the poet themselves but are merely transcribing their sources. Moreover, one does not in the whole of *CIL* 4 and its supplements come across a single example of the hendecasyllable, which is surprising when one considers the obscenities written on the walls of Pompeii and the frequent use made of the meter in the three great repositories of classical Latin scatology, namely Catullus, Martial, and the *Priapea*. We may, before proceeding any farther, dismiss out of hand Copley's notion that our poem owed something to Catullus (of whom there is not a single verbal echo).

To judge by Manilius the last stage in formal instruction implies some elementary verse composition, and it is its mistakes and incoherence and immaturity, so abundantly exemplified in the Pompeian verse graffiti, which enable us to identify 4.5296 as student work.

Let us begin with meter. The metrical form of our poem is confused and fluctuates awkwardly between hexameters and elegiac distichs. No reputable author ever mixed up his hexameters with elegiacs; but in the volumes of *CLE* we find plenty of evidence to support the contention that many an amateur elegist, finding the restrictions of the verse-form impossibly exacting, yielded to the temptation of breaking the strict sequence of hexameter (H) and pentameter (P), so that hexameters clustered together and pentameters were only occasionally inserted, for example:

CLE 880: HP P *CLE* 1058: HP HH HH HP HP

CLE 914: HH HP *CLE* 1253: H HP HP HH

Of these mongrel elegiacs we have one fascinating illustration in classical literature, for in his *Satyrica* Petronius brings on Trimalchio to flaunt a couple of his compositions:

cap.34: H HP cap.55: H HP

Their literary merit splendidly matches the crude quality of the epigraphic material:

eheu, nos miseros, quam totus homuncio nil est!
sic erimus cuncti, postquam nos auferet Orcus.
  ergo uiuamus, dum licet esse bene. (*Sat.*34)

quod non exspectes, ex transuerso fit ubique,
nostra et supra nos Fortuna negotia curat.
  quare da nobis uina Falerna, puer! (*Sat.*55)

Incorrect scansion in our poem is also typical of other Pompeian graffiti. *CIL* 4.813 (grandly inscribed over the entrance to the library of the Warburg Institute, London) otiosis locus hic non est: discede morator

must have the first word scanned as a molossus. Similarly in verse 2 of our poem *oscula* is meant to be scanned as a molossus, for though the verse has clearly been derived from a pentameter which ended *oscula ferre labris*, it is no less clear that the diminutive *labellis*, giving a hexameter clausula, was purposely contrived, to match the diminutives *braciola* and *pupula*: in no way can it have been corrupted from *labris*. The verse is an example of the ruthless violence needed to turn a pentameter into a hexameter, violence which recurs in the hexameter *CIL* 4.1837.3,

ergo coge mori, quem sine te uiuere cogis.

This will only produce a permissible rhythm by scanning *sine te* as a dactyl. Here, however, we know the actual verse it was derived from, a pentameter of Ovid's (*Heroides* 3.140):

quam sine te cogis uiuere, coge mori.

This immature prosodist (or rather pasticheur, for his other verses come from Ovid, *Amores* 3.6.87, Tibullus 2.6.20, and Ovid, *Tristia* 1.2.52), clearly aiming at an elegiac alternation of hexameter and pentameter, realized that a hexameter was needed, and produced a correct beginning and a correct end, but imposed a perverse scansion upon the connecting words. We should not jump to the conclusion that the composers did not know that *oscula* was a dactyl and *sine te* an anapaest, but rather that their compositional technique was so inadequate that they tortured intractable words into the only cola available to them. Not for them the fastidiousness of Ovid, who refused to take liberties with the scansion of *Tuticanus*.

Passing over verses 3 and 4 of our poem, where the mistakes are more likely to be due to carelessness in writing than ignorance of meter, we move on to the next sentence, which reveals the composer's lack of experience in verse-making. To begin with, there is no main verb, unless we take *meditans* (sc. *eram*) to be an equivalent for *meditabar*. Then verse 6, even if we stop our ears and scan *-tuna quos* as a dactyl, cannot be accepted as a hexameter, for it has seven feet. The sense too is confused, though it is clear what was intended: "those whom Fortune has raised aloft one moment, she casts down the next," and we may dimly discern behind the garbled couplet the outline of the model from which it was cobbled:

Fors, quos modo sustulit alte,
hos modo proiectos praecipitesque premit.

Verse 8 is a perfect hexameter, but oddly phrased for its context, which becomes understandable once one realizes that it has been torn from its home in Lucretius (5.962):

et Venus in siluis iungebat corpora amantum.

I have left to the end a feature of the graffito which compels us to ascribe it not to an individual but to two scribes: the epigraphy. The last line, which meter shows to have been organically connected with the one before, was written with the letters A and M formed quite differently from their shapes in previous occurrences. It seems that as the writing approached the bottom of the wall the first of the boys tired, and for the last line a companion took over, possibly a smaller boy who, if more nimble in stooping down, was less accomplished in calligraphy and less orthodox in spelling: the final word, *agam*, he pronounced and consequently spelled without the intervocalic *g* (as one might say "Wha'll I do?"), a feature of colloquial speech inherited by Italian, as we see from *ego* and *magistrum* and *viginti* becoming *io* and *maestro* and *venti* respectively. Two boys then were involved, and possibly more. Let us take a closer look at the final verse.

dividit lux et se | paries quid aam

The last three words appear on a separate line, so let us start with the beginning. We are faced with a problem of scansion: the second syllable of *dividit* ought to be short; however, examples given above make it pretty certain that we are to scan it as a molossus like *oscula* in the second verse. A problem less easy to solve is posed by the word invariably transcribed as *se*. In fact Della Corte's copy reads *:s:e* or rather *:s:ic*, for apart from the dots of interpunction the letters strikingly resemble the *sic* at the beginning of verse 8. Furthermore, *se* "themselves" ought to be reflexive and is not the correct pronoun. Other doubts arise, for the singular verb *dividit* ought not to have two subjects, *lux* and *paries*; and in any case the sophisticated word-order *dividit lux et se paries* is incompatible with the naive style of the rest of the graffito.

Is it possible, as some believe, that the poem breaks off at *se*- and that the last three words are unconnected with it? Hardly. Copley thought that *se*- might be the first letters of *seiunxit*: this would nicely balance *coniunxit*, but is ruled out of court by the fact that the composer preferred *dividit* as well as by the awkward rhythm involved. Moreover, even assuming that at *se* there is a switch in sense, we must recognize that there is no interruption to the meter, since *paries* cannot commence a dactylic line; we are compelled to assume on metrical grounds that the composition is complete: the last verse is patently a pentameter. *Se* being impossible, and *sic* being equally out of the question though closer to the ductus, and *paries* badly needing some justification as a vocative, I incline to the view that the word intended was *dic* (I forget who first made the suggestion, but it was not I; *sic* will then be a *Perseverationsfehler* from the preceding verse). I believe therefore that we should read

dividit lux et dic, paries, quid a'am?

and translate, "dawn separates them and say, wall, what am I to do?" The closure is undeniably abrupt; but a closure it certainly is, and for this a pentameter was appropriate. It must have been obvious to the writers from the start that space did not allow for more.

At any rate the explanation of *CIL* 4.5296 seems to be that some Pompeian youth accompanied by a friend or friends scribbled these verses on the wall as a *jeu d'esprit*. This was something that they had elaborated in idle moments; they may even have chosen a house where there lived a girl who had set their hearts a-flutter. We may compare them with Encolpius and Giton in Petronius' wonderful novel, hardly studious and certainly not scholars, but for all that not uneducated. Roaming the streets and with nothing to do after leaving the establishment of some Agamemnon or eluding the importunities of some Eumolpus, they can have had no idea, when they posted their "paraclausithyron," of the mortal peril from Vesuvius that impended. Still, if they were anything like their Petronian counterparts, they will have been resourceful enough to make good their escape from the holocaust. Dare we indulge our imaginations and fancy that they hurried to the doctor's house to ensure that of their *pupula perdita*?

SERVIUS AS COMMENTATOR ON HORACE

MARIO GEYMONAT

Horatium, cum in Campania otiarer, excepi: thus writes Servius in the letter *Fortunatiano dn.*, prefixed to his catalogue of the nineteen metres of the *Odes* and *Epodes* which Keil published in the editio princeps of 1864 (*GL* IV, 468-72).[1] The word *otiarer* recalls an ironical remark made by Horace in his autobiographical satire addressed to Maecenas (1.6.128: *domesticus otior*), and it is pleasant to imagine that the commentator *par excellence* of the *Aeneid*, the *Eclogues* and the *Georgics* passed his leisure hours in a region close to Rome which was so dear to Virgil.[2] Now at the beginning of the fifth century it was not immediately

---

[1] The title *De metris Horatii* is missing in Parisinus 7530 (ff. 35^v and 37-38^r), a codex in Beneventan script dating from the beginning of the ninth century which is the sole testimony to this and several other important grammatical texts beginning with the *Anecdoton Parisinum* ff. 28-29r (= *GL* VII, 533-37 = *GRF* 54-56 Funaioli). Shortly after the first edition, L. Müller maintained peremptorily ("Sammelsurien," *JKPh* 93 [1866] 564-65) that "der Verfasser des Büchleins unmöglich derselbe sein kann" as the author of the *Centimeter* (or *De centum metris*, *GL* IV, 456-67), a brief tract on metrics transmitted in several codices (including, among others, the Parisinus) under the name of Servius, which certain scholars claim to have been written between 400 and 410 (N. Marinone, "Per la cronologia di Servio," *AAT* 104 [1970] 210-11) and which others consider apocryphal, dating it to the sixth century (G. Brugnoli in *Enciclopedia Virgiliana* IV [1988] 806). In his analysis of the whole manuscript ("Le Parisinus Latinus 7530, synthèse cassinienne des arts libéraux," *Stud. Med* 16 [1975] 115) L. Holtz drew attention to the fact that the short work on the metres of the *Odes* and *Epodes* is "authentifiée par la dédicace", and in *Donat et la tradition de l' enseignement grammatical* (Paris 1981) 223, he put its matter "en rapport avec les exercices scolaires auxquels donne lieu la lecture expliquée des poètes". The *amicus* whom the grammarian is addressing was not ignorant of Greek (*non mihi autem criminandum puto, si in tractatu Latinis Graeca miscuero, cum metrorum ratio huius linguae egeat sollertia et adprime his litteris sit erudita)*, but we cannot with certainty identify this *Fortunatianus d(ominus) n(oster)* either with Clodianus Chirius Fortunatiannus (*PLRE* I.369,3), author of an *Ars rhetorica de statibus* edited by Halm (*RLM* 590-92), or with Atilius Fortunatianus (*PLRE* I.369.2), whose *Ars* (on metrics) is included in Keil's collection (VI, 278-304).

[2] *Illo Vergilium me tempore dulcis alebat | Parthenope studiis florentem ignobilis oti* (*Georg*. 4.563-64), which is further explained: *habuitque domum Romae Esquiliis iuxta hortos Maecenatianos, quamquam secessu Campaniae Siciliaeque plurimum uteretur* (*Vit. Don*. 43-45). That Virgil personally knew the promontory of Misenum, the cave of the Sibyl and the whole Phlegraean region is confirmed by his accurate descriptions in the *Aeneid*. At Atella, in Campania, over a period of four days at the beginning of August 29 B.C., the poet had read (in Maecenas' presence)

clear why a critic of Servius' reputation would turn his attention to Horace.[3]

Exegi monumentum aere perennius: so the poet confidently wrote, and a few stained and grimy busts still linked his name with Virgil's in some schools at the time of Hadrian,[4] but *ille fidicen*, as Servius calls him in the letter to Fortunatianus (468.8), echoing *Carm.* 4.3.23 (*Romanae fidicen lyrae*), had in fact already fallen into disregard by Pliny's days,[5] and Quintilian himself, in making his own assessment of Horace, did not conceal his strong objections on moral grounds.[6] After the middle of the second century A.D. references to Horace are rare even in works of scholarship: Gellius mentions his name once, somewhat imprecisely,[7] though he continually cites Virgil; Macrobius ignores Horace in his commentary on the *Somnium Scipionis* and mentions him only twice in the *Saturnalia*[8] compared with over 800 passages in which

---

the *Georgics* to Augustus on his return from the East shortly before celebrating his triple triumph at Rome (this story apparently originates with Suetonius: *Vit. Don.* 93-97).

[3] The young Servius is already presented as a successful grammarian in the *Saturnalia*: *qui priscos praeceptores doctrina praestat* (1.24.8), *litteratorum omnium longe maximus* (1.24.20), *non solum adulescentium qui tibi aequaeui sunt, sed senum quoque omnium doctissime* (7.11.2).

[4] Juvenal 7.226-27 *cum totus decolor esset* | *Flaccus et haereret nigro fuligo Maroni.*

[5] In the 37 books of the *Naturalis Historia*, in contrast with about 70 references to Homer and 50 to Virgil, Horace is remembered solely for his opinion that elongated eggs are better than those of a rounder shape (10.12.145: *Serm* 2.4.12-13). Moreover Pliny the Younger mentions Horace only once by name (*Epist.* 9.22.2).

[6] *Inst.* 1.8.6: *nam et Graeci licenter multa et Horatium in quibusdam nolim interpretari.* For Horace and the *grammatici* before the III/IV century, see F. Lo Monaco, "Note sull'esegesi oraziana antica," in *Studia classica Iohanni Tarditi oblata* (Milan 1995) 1203-24 and G. Brugnoli, *Da Orazio lirico a Leopardi* (Venosa 1996) 19-22.

[7] At 2.22.25 (*ut est Horatianus quoque ille Atabulus*) he refers to the wind that *torret* the mountains of Apulia in *Serm.* 1.5.78 (which describes Horace's journey to Brindisi in the company of Virgil and Maecenas), but it is not certain that Gellius is speaking of Horace when he says a little earlier *legebatur ergo ibi tunc in carmine Latino 'Iapyx' uentus* (2.22.2: the *Iapyx* is named twice in Horace's *Odes* 1.3.4 and 3.27.20, but is certainly mentioned by other poets, including Virgil, *Aen.* 8.710).

[8] Macrobius censures him (3.18.13: *quam in culpam etiam Horatius potest uideri incidere*) for having derived the name *Tarentum* from *terenus*, which means "tender" or "soft" in the Sabine dialect (though in his use of *Tarentum* in *Serm.* 2.4.34 the poet is referring to the loose morals of the inhabitants rather than to the

Virgil is discussed. The situation begins to change in the schools of Gaul at the time of Ausonius (who advises his nephew to read Horace together with Virgil, Terence and Sallust: *Protrepticus* 56), but Horace's sorry fate is substantially confirmed by the "affligeante médiocrité" of the scholia on his work[9] if we compare them with the very much more extensive and learned commentaries on Virgil and Terence. It should be added that of the third century commentary of Porphyrion we possess only a few *excerpta* dating from the fifth century; from this period survive, too, the more authoritative Horatian scholia which Keller published under the name of pseudo-Acron (Leipzig 1902-4).

In a study written just a few years later, August Langenhorst listed some striking correspondences between redaction *A* of Keller's collection and Servius' Virgil commentary: in many cases the two commentaries offer their readers identical glosses, each referring back to the same verses of both Virgil and Horace.[10] He also drew attention to the number of Virgilian citations (about 800) in the scholia on the *Odes* and *Epodes*, rather more than the references (fewer than 100) in the scholia

etymology of the word). At 5.17.7 Macrobius recalls Horace's opinion that Pindar is inimitable (*quem Flaccus imitationi inaccessum fatetur*): this is not a quotation but merely a summary of the opening words of the ode *Pindarum quisque studet aemulari* (4.2).

[9] J. Schwartz, "L'ombre d'Antoine et les débuts du principat (à propos de commentaires perdus d'Horace)," *MH* 5 (1948) 155. The extent of the ancient commentaries on Horace has been delineated by, among others, R.G.M. Nisbet and M. Hubbard, *A Commentary on Horace Odes Book I* (Oxford 1970) xlvii-li and J.E.G. Zetzel, *Latin Textual Criticism in Antiquity* (New York 1981) 168-70.

[10] Here are a few of the more convincing coincidences: Servius on *Aen.* 1.292 ~ pseudo-Acron on *Carm.* 1.35.22; 2.133 ~ 3.23.4; 3.2 ~ 1.33.10; 3.64 ~ 2.14.23; 3.506 ~ 1.3.20; 4.76 ~ 4.1.36; 4.266 ~ 1.2.20; 5.71 ~ 3.1.2; 6.15 ~ 1.3.35; 9.582 ~ *Epod.* 5.21; 10.727 ~ *Carm.* 3.12.2; 12.527 ~ 1.6.6; on *Ecl.* 3.88 ~ 2.1.15; 8.28 ~ 1.2.9. Langenhorst divides the parallels between the scholia of redaction *A* on the *Odes* and *Epodes* and those of the Servian commentary on Virgil into five categories: "I. scholia, in quibus Vergilii aliquis versus commemoratur, quem explicans Servius Horatii ipsius illius versus mentionem facit; II. scholia, in quae non hoc solum cadit sed quorum etiam interpretatio cum Servii verbis congruit; III. scholia, in quibus Vergilii versus aliquis commemoratur quorumque interpretatio congruit cum verbis, quibus Servius illum versum explicat, Horatii tamen non memor; IV. scholia, quae cum Servii verbis Horatii versum in illis ipsis explicatum commemorantis congruunt nec tamen Vergilii huius versus mentionem faciunt; V. scholia, quae cum Servii verbis congruunt etsi in neutris alterius poetae mentio fit" (*De scholiis Horatianis quae Acronis nomine feruntur quaestiones selectae* [Bonn 1908] 30).

on the *Satires* and *Epistles*, though the latter are as copious.[11] For all these reasons Langenhorst conjectured that the author of redaction *A* of the commentary on Horace which is preserved in the margins of Parisinus 7900/A, and which consists solely, as has been noted, of scholia on the *Odes* and *Epodes*, had been a grammarian of the school of Servius and perhaps even one of his own pupils.[12]

That Servius' commentary on Virgil is "einen festen *terminus post quem*" for this portion of the scholia on Horace is nowadays usually accepted,[13] and I would have no hesitation in attributing to Servius a large part in the revival of interest in Horace at the beginning of the fifth century, even if he is only named once in the Horatian scholia—in a gloss which provides us, moreover, with a precious biographical detail (*sic Servius magister urbis exposuit*).[14] Nor is Donatus cited in the DS

[11] Langenhorst (above, n.10) 42. It is also worth observing that in the scholia on the *Satires* and *Epistles* Virgil is sometimes referred to as "Maro" or "poeta", titles which never occur in the scholia on the *Odes* and *Epodes*. In the latter, too, there are 47 citations of Juvenal (one of the *idonei auctores* of Servius: see notes 14 and 25) while, contrary to what one might expect, there are barely 5 in the scholia on the *Satires* and *Epistles*.

[12] Langenhorst (above, n. 10) thus develops his hypothesis elsewhere: "Itaque hoc statuendum est, antiquissimam scholiorum Horatianorum recensionem ante Servii tempora oriri non potuisse" (29); "Omnibus his locis accuratius examinatis scholiorum ad Horatii carmina et epodos adscriptorum permulta apprime concinere cum Servii commentariis Vergilianis eorumque magnam partem prorsus ab illis pendere facile intellegitur" (41); "Quod commentum qui composuit, non solum Servii commentariis saepius usus est, sed artiore etiam vinculo cum eo videtur coniunctus fuisse. Nam ut ille Vergilio studiosissimus fuit, eorundem potissimum poetarum utitur testimoniis quibus ille, similem atque ille interpretandi rationem multis locis sequitur. et cum ne tantillum quidem talis necessitudinis ei intercedat cum aliorum poetarum commentariis, a vero vix aberres, si eum ex Servii discipulis fuisse conicias" (43). In re-examining this position P. Wessner unnecessarily proposed the existence of a third scholiast as the source in turn of both pseudo-Acron and Servius (*BPhW* 29 [1909] 1112).

[13] This is the opinion of, among others, G. Noske, *Quaestiones Pseudacroneae* (Munich 1969) 275 and S. Borzsák in his note "Esegesi" in the *Enciclopedia Oraziana*, a work in three volumes that will be published in Rome in 1998-99 under the editorship of Scevola Mariotti (there we shall see a number of other entries devoted to the "Fortleben" of Horace in ancient, medieval and modern literature: the note "Servio" will be written by S. Timpanaro).

[14] The detail turns up explaining the word *antestari*, which appears in Horace but was never used by Virgil: *Denuntiantes litem antestatos habebant, quibus praesentibus conueniebant, ita ut aurem illis tertio vellerent. Sic Seruius magister urbis exposuit. Alii sic exponunt* etc. (redaction Γ on *Serm.* 1.9.76: Keller II.104). The

scholia on Virgil, yet it is commonly accepted that these are in large part derived from his work.[15]

In commenting on either Virgil or Horace Servius had an outstanding predecessor in Valerius Probus who, at the end of the first century, had employed diacritical marks *in Vergilio et Horatio et Lucretio, ut Homero Aristarchus*.[16] The views of the grammarian who came to Rome from Beirut (*Berytus*) on the subject of the textual variants and interpretation of Virgil are treated by Servius with great respect,[17] but the Scholia Veronensia on *Aen.* 9.373 confirm that some specific comparisons with verses of Horace can go back to Probus:[18] *SVBLVSTRI. Asp.: utrum 'non nubila inlustrique'? Nam 'sub' pro 'parum' ponitur. Lucil.: 'Facti subpudet, ut di...'* (fr. 1171 Marx). *An pro 'sub inlustri' positum? ut* (*Aen.* 1.453-54) *'Namque sub ingenti lustrat dum singula templo | r.o.' Prob. hic posuit aptissimum hoc ex-*

---

title *magister urbis* is consistent with what we know from other sources: Servius is a character in the *Saturnalia* of Macrobius, where he is represented as having received a letter from Symmachus, *praefectus urbi* at the end of the fourth century (8.60); and Juvenal, an author whose work has much in common with Horace, was read at his school in Rome (see notes 11 and 25). Other evidence that Servius was active in Rome is assembled by G. Thilo in the preface to his edition (I, 1881, lxxiii: in particular note the scholia on *Aen.* 3.444 *signis, id est quibusdam notis, ut in obelisco Romae videmus*, and on *Aen.* 8.271 *ingens enim est ara Herculis, sicut uidemus hodieque*).

[15] Donatus clearly never mentioned his own name in his commentary on Virgil. It is for this reason that it was not transmitted to the DS; cf. G.P. Goold, "Servius and the Helen Episode," *HSCP* 74 (1970) 110 and S. Timpanaro, *Per la storia della filologia virgiliana antica* (Rome 1986) 130.

[16] The evidence is in the *Anecdoton Parisinum*, a text very close to Probus who is mentioned on four occasions, each of a few lines (54-55 Funaioli: 10, 26, 30, 45).

[17] See the scholia of Servius on *Aen.* 1.194, 1.441; 3.3; 6 praef., 6.177, 6.473, 6.782; 7.421, 7.543, 7.773; 8.406; 10.18, 10.33, 10.539 and on *Ecl.* 6.76.

[18] The evidence escaped L. Lehnus, who believes that there are no surviving traces of Probus' commentary on Horace (*Enciclopedia Virgiliana* IV [1988] 285). Moreover the note immediately preceding in the *Scholia Veronensia* (on *Aen.* 9.369) says of Probus and Sulpicius Apollinarus (the master of Gellius) that *hoc loco adnotant*, because it contradicts what Virgil had written (*Aen.* 7.600) on the attitude of King Latinus to the outbreak of hostilities. Against this supposed contradiction, though without naming the grammarians among whom it arose, Servius clearly takes a polemical position at *Aen.* 9.367: *VRBE LATINA non est contrarium illi loco, ubi ait* (7.600) *'saepsit se tectis rerumque reliquit habenas', quod modo a Latina urbe auxilia uenire commemorat: intellegimus enim Latinum in principio discordiae et tumultus paululum se abstinuisse, postea tamen nec suorum copias nec propria denegasse consilia; nam eum et coetui et foederibus interfuisse dicturus est.*

emplum ex Horatio (*Carm.* 3.27.31-32) *'Nocte sublustri nihil astra praeter | uidit et undas'.*[19] Servius' note is rather briefer, omits Asper's opinion, passes over the fragment of Lucilius and suppresses the very name of Probus, but coincides in its citation from the Horatian ode: *SVBLVSTRI NOCTIS IN VMBRA sublustris nox est habens aliquid lucis: Horatius 'Nocte sublustri nihil astra praeter | uidit et undas'.* Servius and Probus share a similar interest in Horace in the scholium on *Aen.* 10.444: *AEQVORE IVSSO pro 'ipsi iussi'. et est usurpatum participium: nam 'iubeor' non dicimus, unde potest uenire 'iussus'. sic ergo hic participium usurpauit, ut Horatius verbum, dicens* (*Epist.* 1.5.21-22) *'Haec ego procurare et idoneus imperor et non | inuitus'. ergo satis licenter dictum est, adeo ut huic loco Probus alogum posuerit.*[20]

Servius' familiarity with the poems of Horace is revealed, certainly, in the catalogue of the metres of the *Odes* and *Epodes* which we discussed above, but more particularly in the number of citations which are more numerous than those of any other Greek or Latin author except Virgil. [21] For example, in his commentary on the *Aeneid*, the *Eclogues* and the *Georgics* Servius has 251 explicit references to Horace,[22] more than half of which refer to the *Odes* (137 in all: 63 to Book I, 29 to Book II, 30 to Book III and 15 to Book IV), while in the DS scholia, of

[19] The citation of Horace is omitted in S. Borzsák's otherwise full apparatus, though he includes the others from the Scholia Veronensia, which are independent of Servius and of the scholia of both Porphyrion and pseudo-Acron (*Carm.* 4.9.9 and *Serm.* 1.9.1-2 to *Ecl.* 6.1; *Epist.* 1.1.12 to *Ecl.* 6.7).

[20] *Probus alogum posuerit* is, I believe, the correct reading of the Leiden codex, but the passage is corrupt in the other manuscripts. There is an examination of the entire scholium, though without reference to the Horatian citation, in R.A. Kaster, *Guardians of Language: The Grammarian and Society in Late Antiquity* (Berkeley 1988) 190-92.

[21] In the *Commentarius in artem Donati* (*GL* IV, 403-48), another work probably by Servius, Horace is named 7 times, or as often as Donatus himself: only Virgil (63) and Cicero (9) receive more notice, while Probus is mentioned 6 times, Sallust 5, Plautus 3, Persius twice. There is one reference each to Terence, Lucretius, Varro, Ovid, Petronius and Terentianus Maurus. It is worthy of note that three of the references to Horace occur in just seven lines, at 432.18-24: these are *Epist.* 1.6.7, *Carm.* 2.7.3 and *Epod.* 17.48 (this last verse is also cited by the scholia on *Aen.* 5.64).

[22] A complete list is provided by P. Santini, *L'auctoritas linguistica di Orazio nel commento di Servio a Virgilio* (Florence 1979): the term *auctoritas* is employed in reference to Horace in the scholion to *Aen.* 2.272.

a slightly earlier date, there are only 20 citations of Horace, 10 being from the *Odes*.[23] The comparison is striking if one examines individual works: the *Ars Poetica* is cited 23 times by Servius but ignored by DS, the *Satires* 43 times compared with 4 in DS, the *Epistles* 25 as against once, the *Epodes* 17 as against once.[24] Among other authors, Homer comes next after Horace: there are 199 citations in Servius and 59 in DS. Then follow Sallust (182 and 74), Cicero (169 and 75), Lucan (151 and 5), Terence (140 and 84), Juvenal (93 and 4), Statius (83 and 1), Ennius (81 and 57), Lucretius (66 and 12) and so on.[25]

Apart from their frequency, the Servian citations are important for their critical method, for the ease with which the grammarian, in commenting on Virgil, introduces material from Horace and passes judgement upon it. There is the case of the scholium on *Aen.* 8.83: *CONSPICITVR SVS Horatius et 'amica luto sus'* (*Epist.* 1.2.26). *Sciendum tamen hoc esse uitiosum, monosyllabo finiri uersum, nisi forte ipso monosyllabo minora explicentur animalia, ut 'parturient montes, nascetur ridiculus mus'* (*Ars Poetica* 139): *gratiores enim uersus isti sunt secundum Lucilium.* Virgil's *sus* could easily have been compared

[23] We cannot exclude the possibility that in the commentary on Virgil (probably the work of Donatus) from which the DS scholia were derived there might originally have been additional Horatian citations, and that the compiler may have introduced only those of them which were not available in the Servian text. Of these I would cite at least the variant *crepet* in the citation of *Carm.* 1.18.5 at *Aen.* 1.738: it is unfortunate that Borzsák's apparatus fails to distinguish between Servius and DS. In the (acknowledged) works of Donatus, however, a specific interest in Horace is not observable: in the *Life of Virgil* and in the preface to his commentary on the *Eclogues* Horace is never mentioned, while one verse only is cited in the *Ars maior* (as an example of "tapinosis", and without actually naming him: *Carm.* 1.6.6 in *GL* IV, 395,17). More numerous, and not limited to the *Satires*, are the Horatian citations in Donatus' commentary on Terence, but it is not easy to establish how much of this material is actually derived from Donatus; not one of the 22 instances in the index of Mountford and Schultz appears to refer to Horace as its primary purpose. On the other hand it is significant that Jerome, who studied directly under Donatus, is clearly interested in the second-ranking Augustan poet; he offers about 65 citations in works written at different times, an indication that the study of this *uir acutus et doctus* (*Epist.* 57.5.5) formed part of his literary education and was not a discovery late in life (H. Hagendahl, *Latin Fathers and the Classics* [Göteborg 1958] 281-83).

[24] There are moreover 5 citations of the *Carmen Saeculare* in Servius compared with 2 in DS; there is one general reference to Horace in Servius but 2 in DS.

[25] In the commentary of Servius Ovid is cited only 24 times and 3 times in DS, Lucilius 16 and 23, Catullus 6 times in Servius and never in DS.

with the frequent use by Ennius of monosyllables at the end of his verses (there are 7 instances in the fragments of the first book of the *Annales*) or with *uulnificus sus* in Ovid's account of the hunting of the Caledonian Boar (*Met.* 8.359), but Horace occurred first of all to Servius' mind, and to justify his style he resorted to Lucilius.

In regard to Horace's language there is naturally no shortage of criticism, as in the scholium on *Aen.* 2.554: *HAEC FINIS omnia Latina nomina inanima, simplicia, a uerbo non uenientia, 'nis' syllaba terminata, masculina sunt. inanima propter 'canis', simplicia propter 'bipennis', a uerbo non uenientia propter 'finis'; ergo 'clunis' Iuuenalis bene dixit 'tremulo descendant clune puellae', Horatius male 'quod pulchrae clunes'*. It is odd, however, that in order to explain the feminine *haec finis*, which tragically signalled in the *Aeneid* both Priam's death and the fall of Troy, the grammarian could find nothing better to rely upon than *clunis*, masculine in Juvenal for the buttocks of lascivious girls (11.164), feminine in Horace for the rumps of horses in the marketplace (*Serm.* 1.2.89). Servius had himself promoted Juvenal to be an *idoneus auctor* for the explication of Virgil, and with the *Satires* he was particularly engaged in teaching his pupils.[26] As for Horace, the grammarian had also criticized him elsewhere for his use of gender,[27] and for the rest the equivocal *pulchrae clunes* had already offered Nonius Marcellus an opportunity for one of just five Horatian citations in his *De compendiosa doctrina*.[28] But in this case we possess a passage

[26] Confirmation of this comes from the colophon of a tenth-century codex of Juvenal, Leiden 82: *Legi ego Niceus Romae apud Seruium magistrum et emendaui* (also, *Legi ego Niceus apud M. Serbium Romae et emendaui* appears in the eleventh-century Florence Laur. XXXIV.42). In his article, "Servius and *idonei auctores*," *AJP* 99 (1978) 181-209, R.A. Kaster did not pay sufficient attention to the *auctores antiqui*, Terence, Sallust and Horace, while diligently tracing citations of the so-called *neoterici*: Lucan, Statius and Juvenal.

[27] See particularly the Servian scholia on *Aen.* 8.571: *VIDVASSET CIVIBVS VRBEM scilicet uiris fortibus peremptis. et proprie 'uiduasset' dixit, quia urbs generis est feminini: Horatius abusiue et satis incongrue in genere masculino posuit, dicens 'uiduus pharetra risit Apollo'* (*Carm.* 1.10.11-12); and on *Aen.* 12.208: *IMO DE STIRPE ideo genere masculino est, quia de arboribus loquitur: nam de hominibus genere feminino dicimus, ut* (*Aen.* 7.293) *'heu stirpem inuisam'; dixit tamen, sed usurpatiue, Horatius etiam de arboribus, ut 'stirpesque raptas et pecus et domos'* (*Carm.* 3.29.37).

[28] *CLVNES feminino Horatius: 'quod pulchrae clunes, breue quod caput, ardua ceruix'. Masculino Plautus Agroico: 'quasi lupus ab armis ualeo, clunis infractos*

of Charisius (101 K, 128 Barwick) which allows us to connect the argument with the lighthearted temperament of Roman poets and scholars of the first century B.C.: *CLVNES feminino genere dixit Melissus* (fr. 4 F) *et habet auctorem Laberium, qui in Ariete sic ait* (v. 7 R^3) *'uix sustineo clunes*'*, *et Horatius* (*Serm.* 1.2.89) *'<pulchrae clunes>'*, *et Scaevola* (fr. 2 B) *'lassas clunes'. Sed Verrius Flaccus* (fr. 15 F) *masculino genere dici probat, quoniam 'nis' syllaba terminata anima carentia nominatiuo singulari masculina sunt, ut 'panis', 'cinis', 'crinis' et similia.*

The easy familiarity of Augustus and Maecenas with the poets and *litterati* of their day appears again in the aristocratic circles of Rome in late antiquity with the grammarians whose task it was to defend the Latin language and the memory of the past.[29] The analogous political function which the poets exercised in former times and grammarians generations later was by no means a secondary reason, in my opinion, for the fascination which Virgil and Horace held for Servius[30] and his colleagues.

In successive stages the commentaries on the *Eclogues*, the *Georgics* and the *Aeneid* represented a collection of the critical exercises of earlier generations. They are for example enriched by well chosen comparisons with Sallust and Terence when Asper and Donatus, the greatest grammarians of the third and fourth centuries, extend their treatment beyond Virgil to turn their attention to those authors: even in the commen-

fero' (289 L). The fragment of Plautus (5 L) has survived, on this occasion not linked with Horace, in the entry on *clunes* in Paul the Deacon's epitome of Festus (54 L).

[29] Consider the opening remarks of Macrobius: *Saturnalibus apud Vettium Praetextatum Romanae nobilitatis proceres doctique alii congregantur et tempus sollemniter feriatum deputant colloquio liberali, conuiuia quoque sibi mutua comitate praebentes, nec discedentes a se nisi ad nocturnam quietem* (1.1.1). On the function of grammarians in late antique society Kaster (above, n. 20) is valuable in its entirety.

[30] It is noteworthy that the *uerecundia* valued in Horace (*Epist.* 1.7.37 *saepe verecundum laudasti*) by the powerful Maecenas came to be attributed in the *Saturnalia* to Virgil as well (1.16.44 *poeta doctrina ac uerecundia iuxta nobilis*), and this is above all the characteristic of the young Servius: *Hos Seruius inter grammaticos doctorem recens professus, iuxta doctrina mirabilis at amabilis verecundia* (1.2.15), *Inter haec cum Seruius ordine se uocante per uerecundiam sileret* (2.2.12), *His dictis cum ad interrogandum ordo Seruium iam vocaret, naturali pressus ille uerecundia usque ad proditionem coloris erubuit* (7.11.1). See R.A. Kaster, "Macrobius and Servius: Verecundia and the Grammarian's Function," *HSCP* 84 (1980) 219-62, reproduced with revisions in the volume cited above, n. 20.

tary of Servius which has come down to us, as I have noted, there remain 182 citations of Sallust and 140 of Terence; in the DS there are 74 and 84 respectively. That Servius was one of the most authoritative promoters of the resurgent interest in Horace at the beginning of the fifth century can, I believe, be confirmed by those citations in the grammarians of a later age which coincide with the ones used by Servius: in the *Institutiones* of Priscian, to give just one example, Servius' note on *Aen.* 2.554 is repeated almost verbatim, with the discussion on the gender of *clunis* and the verses, one after the other, of Juvenal and Horace.[31]

Horace was soon to fall into profound oblivion until the ninth century, the period of our oldest manuscripts, when the *Odes*, the *Epodes*, the *Satires* and the *Epistles* became a learned fashion: Alcuin, the greatest scholar of the Carolingian age, perhaps ironically gave himself the nickname "Flaccus". Throughout this long eclipse, however, Horace was always kept alive in the scholia on Virgil, some of which seem to have been written to interpret both poets.[32]

[31] On the other hand nothing is said about the Virgilian expression *haec finis*: *'finis' quoque et 'clunis' tam masculini quam feminini generis usurpauit auctoritas in una eademque significatione. Iuvenalis in IIII 'Ad terram tremulo descendant clune puellae'. Horatius in I sermonum 'quod pulchrae clunes, breue quod caput, ardua ceruix'* (*GL* II, 160.10-15).

[32] As an example, the Bern scholium on *Ecl.* 7.29 (in the right margin of MS Bern 172 and almost identical to the note in *Explanatio I* of Philargyrius) refers as much to Horace as it does to Virgil: *PARVVS, Micon puer eius vel, ut Iunilius dicit, 'pauper', dicens ut Horatius: 'parui properemus'; et e contrario 'ampli' diuites dicuntur.* Here the final words refer to *hoc opus, hoc studium parui properemus et ampli*, a verse of Horace (*Epist.* 1.3.28) on which no other scholia have survived. Servius for his part ignores *Ecl.* 7.29 and DS merely remarks: *PARVVS vel humilis, uel pauper, uel minor aetate*; but it may be recalled that Virgil, speaking in the person of Thyrsis, took from the young Horace some of his stylistic features (see my "Tirsi critico di Lucilio nella settima egloga virgiliana," *Orpheus* 2 [1981] 366-70).

THE FIRST PERSON SINGULAR IN HORACE'S *CARMINA*

NICHOLAS HORSFALL

The biographical tradition about Greek authors has for over twenty years been a battleground: engrossing and highly instructive if you have been able to watch from a distance.[1] No less fierce (and a bit more complicated) is the debate on the first person singular in Pindar.[2] It is then a bit of a shock to come down to earth again in Latin studies, where many authors still seem to have biographies[3] and where serious scholars still write at length about Horace's life.[4] "Horace never lies," uttered Eduard Fraenkel, notoriously.[5] Were that quite true, it might be very useful to us, for we know strikingly little else about him.[6] Not that that matters a scrap (or hardly so: it would indeed help with *Epod.* 9 if Dio told us the poet had been at Actium!) to how we interpret the poems but *de facto* it is still important to the way people write about Horace, and the hunt for biographical "facts" continues undeterred.[7] The contrast with the bulk of

---

[1] See, for example, M.R. Lefkowitz, *Lives of the Greek Poets* (London 1981); J. Fairweather, "Fiction in the Biographies of Ancient Writers," *Anc. Soc.* 5 (1974) 231-75 and "Traditional Narrative, Inference and Truth in the *Lives* of Greek Poets," *PLLS* 4 (1983) 315-69; G. Brugnoli, *Enciclopedia Virgiliana* 5*.575-85.

[2] See, for example, G.B. D'Alessio, "First-person Problems in Pindar," *BICS* 41 (1994) 117-39, M.R. Lefkowitz, *First-Person Fictions* (Oxford 1991).

[3] Despite the efforts of (e.g.) E. Courtney, *A Commentary on the Satires of Juvenal* (London 1980) 1-10, A.S. Gratwick in E.J. Kenney and W.V. Clausen, edd., *Cambridge History of Classical Literatur, Vol. II* (Cambridge 1982) 814-6 on Terence, and both H. Naumann (e.g.) "Suetonius' life of Virgil," *HSCP* 85 (1981), 185-7 and G. Brugnoli (above, n. 1) *passim* on Virgil.

[4] F. Della Corte, *Enciclopedia virgiliana* 5**, 2-97, for an extreme case.

[5] *Horace* (Oxford 1957), index, s.v.; cf. G.W. Williams, "Libertino patre natus," in S.J. Harrison, ed., *Homage to Horace. A bimillenary celebration* (Oxford 1995) 312.

[6] *ILS* 5050.149, Ov. *Tr.* 4.10.49, Aug. *Ep.* frr. xxxiii, xli Malcovati, Maec. fr. 3 André, and his villa—if it really is his villa!

[7] *Homage to Horace* is not immune; nor is the *Enciclopedia Oraziana*, to judge from the elegant "Estratto" entitled "Biografia" (Roma 1992).

contemporary testimony about Virgil is striking, little though that testimony really tells us about Virgil the Man, or Virgil the Poet.[8]

Not long ago, I spent some time on a brief analysis of Suetonius' methods of literary biography, and on a much fuller and severely negative study of the biographical tradition about Virgil;[9] a touching faith in what the *Vita Suetonii–Donati* tells us about him has lingered on long after Fraenkel, in the first chapter of his *Horace*, destroyed with exemplary vigor any sense of the historical reliability of the Suetonian *Vita Horati.*[10] At that point, having taken a pick to what was left of Virgil's farm,[11] I began, almost inevitably, to worry about the continued and almost unchallenged use of Horace's text as evidence for his life: by chopping wood, you learn about axes, wedges, mauls and the like! In other words, once you realise that both biographical statements about poets and likewise poets' own apparent 'autobiographical' statements need not always be literally true,[12] you begin to see more readily that scepticism about poets' lives is not just a tic, or a fashion, or an aberration, but the only way to read the text with critical intelligence, essential though 'biographical' readings are for the reception-history of the poems.[13] Whether what can be done towards scrubbing the historical accretions of biographical interpretation off the text of Horace has any relevance to the way in which Catullus may or should be read is not a question which I wish to address here, though the restoration of

---

[8] Cf. G. Brugnoli and F. Stok in *Enciclopedia Virgiliana* 5**.429-36, Horsfall (next note).

[9] "Problemi della biografia letteraria," *Atti Accademia peloritana dei Pericolanti* 68 (1992) 41-53; N. Horsfall, ed., *A Companion to the Study of Virgil* (Leiden 1995) 1-25.

[10] P. White, *Promised Verse* (Cambridge, Mass. 1992) 124-25, 127-32 still convinces me less than Fraenkel (above, n. 5) 364-65 on the genesis of *Carm.* 4. Cf. below, n. 17.

[11] Cf. n. 9 above (1995); Wendell Clausen does the same himself gently but lethally: *A Commentary on Virgil: Eclogues* (Oxford 1994) 267, n. 9.

[12] Our honorand will not be surprised that I took care to discuss fruitfully, as ever, the elusiveness of such statements with a dear common friend of ours, the poetess Julia Budenz.

[13] Cf. I. Dionigi, "Interpreti recenti di Orazio," *Atti dei convegni di Venosa, Napoli Roma* (Venosa 1994) 273-285, W. M. Calder, "F.G. Welcker's *Sapphobild* and its reception in Wilamowitz," in *Friedrich Gottlieb Welcker, Werk und Wirkung,* Hermes Einzelschriften 49 (Stuttgart 1986) 131-56.

sequence, if not of respectability to his private life has recently returned to favor.[14]

That should not be taken as the prelude to an outburst of destructive violence for its own sake: I start rather from a point, so far as I can see, objectively present in the text and largely uncontroversial:[15] the presence of recurrent themes in the *Carmina* and their role in the creation of some sort of "architecture" in the collections. Particularly, the theme of poetry itself. What I say is the result not of bibliographical enquiry, but of reading the poems straight through in the summer of 1994 (four books: four days). The *Carmina* turn out to be at least as heavily metapoetic as what I still call the *Bucolics*.[16] Notably, in *Carm.* 4,[17] a collection particularly rich in poetry about poetry;[18] not just the commissioned bombast of Horace's sold-out senility, as the heirs of Timagenes prefer to believe! In Book 4, odes 2, 3, 6, 8, 9, 11 and 15 are concerned explicitly with poets and poetry; even those unloved patriotic odes (and it is a pleasure to find Du Quesnay [above, n. 17] 128-87 championing their merits) are themselves very much a statement about the poet in the *res publica* (cf. in particular 4.15.1-4, 32). The collection *Carm.* 1-3 has (typically) a prominent metapoetical proemium and sphragis. Horace writes towards his recognition as an acknowledged "voice" of Rome and Italy (undeniably so, at the level of the explicit rhetorical strategy of the texts themselves!) and it is hardly necessary to argue that the Sabine villa[19] and the surrounding country (note 3.29.6-8) is just as fundamental to Horace *qua* poet as was Ascra to Hesiod, or that Horace's sense of his roots between Lucania and Apulia is an element integral to

---

[14] T.P. Wiseman, rev. of F. Stoessl, *C. Valerius Catullus. Mensch, Leben, Dichtung* (Meisenhein am Glan 1977) in *JRS* 69 (1979) 161-8.

[15] Curiously, H. Dettmer, *Horace: A Study in Architecture* (Hildesheim 1983) leaves *Carm.* 4 out of consideration.

[16] Cf. Clausen (above, n. 11) xxiii-xxv.

[17] On the book's genesis, cf. I.M. LeM. Du Quesnay, "Horace *Odes* 4.5," in Harrison, ed. (above, n. 5) 131-5 and White (above, n. 10) 127-32.

[18] Cf. S.J. Harrison, "Horace, *Odes* 4.2" in Harrison, ed. (above, n. 5) 108-27. Note that the bibliography offered here is deliberately minimalist.

[19] Cf. S. Quilici Gilgi, "Biografia; la villa in Sabina," in *Enciclopedia Oraziana* ("Estratto" [Roma 1992]) 37-41; *Atti del convegno di Licenza 1993* (Venosa 1994) *passim*.

his poetical self-expression—just as Boeotia was to Pindar.[20] Both geography and autobiography have therefore for Horace a metapoetic role and geography repeatedly provides the context for Horace's expression of a sense of his own originality.[21] Rome herself is not "mere" setting or scenery, but the essential and inescapable frame for Horace's view of himself as public, ceremonial poet.[22] The topography of the city[23] is both the visual, evocative context of numerous poems and the guarantor, in tufa, marble, and history of the poet's fame. Horace's recognition of public life as the central topic of serious lyric belongs very precisely in the restored monumental city: it is there that he envisions his numerous poetic *contiones* to the assembled citizens.[24] The public voice of the city (and of Italy at large) thereby reacquires its lyric expression. Horace's poetic *iter* from his nurse's *limen* (whatever she was called) to the Capitolium is of course both the poetical representation of a real, biographical *iter* and a poetic journey built in part round certain biographical elements. At that point it is really no longer possible to separate apparent biographical statements in the *Odes* from vastly complex questions raised by apparent autobiography in Greek lyric and from the biographical tradition (and Horace's knowledge of it!) about the Greek lyric poets.

When Horace exclaims *Quo me, Bacche rapis | tui plenum* (3.25.1f.), when he visualises himself as undergoing metamorphosis into a swan (2.20), when he claims actually to have seen Bacchus (2.19.1), when he tells us of the miracle of the doves who protect his infant slumbers (3.3.9ff.), the reader who reacts with mirth or mockery should change career (or at least author); our first duty is still to try to understand something of the intellectual context from which these dis-

---

[20] D. Gagliardi "La Lucania nella poesia di Orazio," *Atti del convegno di Venosa* (Venosa 1993) 181-92. Pindar: Lefkowitz (above, n. 1) 65; cf. *Ol.* 6.90, *Pyth.* 5.75f., 9, 11, fr. 83.

[21] At this point in the argument, it would be easy to deck the notes with abundant references to the text of Horace; I do not suppose that readers who have got this far, let alone the honorand, actually need them.

[22] Thumb and all, Fraenkel (above, n. 5) 403.

[23] Cf. "Roma; città augustea e luoghi di Orazio," in *Enciclopedia Oraziana* ("Estratto", Roma 1993), 5-14, with map.

[24] *Carm.* 3.6.2, 3.14.1, 4.2.45ff.

concerting claims emerged. They may seem silly to us at first sight, but if we stop there, we have not begun to contemplate a serious critical reading of the texts! Such claims are entirely traditional in character: that is not to say that they could not in some sense have 'happened' (whatever that means!), but, more to the point, these asseverations open the door to the possibility that stories told by the poet about himself can serve as tropes that contribute to the creation of the poet's particularly complex *persona* in *Carm.* The serious student of 'infancy miracles' will read 3.4.9ff. alongside the stories collected in the VSD of Virgil (chs. 3-5). Likewise, the *Vita Pindari* 2 tells of a bee that landed on the boy's lips when he slept exhausted near Mount Helicon, and built a honeycomb in his mouth. The same story is told of Plato.[25] I can offer no novel defence of the feathered Flaccus (2.20), but to comprehend something of this (to us) rebarbative text you need to remember (and most of the material is there to hand in the Nisbet-Hubbard commentary) the importance of: (i) the swan as metaphor for the poet;[26] (ii) of flight as a metaphor for fame; (iii) of Euripides fr. 911N, as interpreted by Satyrus[27] as referring not to avian flight, but to hasty, human flight from Athens; (iv) of Ennius' epitaphic claim *uolito uiuo' per ora uirum*;[28] (v) of Homer's metempsychosis—into Ennius;[29] and lastly of (vi) claims that might be made in the closure-poems of poetical books.[30]

I offer no more borrowed plumage, so to speak, but that should be enough to show that Horace was hardly straining the credulity or good-will of the educated ancient reader, little though we may share that toler-

---

[25] Cic. *Div.* 1.78, A.S. Riginos, *Platonica* (Leiden 1976), 17-21, Lefkowitz (above, n. 1) 59, Soph. T. 108 Radt.

[26] R.G.M. Nisbet and M. Hubbard, *A Commentary on Horace: Odes, Book II* (Oxford 1978) 322.

[27] fr. 39; Lefkowitz (above, n. 1) 169, n. 13; Nisbet and Hubbard (above, n. 26) 335.

[28] *Var.* 17 f., Nisbet and Hubbard (above, n. 26) 336, O. Skutsch, "On the Epigrams of Ennius," *LCM* 10 (1985), 146f.; OS it was who introduced me to Wendell Clausen; *uolitat uiuos et ille per ora uirum.*

[29] Nisbet and Hubbard (above, n. 26) 336.

[30] On the sphragis, cf. H. Lloyd-Jones, "The Seal of Posidippus," *JHS* 83 (1963) 75-99 (= *Greek Comedy, Hellenistic Literature, Greek Religion, and Miscellanea: The Academic Papers of Sir Hugh Lloyd-Jones* [Oxford 1990] 158-94), D. Fowler, "On Closure," *MD* 22 (1989) 75-122.

ance. A figure is pushed, we may admit, as far as it could readily go (or a little further, but metamorphosis here is not biographical but metaphorical).

The vision of Bacchus enjoins similar treatment: *uidi* is conventional language in the description of a theophany,[31] conventional and treacherous, just as it is in didactic, in *Buc.* 6.65 (of Cornelius Gallus), and in Ennius. 'Bacchus' is a way or writing about that difficult topic, inspiration;[32] not liquid indeed, but poetic. The Augustan poets knew what it was (and real poets do still); they had felt it, but knew that they were compelled to use traditional figured language in explaining what it was to readers who, in most cases, never had.[33] *Credite posteri* envisages a reaction of perplexity, or scepticism, but Horace has no intention of reneging on his sense of writing as an inspired *uates*. If there seems rather a lot of physiology in Horace, I venture to cite in support A. E. Housman, writing as poet:[34] "experience has taught me, when I am shaving of a morning, to keep watch over my thoughts, because, if a line of poetry strays into my memory, my skin bristles so that the razor ceases to act." Bristles, like feathers, help confirm inspiration as physiologically real: seven centuries after Hesiod's encounter on Mount Helicon, Horace has to express his unaltered sense of the divine (in some sense) origin of poetry as best he can. It has not changed, but progress has brought scepticism (as it has, conventional adherence) and scepticism has driven Horace to pilfer his predecessors for elaborate tropes in which to express an experience both profoundly simple and incomprehensibly complex.

Horace's status as an inspired poet, under the gods' special favor (*non sine dis animosus infans*, 3.4.20) and protection depends not only on the direct 'evidence' so far cited but also upon a series of famous, and likewise exemplary, anecdotes that he quotes in illustration of his sense of a privileged status. They are 'evidence' after the manner of the

---

[31] Nisbet and Hubbard (above, n. 26) 315.

[32] Nisbet and Hubbard (above, n. 26) 316.

[33] Nisbet and Hubbard on 2.19.6 show much learning but perhaps uncharacteristic lack of understanding for that impalpable, external, even 'divine' element in the composition of poetry.

[34] A.E. Housman, *Selected Prose* ed. J. Carter (Cambridge 1961) 193.

miracles in the lives of Homer,[35] Pindar and Virgil (cf. p. 44 above): the paradeigmatic role of the miracles in (e.g.) Sulpicius Severus' life of St. Martin of Tours is hardly very different.

Di me tuentur, dis pietas mea | et Musa cordi est (1.17.3-4): the gods love Horace because Horace is a poet who takes that calling (or vocation) seriously and it is above all Mercury who cares for him[36] and saved him on the stricken field of Philippi (2.7.13f.), *denso pauentem sustulit aere* indeed, after the manner of an Homeric deity, though the poet had "thrown away his shield", *relicta non bene parmula* (2.7.10), in the best tradition, after Archilochus, Alcaeus, and perhaps Anacreon too.[37] No need whatever to question that Horace fought with Brutus, had been present at Philippi (one battle or both? We have no idea),[38] realised he was on the losing side, and fled. Mercury's intervention and the jettisoned shield are narrative elements perfectly calculated to throw an elegant smoke screen of inherited allusive density over "what actually happened". At 3.4.25ff. Horace describes himself as *uestris amicum fontibus et choris*: the proof of this devotion is seen in three exemplary episodes, cited in regular manner: (i) Philippi (v. 26), (ii) the fall of a tree (see below), and (iii) the terrible storm off Cape Palinurus in 36, during the campaign against Sex. Pompeius.[39]

It may be that 3.27.19ff. *ego quid sit ater | Hadriae noui sinus et quid albus | peccet Iapyx* refers to further perils during the poet's travels. All, though, overcome. The fall of the tree Horace says he commemorated on the 1st of March (3.8.1) and he associates the event with Maecenas' recovery from illness (2.17.27ff.). A fourth reference

---

[35] Lefkowitz (above, n. 1) 12-24, *passim*.

[36] R.G.M Nisbet and M. Hubbard, *A Commentary on Horace: Odes, Book 1* (Oxford 1970) 127-8.

[37] Nisbet and Hubbard (above, n. 36) 113

[38] Nisbet and Hubbard (aboove, n. 36) 113, M. Malavolta, "Le battaglie di Philippi," in *Enciclopedia Oraziana* (above, n. 7) 26-32, G. Davis, *Polyhymnia* (Berkeley 1991) 91-3, H. Krasser, *Horazische Denkfiguren* (Göttingen 1995), 71-7, D.A. Russell, "Self-definition in Plutarch and Horace," in *Philanthropia kai eusebeia. Festschr. A. Dihle* (Göttingen 1994) 433.

[39] Russell, *loc. cit.*, R.G.M. Nisbet in *Enciclopedia Oraziana* (n. 7), 3, T. Rice Holmes, *The Architect of the Roman Empire* I (Oxford 1928) 114.

(3.8.8) encourages (e.g.) Nisbet and Hubbard to assert:[40] "the historicity of the experience should not be doubted". Not an unusual event, and one that the poet might fairly cite as an *exemplum* of divine benevolence, elaborated in the traditional manner of literary accounts of death (or escapes therefrom) from falling objects. I only urge the presence of various literary 'filters', (at whose nature the text here just hints in passing) between 'event' (or event) and account. Such filters are just as densely present in the case of the wolf Horace encounters in one of his lightest and funniest odes (1.22), *silua ... in Sabina*, a veritable *portentum*: nothing like it even in Mauretania, home of lions, discussed in King Juba's recent *Histories*.[41] No zoological implausibilities in the text, though;[42] just a lot of fun, which does not make the episode in itself any less (or more) plausible. Poetry (*dum meam canto Lalagen*, 1.22.10) again is Horace's salvation and the wolf turns tail.[43] Whether pedants too come under the protection of some minor demon is not at all clear: the writer's own refuge is some twenty miles beyond Horace's into the mountains. A remote spot, but renewed study of *Carm.* 1.22 hardly suggest a need for *uallum* and firearms! Citation as a poetic *exemplum* and historicity are not incompatible; rather, they are distinct issues. The recent sight of a fiery salamander near the gate of our house is, as it happens, fact; what I might make it in a poem does not diminish the historicity of the encounter (a human fled; the reptile did not!), but nothing guarantees that the account in my hypothetical poem would be historically precise. So too Horace's wolf; one wonders whether, in the guise of poet of love, he might be thought to exercise Orpheus' power over the animal kingdom.

One last miracle remains: that of Diespiter *qui per purum tonantis* | *egit equos* (1.34.7f.) and drove Horace "to abandon Epicureanism". This is at a quite different level, for thunder in a clear sky is a familiar absurdity in solemn lists of portents, a notorious impossibility already to

---

[40] (above, n. 36) 201.

[41] *FGH* 275 F 55, Nisbet and Hubbard (above, n. 26) 270.

[42] Nisbet and Hubbard (above, n.26) 268.

[43] Cf. Fraenkel (above, n. 5) 187, G. Davis, "The Literary-Generic Dimension of Hor. *C.* 1.22," *QUCC* 27 (1987) 68.

Epicurus,[44] essential to the intellectual structure of the poem, as has long been recognised. But the whole point is, clearly, that Horace did not hear, and knew he could not have heard, thunder from a cloudless sky.

At that point, we can hardly avoid a skirmish with the problems of Horace's private life. Thus Suetonius: *ad res uenereas intemperantior traditur*: that *traditur*, as we have learned,[45] proves nothing and might even mean no more than 'as his poems make clear'. That, after all, is how the biographical criticism of ancient literature worked: according to Quintilian (10.1.100) there were so many homosexual themes in Afranius—because of course the author himself was homosexual. Suetonius' text continues with a glittering jewel: Horace was said (*dicitur* this time) to consummate his amours in a room with a mirror on the ceiling. "This one extraordinary fact that has perplexed and disturbed classicists for centuries";[46] neither unknown nor impossible, given that the younger Seneca offers another instance (*NQ* 1.16.2). But Suetonius had little choice: a paragraph on the poet's sexual habits was almost obligatory;[47] *dicitur* is worth exactly as much as *traditur* (cf. n. 36), a detail possibly transmitted rather than invented, but not for that reason of any evidential value whatever. While it is still possible to read ancient literary biography with a straight face and a credulous pen, the serious reader will apply to such texts the critical standards which, in biblical and hagiographic scholarship, have been current for close on four centuries.[48]

There are four homosexual poems in Horace: their very existence does not necessarily make the poet bisexual in tastes or habits—not that there was anything unusual about that at Rome.[49] If one looks at the two Ligurinus poems, the second (4.10), on the time when Ligurinus'

[44] Nisbet and Hubbard (above, n. 36) 377, Fraenkel (above, n. 5) 255, Lucr. 6.99, 247ff., 400ff.

[45] Cf. my remarks in Horsfall, ed. (above, n. 9) 3.

[46] W.S. Anderson, in G. K. Galinsky, ed., *Perspectives of Roman Poetry* (Austin 1974) 34.

[47] Cf. J.A. Fairweather (above, n. 1) 263-4.

[48] For an extreme case of the straight face, cf. Della Corte, cit., and, for a very different view of Virgil's life, ch. 1 of Horsfall, ed. (above, n. 9).

[49] On Roman bisexuality, cf. E. Cantarella, *Secondo natura* (Roma 1988) 157-98.

beauty and the *superbia* it inspired will pass away, is narrowly Hellenistic in inspiration,[50] while the first (4.1) lists the symptoms of love very much 'by the book'[51] and sets them in the traditional palaestra-landscape of the Campus Martius, as in 1.8, 3.7, or 3.12. There are indeed homosexual elements elsewhere in the *Odes.*[52] Not many, though; that reflects rather Roman life than the tradition of Hellenistic epigram,[53] though distinction between the two is rarely possible, whether in general or in the particular case of Horace.

It does not help much if we change gender. The girls mentioned in Horace's poems are indeed very numerous[54] and that already distinguishes him from Catullus and from the elegists. We may add a further distinction which counts, I hope, not as a misguided revival of the biographical fallacy but as the sensible application of social history: Horace's own modest beginnings, experience in Brutus' army and service as a *scriba quaestorius*. He never, that is, shared as an equal in the pleasures of his more moneyed contemporaries at Rome, as Catullus and the elegists far more clearly had done,[55] though we cannot exclude something of the kind during his studies in Athens or while on Brutus' staff.[56] In the *amicorum numerus* of Maecenas, it is my impression that certain unwritten laws are likely to have obtained, under which a rich,

---

[50] See G. Pasquali, *Orazio lirico* (ed. 2 Firenze 1966) 460-3.

[51] Cf. Nisbet and Hubbard on *Carm.* 1.13.

[52] S. Lilja, *Homosexuality in Republican and Augustan Rome*, Comm. Hum. Litt. 74 (Helsinki 1983) 70-4.

[53] F. Buffière, *Eros Adolescent* (Paris 1980) 293-5, K.J. Dover, *Greek Homosexuality* (New York 1978) 171.

[54] S. Treggiari, *Roman marriage* (Oxford 1991) 302. P. Veyne, *L'élégie érotique romaine* (Paris 1983) annoys and informs in pretty equal measure, while T. Zielinski, *Horace et la société romaine* (Paris 1938) 157-84 is still of more than antiquarian value.

[55] T. McGinn, *Prostitution and Julio-Claudian Legislation* (Diss. Univ. of Michigan 1986) Ch. 1; Veyne (above, n. 54) 78-96, H. Herter, "Die Soziologie der antiken Prostitution," *JhbAC* 3 (1960) 70-111.

[56] T.P. Wiseman, *Catullus and his World* (Oxford 1985), 10-4, *et passim*, J. Griffin, "Augustan Poetry and the Life of Luxury," *JRS* 66 (1976) 87-105 (= *Latin poets and Roman life* [London 1985] 1-31).

varied, and even notorious private life might not have been encouraged.[57]

Horace's bed, however, continues to fascinate even the soberest Latinists: "And Pyrrha herself is the wayward beauty of fiction, quite unlike the compliant *scorta* of Horace's own temporary affairs".[58] Possibly the distinguished editors of *Carm.* 1 and 2 had the fiasco in the inn at Trivicum in mind (*Serm.* 1.5.82-5). I cannot say, but I do wonder how Nisbet and Hubbard could possibly have acquired the certainty which their epigram implies; such observations tell us more about the scholars themselves than about the object of their enquiries! Pyrrha, to be sure, is as Hellenistic and conventional as they come, but it is those compliant *scorta* which perplex me deeply. We know far more than we did about the range of female sexual partners available at Rome, from Volumnia Cytheris herself, to *prostibula* in their *angiportus*—note the Lydia of 1.25. Of this range, Horace himself is to some degree explicitly aware: note Lyde, the *deuium scortum* who arrives with her music (1.25.10), the Myrtale *libertina* of 1.33.14f., the *ancilla* of 2.4.1, even the inebriated Damalis of 1.36.13 and perhaps above all the contemptuous (and in some ways markedly Ovidian) description (3.6.25-32) of undiscriminating adultery, with the husband's consent, born over the wine and consummated (what crushing disdain!) with an *institor* or *nauis Hispanae magister.* It is not much to go by: we might even conclude that Horace had no special love of contemporary detail, telling but also tangible, in his amatory poetry. Though we may seek (vainly!) to distinguish between literature and life (cf. n. 41), between Fraenkel's Horace who never tells a lie and the Horace of Nisbet and Hubbard, who transcribes with flawless beauty Hellenistic conventions (I exaggerate, of course, a little), between the sincere and the banal, between the Greek and the Roman, that bed adamantly preserves its secrets.

[57] What Maecenas might allow himself and what could be tolerated (or, more important, seen to be tolerated) in his protégés were two quite different matters! Cf. I.M. Le M. du Quesnay, "Horace and Maecenas" in A.J. Woodman and D. West, edd., *Poetry and Politics in the Age of Augustus* (Cambridge 1984) 19-58 and my note on *Ep.* 1.7.37 in *La villa sabina di Orazio* (Venosa 1993). Of this almost unobtainable little book S. Borszak has published an extraordinary account in *Gymn.* 101 (1994) 463-64, rich in confusion and inaccuracy; for a briefer, but entirely correct statement of how I read the poem, I am obliged to L. Deschamps, *REL* 71 (1993) 276-8.

[58] Nisbet and Hubbard (above, n. 26) 73

There is at least one external reason for this: the shortage of good independent anecdotal evidence for the Augustan period (I except some stories in Seneca Rhetor, in Suetonius, in Macrobius).[59] Where we have Cicero's letters and a multiplicity of biographies from Plutarch, things are very different. Reliability is relatively unimportant; even gossip gives us a "feel" for the character, the "tone" of sexual mores which we simply do not have—outside the imperial family—for the Augustan year.

The girls have Greek names. But were they Greek? From Greece or from Greek literature? We have no idea. Their very multiplicity is a literary inheritance, as a glance at Meleager's addressees will show. Horace as lover is altogether irrelevant to our understanding of Horace as poet of love. Mirrors have as little to do with "what actually happened" as do "compliant *scorta*". That is why nutrient doves, stray wolves and falling trees are important: they establish Horace's character as inspired and privileged poet; Lalage, Chloe, Lydia and all the rest tell us not about love, but about love–poetry.

Let us turn very briefly to Horace's laments, of wide distribution and prolonged duration, upon his increasing years and declining health.[60] I do not mean to suggest that he was in fact blessed with eternal youth and unbounded irony, but the evidence needs to be seen in some sort of context if it is to be evaluated sensibly: in *Carm.* 3.26, which cannot be later than 23 (when the poet was 41), he declares, as a general principle:[61] *uixi puellis nuper idoneus | et militaui non sine gloria.* Not only, that is, Pyrrha, thankfully eliminated in 1.5, but *puellae* in general; as a literary lament, extraordinarily common.[62] The theme occurs with particular frequency in Book 4 (from the first poem on), a collection issued ten years later, though perhaps accumulated over the entire intervening period. Cinara seems to be the name (I forbear to say "girl") that focuses the issue: "Horatii amica mortua", according to the index of the

---

[59] Cf. my remarks in Horsfall, ed. (above, n. 9) 3.

[60] Cf. my note on *Ep.* 1.7.4f., J. F. D'Alton, *Horace and his Age* (London 1917) 140-2.

[61] Shackleton Bailey: *uixi, puellis.*

[62] Cf. *C.* 2.4.21-4 (necessarily close to 3.26 in date). "Too old": old hat; cf. the parallels collected by Nisbet an Hubbard (above, n. 26) 72, F. Cairns, *Generic Composition* (Edinburgh 1972) 80.

latest Teubner edition (Shackleton Bailey). Now Cinara appears not only in Book 4 (1.4; 13, 21f.), but also in *Ep.* 1 (7.27, 14.33), published six years before, Cinara appears consistently (and exclusively) as a love long gone and never "in play". Her textual and structural function is therefore solidly symbolic; the name, in Greek, suggests "artichoke" or, perhaps more significantly, "plaintive". She is dead (4.13.23): had she in truth been Horace's great love, we might wonder at the marked lack of specificity in the texts. In other words, in the case of a real girl, really bedded, would not a great poet have found a way of—well, bringing her back to life?

Horace's declining health is a recurrent motif in *Ep.* 1,[63] after whose publication, he lives for another eleven years. Should that make one think that Horace's carefully created *persona* in *Ep.* 1.7 is that of the valetudinarian old buffer? Of course I do not pretend to know, but the concentration of the 'evidence' does suggest to me a function more poetical than autobiographical. That eyes[53] and stomach mutinied during the *iter Brundisinum* is surely irrelevant to this argument.

I did not set out to address myself to the problems of Horace's self-representation outside the *Odes*, but the issues raised are so strikingly different that that very difference merits a paragraph.[64]

The hexameter poems are not necessarily more secure and more reliable: that becomes clear to anyone who looks carefully at the scene-setting for *Ep.* 1.7 and watches the 'evidence' for a tiff between Horace and Maeceanas crumble before their eyes.[65] We do not in any way impugn the historicity of the *iter Brundisinum* when we recall that we must read it, where we can, through the filter of Lucilius' *iter Siculum* (Books

---

[63] Cf. n. 59, *Ep.* 1.12.5ff, 14.35, 15 *passim*, 16.14.

[53] Cf. my "Rome Without Spectacles," *G&R* 42 (1995) 49-56.

[64] *Epodes* are peculiarly difficult: the first poem *may* be about Actium (cf. Nisbet in *Enciclopedia Oraziana* (n. 7), 4): I remain one of the few who is not at all sure that the ninth shows Hor. to have been present at Actium: *CR* 35 (1985) 52. For Dr. L. Watson's commentary on *Epd.* we have been waiting almost as long as for mine on *Aen.* 7!

[65] Cf. now my commentary, pp. 37-8, after (e.g. G. W. Williams, *Tradition and Originality in Roman Poetry* (Oxford 1968) 19-22, R. Kilpatrick, *The Poetry of Friendship* (Edmonton 1986) 8.

3-4).[66] More destabilizing is the realization that Horace's portrait of his father in *Serm.* 1.6, to Fraenkel sacred,[67] and undeniably brilliant and memorable to a high degree, is itself the literary offspring of Micio's self-portait in Terence's *Adelphi.*[68] And the education Horace received from his father has itself a rhetorical and structural function, as we realise in retrospect, when we look at the way in which the ethical foundations of the poet's friendship with Maecenas are presented in the text:[69] it is the poet's *uerecundia* that his patron, significantly called *pater*, admired (*Ep.* 1.7.37): it was *pudicitia* that his freedman father had taught him (*Serm.* 1.6.57)[70] and *pudor* that stayed his tongue during the first interview with Maecenas (*Serm.* 1.6.57). In *Serm.* 1.6, Maecenas *turpi secernis honestum* (v. 64); in *Ep.* 1.7 he likewise *dignis ait esse paratus* (v. 22). The poems, and the relationships, are intimately linked. It is very easy to be charmed by Horace's portrait of life round about Maecenas; we are after all meant to be (cf. n. 59). Such episodes are not pages *ex commentariis alterius Sam. Pepys.* Such vignettes from the poet's life are best compared not to photography, but to painting, impressionist, even!

If we remove the whole elaborate scaffolding of pseudo-biography built up around the person (and text) of Horace, the extraordinary thing is, that nothing happens. When I undertook the same operation in respect of the "Life of Virgil" (cf. n. 9), the sky did not fall in; that emboldened me to try it again! Horace's poems actually increase, rather than decline, in interest. To the reader of the *Odes* (and the case of *Serm.* and *Ep.* is if anything rather more complicated), the poet is his text. Something of the kind has been said before;[71] few readers of Horace were convinced, and that emboldened me to try Wendell Clausen's

---

[66] Cf. (e.g.) Fraenkel (above, n. 5) 105-7.

[67] "A reader who cannot afford the time to read it at leisure ... had better leave Horace alone", Fraenkel (above, n. 5) 5.

[68] Ter. *Ad.* 48-58.

[69] Cf. above all Du Quesnay, cit. (1984).

[70] Cf. Williams (above, n. 5) 296-313.

[71] H. Cherniss, "The Biographical Fallacy in Literary Criticism," *UCPCP* 12 (1943), 279-92, *passim* on the battle between Wilamowitz and Gundolf: this splendid paper is repr. in *Critical Essays on Roman Literature: Elegy and Lyric* ed. J. P. Sullivan (London 1962) 15-30.

good nature with a brisk restatement of an approach to Horace not, I think false just because it is old and (inexplicably!) rather out of favor.[72]

[72] Gregson Davis was amused to find that I had unknowingly reached—more or less—the conclusions of his *Polyhymnia* (above, n. 38) 78-114. I blush at the incompleteness of my reading in modern Horatian studies, but am delighted both to discover a distinguished ally and to find that (relatively) traditional and (comprehensibly) modern methods can quite independently reach the same conclusion. And after I had finished typing the very last draft of this paper, there came into my hands a copy of Krasser (above, n. 39), with pp. 71-7 of which I found myself in very substantial agreement. It is tiresome but consoling to discover that what started out as audacious polemic has apparently become, in little over a year, a sort of orthodoxy! The *nouerca* of this paper was "The Swiss office for in-service training of upper secondary teachers" (their terminology), whom Prof. F. Graf invited me to address in Sept. 1994. They were a delightful audience; "'Je' et Jeux dans les Odes d'Horace" enraged them (it was meant to). That encouraged me to hope that Wendell Clausen, who has viewed my efforts with gentle but increasingly generous humor for just on twenty years, might be tickled to see a confirmed and crusty pedant turn and offer not an ultra-literalist but a—dare I say—metapoetic reading of a series of familiar Horatian texts.

THE METAMORPHOSIS OF HERO

E.J. KENNEY

I

The hazards of any attempt at reconstructing a lost Hellenistic original from coincidences between Ovid and a late Greek poet have been well demonstrated by Peter Knox.[1] On Hero and Leander Neil Hopkinson is austerely discouraging: "the standpoint from which [the lovers] write is so far from that of the putative source as to make reconstruction of it impossible".[2] Nothing venture, nothing win; though such a reconstruction even in outline may be chimerical, I believe that it is possible at least to identify with a fair degree of probability some of the elements in Musaeus' version that were *not* in the putative original. That may in turn enable us to appreciate more clearly the character and quality of the creative transformation wrought by Ovid on the material of what was to become one of the great romantic love-stories of all time. Comparison of Ovid with Musaeus, which is inevitably the starting-point of the investigation, serves only to enhance that appreciation. Though Musaeus' connected narrative has probably exerted more influence over the later literary tradition than Ovid's epistolary vignettes,[3] there can be no argument about who is the greater artist. Musaeus' characterization of Hero, critically considered, emerges as a tissue of inconsistencies verging at times on outright absurdity. Ovid's portrait is of a piece, among the most poignant and moving in the long gallery of his suffering heroines.

---

[1] P.E. Knox, "Phaethon in Ovid and Nonnus," *CQ* 38 (1988) 536-7.

[2] N. Hopkinson, ed., *Greek Poetry of the Imperial Period* (Cambridge 1994) 137.

[3] K. Kost, ed., *Musaios. Hero und Leander*, Abh. zur Kunst-, Musik-, und Literaturwissenschaft 88 (Bonn 1971) 69-85, esp. 74-5, Hopkinson (above, n. 2) 139-40. L. Lerner, "Ovid and the Elizabethans," in C. Martindale, ed., *Ovid Renewed. Ovidian Influences on Literature and Art from the Middle Ages to the Twentieth Century* (Cambridge 1988) 128-31, discusses Marlowe's *Hero and Leander* without mentioning Musaeus.

II

Before turning to Ovid and Musaeus, however, it is appropriate to notice the earliest certain appearance of the story in European literature, in Virgil's *Georgics* (3. 258-63):

> quid iuuenis, magnum cui uersat in ossibus ignem
> durus amor? nempe abruptis turbata procellis
> nocte natat caeca serus freta, quem super ingens
> porta tonat caeli, et scopulis illisa reclamant
> aequora; nec miseri possunt reuocare parentes,
> nec moritura super crudeli funere uirgo.

From this a number of points emerge:

1. Virgil does not name the protagonists or fill in all the details.

(a) There is no explicit mention of the lamp which plays a key role in other treatments of the story, except in so far as its absence on the night of Leander's death is implied by *nocte ... caeca.*

(b) There is no reference to a tower.

(c) The reference to Leander's parents is both inexplicit and ambiguous. Does *reuocare* mean (i) "call him back", sc. from the dead or (ii) "restrain him from his purpose" (*OLD* s.v. 8b)? If the latter, then *miseri ... parentes* must be taken with Mynors as "not his parents but the thought of them".[4] What part, if any, they may have played in the motivation of the story is left unclear.

(d) The reference to the deaths of the lovers is brief and inexplicit. The words *moritura super crudeli funere* can mean both "to die as well by a cruel death" and "to die on his cruelly-slain corpse";[5] the ambiguity is neatly caught by Jackson's "who shall die, as it is bitterest to die, on her lover's corpse".[6] That nuance could be savoured only by a reader who already knew the denouement of the story, Hero's leap from the

---

[4] Mynors *ad loc.*; but his gloss "as the next line shows" perhaps invites reservations. With reference to Hero sense (i) seems more appropriate than (ii), given that Leander was drowned on his way to her, not on the way back.

[5] Thomas and Mynors *ad loc.*

[6] J. Jackson, trans., *Virgil* (Oxford 1908) 75.

tower to where the dead Leander lay below.[7] This feature, which must have been original, is spelled out in Servius' note on the passage: *Leander et Hero, Abydenus et Sestias, fuerunt inuicem se amantes. sed Leander natatu ad Hero ire consueuerat per fretum Hellesponticum, quod Seston et Abydon ciuitates interfluit. cum igitur iuuenis oppressi tempestate cadauer ad puellam delatum fuisset, illa se praecipitauit e turri.*[8]

2. The emphasis in Virgil's lines is on Leander himself and on the storm in which he perished; Hero is disposed of in two words. Both the context of the passage, the great excursus on the power of sexual desire over the animal kingdom, and the short but graphic description of the power of the elements against which Leander was contending, imply reckless folly on his part rather than heroic courage. Whether or not that was the original emphasis of the story, it agrees with Ovid's characterization.

It is likely *a priori* that Virgil's allusive summary[9] did not misrepresent, even if it did not fully represent, the basic data of the story; that is to say, whatever he left out, he will not have added anything. The omission of any explicit reference to the lamp is striking; but though this was clearly, as the references to it by Ovid, Antipater of Thessalonica,[10] and Musaeus show, an integral and original element, it was irrelevant to Virgil's purpose and called for no more than the throwaway hint of *nocte ... caeca.* That, however, would convey nothing to readers who did not know the story already; it would seem

[7] Musaeus 341-2; T.D. Papanghelis, *Propertius: A Hellenistic Poet on Love and Death* (Cambridge 1987) 110.

[8] Servius or his source could have read about the tower in Ovid; his explicit reference to Hero's death-leap seems to be paralleled in Latin authors only at Ausonius (*Cupido cruciatus*) 19.22-3 Green *fert fumida testae | lumina Sestiaca praeceps de turre puella.* The agreement of Servius with Ausonius suggests that this feature of the story was more familiar than might seem to be implied by the silence of the other poets, whose interest focusses on Leander; cf. Fronto on *fabula histrionibus celebrata* (*Epp.* 14.4, p. 47 van den Hout), though his reference is to Hero's vigils, not to her death.

[9] "Die ganze Sage in sechs Versen" (E. Norden, *Orpheus und Eurydike. Ein nachträgliches Gedenkblatt für Vergil*, Sitzb. d. Preuss. Akad. d. Wiss. Philos.-hist. Kl. [Berlin 1934] 664 = *Kleine Schriften zum klassischen Altertum* [Berlin 1966] 511).

[10] *A.P.* 7.666.4 (= *GP* 132).

that Virgil took it for granted that they did. *Leandri nomen occultauit quia cognita erat fabula*, says Servius auctus; but that need be no more than the usual scholiast's inference from the text itself.[11] How familiar, one wonders, can it have been? Propertius' exploitation of the story, though in the light of hindsight unmistakable,[12] is so indirect that an uninstructed contemporary could hardly have been expected to pick up the allusion. Another evidently original feature, the tower, was like the lamp irrelevant to Virgil for the purpose in hand; but Horace's casual allusion to the Hellespont as *freta uicinas inter currentia turres*[13] seems to assume that his readers knew about it and also the matching tower associated with Leander, otherwise mentioned only by Strabo.[14] Whether all this adds up to a case for believing that the story was indeed well known to Virgil's contemporaries[15] perhaps remains a moot point. How many *docti lectores* actually measured up to the standards set by the *doctus poeta*?

III

It was characteristic of Roman Alexandrianism to select a particular incident or motif from a Greek model for intensive rather than extensive exploitation: the washing ashore of Leander's corpse or the lovesick Acontius' sylvan soliloquy.[16] Ovid tended to go the whole hog.[17] In the single *Heroides* he had created a whole new genre[18] from the isolated experiment of Propertius' "Arethusa". In the double letters he further

---

[11] "Die Bemerkung ... ist gar zu trivial, um als 'richtig' weitergegeben zu werden" (Norden [above, n. 9] 664 = 511, n. 88).

[12] Papanghelis (above, n. 7) 103-10.

[13] *Epp.* 1.3.4; see Mayer (1994) *ad loc.*

[14] Strabo 13.591; but cf. Lucan 9.954-5 *amore notatum | aequor et Heroas lacrimoso litore turres.*

[15] As maintained by e.g. Kost (above, n. 3) 18, Papanghelis (above, n. 7) 103; cf. id. 109 "a famous Hellenistic treatment".

[16] E.J. Kenney, "Virgil and the Elegiac Sensibility," *ICS* 8 (1985) 45-7.

[17] On his transformation of the Acontius and Cydippe story see E. J. Kenney, ed., *Ovid. Heroides XVI-XXI* (Cambridge 1996) 17-18.

[18] *A.A.* 3.345 *ignotum hoc aliis ille nouauit opus*—an adroitly ambiguous claim.

expanded the genre, conjuring a unified triptych of character-studies from extremely diverse materials.[19] The treatment of the stories is necessarily oblique and allusive, being governed (though Ovid of course permits himself much latitude) by the epistolary conventions and, especially in the case of Hero and Leander, the fact that his protagonists are still alive and ignorant of what fate has in store for them. In the course of their correspondence certain features not mentioned, or not expressly mentioned, by Virgil, emerge:

1. The protagonists and the places are identified.

2. Hero lives in a tower with only a nurse for company.

3. The lamp is referred to, though its role is in certain respects somewhat played down.[20]

4. The affair is clandestine because the lovers are of unequal social status. Leander's parents must not know of it (Hero's are not mentioned) and public opinion would evidently be against it; Hero wonders if she is good enough for him.

Of these features the first three are clearly original and fundamental to the story. Parental disapproval as the motivation for a story of thwarted love is familiar in all ages; an obvious parallel is provided by Pyramus and Thisbe.[21] As will be seen, Musaeus' treatment of this mo-

---

[19] Kenney (above, n. 17) 18-20.

[20] At least, compared with its symbolic role in Musaeus' treatment (Kenney [above, n. 17] 9, n. 31). In that connexion I ought in my commentary on *Her.* 19.35 to have pointed out that *in summo ... tecto* does not mean 'auf dem Dach' (Kost [above, n. 3] 172, 464) but "in the top (room) of my abode." Kost's interpretation is refuted by Leander's words (18.105-6) *cetera nox et nos et turris conscia nouit | quodque mihi lumen per uada monstrat iter* (Kenney [above, n. 17] *ad loc.*). If the signal lamp and the lamp in the boudoir were not one and the same the irony of the ambiguous omen given by its spluttering would be lost—another point which I ought perhaps to have spelled out. (Reading *in summa ... turre* [=18.31], as preferred by e.g. Dörrie, makes no difference to the sense.)

[21] H. Bopp, *Inscia capta puella. Akontios und Kydippe bei Kallimachos und bei Ovid* (Diss. Münster 1966) 8-12, P.E. Knox, *Ovid's* Metamorphoses *and the Traditions of Augustan Poetry* (Cambridge 1986) 35-6; cf. id., "Pyramus and Thisbe in Cyprus," *HSCP* 92 (1989) 315-28. What might be termed "bourgeois" motivation was characteristic of the Hellenistic poets, who "loved to picture plain, ordinary people embroiled in extraordinary circumstances" (H. Fränkel, *Ovid. A Poet Between Two Worlds* [Berkeley 1945] 52). For social disparity as a factor in the plots of New

tif is somewhat different, but Virgil's reference to Leander's *miseri parentes* suggests that it is Musaeus and not Ovid who has here departed from the original.

IV

In the long interval separating Musaeus from Ovid there are scattered references to the story in various Latin writers which add nothing of moment.[22] Its subsequent popularity and diffusion since the Renaissance are largely due to Musaeus, whose treatment may fairly be described as enterprising. His narrative begins in high epic style with an invocation of the Muse (1-15), in which the emphasis falls heavily on the lamp and its multiple role as confidant, beacon, and symbol of Leander's life. The salient points of the plot as Musaeus develops it are these:

1. The lovers are equally matched in beauty, the pride of their respective cities, Sestos and Abydos.

2. Hero is of divine descent, a virgin priestess of Aphrodite.

3. She lives in a tower by the sea, described as πύργον ἀπὸ προγόνων (32), a characteristically vague expression[23] that has occasioned some debate. Does it mean a tower that had been in her family for generations or (so Kost) a sort of tied cottage traditionally assigned to the incumbent priestess?

4. She lives there with a single attendant (188), στυγεραῖς βουλῆισι τοκήων (190), which at first sight might seem to favour the first of the above alternatives; but the hateful decision might have been that to apprentice her to the service of the goddess, which would automatically entail banishment to the tower.

Comedy, see M. Willcock, "Plautus and the *Epidicus*," *Papers of the Leeds International Latin Seminar* 8 (1995) 20-1

[22] A. Palmer, ed., *P. Ovidi Nasonis Heroides with the Greek Translation of Planudes* (Oxford 1898) 454-5, E. Rohde, *Der griechische Roman und seine Vorläufer*, 3rd ed. (Leipzig 1914) 144, n. 1; cf. Kost (above, n. 3) 571, n. 32, T. Gelzer, *Musaeus, Hero and Leander* in Callimachus, *Fragments*, ed. C. Trypanis (Cambridge, Mass. 1975) 306-7. On Servius see above, n. 8.

[23] On Musaeus' tendency to subordinate sense to sound and decorative effect see Gelzer (above, n. 22) 314-15.

5. The impediment to marriage with Leander is constituted by her parents, not his. As soon as he accosts her, she warns him to beware of their anger,[24] adding "it is unseemly to touch the priestess of Cypris; it is impossible to approach the bed of a virgin" (125-7). Marriage is out of the question, οὐ γὰρ ἐμοῖς τοκεέσσιν ἐπευάδεν (180), another ambiguous phrase which makes sense in the context only as connoting an objection to marriage as such, not to Leander in particular, of whom they know nothing: i.e. they ruled out marriage as a way of life for her when they made her a priestess.

6. Of the 343 verses of the poem as we have it rather more than half (42-220) are taken up with an account of the setting and circumstances of the first meeting of the lovers. Certain features of this section call for remark:

(a) The meeting takes place at a great festival of Aphrodite, to which worshippers, especially women, had flocked from many different places (42-50).

(b) During the festival Hero is described as "pacing" (55) or "wandering" (71) about the temple, followed and admired by the throngs of young men who had come to Sestos in pursuit of girls (51-4, 69-85)—Hero, who has previously been described as modest, chaste, reclusive, devoted to her duties, and fearful of Eros (33-40).

(c) A single passionate glance from Leander suffices to sweep away these defences. Hero's response is immediate (101-8), and she allows herself to be cornered, wooed, and won with hardly a show of resistance.[25]

---

[24] On the textual problem see Kost (above, n. 3), Hopkinson (above, n. 2) *ad loc.*

[25] A benevolent critic might suggest that the *volte-face* by which her defences almost instantly crumble, and what had earlier been depicted by the poet as selfless dedication is now seen by Hero herself as a kind of durance vile, represents a fine stroke of psychology: her reaction to Leander's wooing forces her to admit that all this time she has been living a lie—*eripitur persona, manet res*. For a defence of Musaeus' powers of construction see O. Schönberger, "Zum Aufbau von Musaios' 'Hero und Leander'," *RhM* 121 (1978) 255-9; but care in small matters can, and in this case does, easily consist with negligence in large ones. An amusing instance of Musaeus' occasional concern for detail is his explanation of Leander's decision to return to Abydos by swimming so as to make sure of his bearings for future trips (227-9); see Kost (above, n. 3) 429-31. The reductively-inclined might mischievously

7. In the subsequent narrative of Leander's first nocturnal visit the union of the lovers is described as a wedding, though one without any of the proper concomitants (274-84);[26] and at the end of their first interview he had already described himself as her husband (220 πόσις).

8. The last part of the poem describes the storm in which Leander dies, Hero's discovery of his body, and her own death (293-343). For obvious reasons Ovid could refer to this part of the story only by way of dramatic irony.[27] As we have noted, the storm figures prominently in Virgil's allusion.

V

That Musaeus and Ovid were independently indebted to a common source is generally agreed, and the matter need not detain us here.[28] What it would be instructive to establish is which elements of Musaeus' treatment are likely to have been original and which represent accretions or embellishments of his own. *Prima facie*, coincidences with Virgil and Ovid identify as original features in the above list (1) the names and places, (3) the tower, (4) an obstructive role played by parents, (5, 7) an emphasis on marriage and incompatibility in that respect, and (8) the storm and the deaths of the lovers. Turning to the remaining heads one

enquire why he had to go back home at all at this point when he could have paid his first visit that evening—as it seems that he did, though when it comes to the point Musaeus is studiously vague about timing. As we learn later (251-2), he had the problem of what to do with his clothes, so to say, buttoned up. Ovid of course could bypass the mechanics of the plot to concentrate on the emotions of the lovers.

[26] All this is interpreted allegorically by Gelzer (above, n. 22) 321-2. I think it far more probable that Musaeus is simply showing off. A closer parallel for this type of negative enumeration than any cited by Kost (above, n. 3) 483-4 is Lucan's description of the "non-wedding" of Cato and Marcia (*B.C.* 2.354-71 and Fantham *ad loc.*). One curious feature is the absence of any mention of the "single attendant": the words Σιγὴ παστὸν ἔπηξεν, ἐνυμφοκόμησε δ' Ὀμίχλη (280) seem to suggest that there was no human attendance at the "wedding". In view of the usual role of the *nutrix* as confidante and accessory in such stories of clandestine love and her presence in Ovid's epistles it seems reasonable to guess (*pace* Kost [above, n. 3] 384) that in the original she had more of a part to play. The point is worth mentioning as further evidence of Musaeus' indifference to consistency and plausibility in details.

[27] Kenney (above, n. 17) 13, 148-9, 151.

[28] Kost (above, n. 3) 21-23, Gelzer (above, n. 22) 304-6, Kenney (above, n. 17) 10-11.

is immediately struck by the long, elaborate and circumstantial account (6) of the first meeting of the lovers. Such meetings and the *coup de foudre* which ensues are a traditional feature of romantic love from the Hellenistic poets on;[29] and it is indeed possible that such a scene occurred in the original. Ovid, however, says nothing of the circumstances in which his lovers first met; if he found any such incident in his source, he silently transformed it, so far as Hero and Leander were concerned, to that of first rendezvous.[30] Coincidences between Musaeus and the epistles of Paris and Helen may lend some colour to the suggestion that, practising the same economy with this part of his material as with the storm (see below, p. 68), he also made use of the first meeting theme by transferring it to another pair of lovers.[31] However, the list of coincidences contains nothing that could not at a pinch be put down to the commonplace nature of the ideas in question.[32]

Even more striking is Hero's comportment at the festival. This was surely an occasion when common sense suggests that she ought to have been exclusively occupied with her duties rather than exhibiting her charms up and down the aisles and flirting with strangers in the vestry (6 (b)). This behaviour, so inappropriate in her case, would be completely characteristic of a visiting pilgrim. Such festivals offered one of the rare opportunities available to Greek girls of marriageable age to mingle freely with crowds in public and to be seen by strange young men. This was a stock literary motif, as evidenced for instance by the *pannychides*, with their attendant chances of seduction, that play a part in many comic plots; and one remembers the meeting of Simaetha and Delphis in the second Idyll of Theocritus,[33] and of course Acontius' first sight of Cydippe in the sanctuary of Artemis.[34] Just as Musaeus' Hero

[29] Jason and Medea, Simaetha and Delphis, etc.; the point hardly requires illustration. On Acontius and Cydippe cf. Kenney (above, n. 17) 211.

[30] Kenney (above, n. 17) 5, 19.

[31] Musaeus 68 ~ *Her.* 16.137-40, 334; 74-6 ~ 16.149-52; 86 ~ 16.135; 89 ~ 16.275-6; 101-2 ~ 17.77-8; 107-8 ~ 16.239; 115 ~ 16.229; 177-8 ~ 17.3-4, 10, 191-2.

[32] See Gelzer (above, n. 22) 310-11, noting Musaeus' debt to the meeting of Odysseus and Nausicaa in *Odyssey* 6 and Plato's *Phaedrus*.

[33] Theocr. 2.66-8 and Gow *ad loc.*

[34] Ov. *Her.* 20.5, 21.105; Aristaen. 1.10.24-5 M.

wanders about the temple of Aphrodite (71 κατὰ νηὸν ἀλᾶτο), so Ovid's Cydippe strolls round the Delian precincts (*Her.* 21.96 *erramus ... uago per loca sacra pede*). Suspicion that Musaeus borrowed this trait from Callimachus hardens into certainty when one turns back to the beginning of the poem and reads that the lovers were "the fairest stars" of their two cities (21-2):

ἡ μὲν Σηστὸν ἔναιεν, ὁ δὲ πτολίεθρον Ἀβύδου,
ἀμφοτέρων πολίων περικαλλέες ἀστέρες ἄμφω.

This is lifted straight from Callimachus (fr. 67.5-8 Pf.):

ἦ γάρ, ἄναξ, ὁ μὲν ἦλθεν Ἰουλίδος ἡ δ' ἀπὸ Νάξου,
Κύνθιε, τὴν Δήλωι σὴν ἐπὶ βουφονίην,
αἷμα τὸ μὲν γενεῆς Εὐξαντίδος, ἡ δὲ Προμηθίς,
καλοὶ νησάων ἀστέρες ἀμφότεροι.

The *furtum* could hardly be more obvious.[35] Similarly with the occasion itself. The commentators offer no evidence for the existence of a famous festival of Aphrodite, such as would attract worshippers from all over the Greek world, at Sestos. Musaeus' tale of the pilgrims begins with "those dwelling in the foothills of the sea-girt islands", ὅσσοι ναιετάεσκον ἁλιστεφέων σφυρὰ νήσων (45); and the only identifiable islands actually mentioned (though editors postulate a lacuna in the text hereabouts) are Cyprus and Cythera, both of course important centres of the cult of Aphrodite, but nowhere near Sestos. All this, as was long ago pointed out by Klemm,[36] would be much more suitable to a festival at Delos. Other coincidences with the extant fragments of Callimachus and with Aristaenetus[37] reinforce the conclusion that Musaeus

[35] "e Call. Musaeus" is the terse annotation of Pfeiffer *ad loc.*; cf. Kost (above, n. 3), Gelzer (above, n. 22) on Musaeus *ad loc.* This emphasis on the matching qualities of the two, which in Callimachus does not conflict with the motivation of the plot, is inconsistent with the implication of Hero's initial rejection in Musaeus of Leander's suit on the grounds, not clearly formulated, of his general unsuitability.

[36] J. Klemm, *De fabulae quae est de Herus et Leandri amoribus fonte et auctoritate* (Diss. Leipzig 1889) 13-23; cf. Thuc. 3.104.3 and the other passages cited by Pfeiffer on Call. fr. 71. This was of course before the publication of *P. Oxy.* 2211, hence Klemm's and G. Knaack's ("Hero und Leander. In honorem Fr. Susemihl," in *Festgabe für Franz Susemihl zur gr. Wissenschaft u. Dichtung* [Leipzig 1898] 46-82) misguided attempts to identify the author of the postulated Hellenistic original as Callimachus (Gelzer [above, n. 22] 304, n. *c*).

[37] Kost (above, n. 3) 36-43.

drew on the *Aetia* for the scene and circumstances of the lovers' first meeting.

None of this necessarily entails that in the original version the meeting did not figure. If the original, as seems to be more likely than not, dates from the later Hellenistic period,[38] the poet might have deliberately adapted this motif from Callimachus' *Acontius and Cydippe*.[39] If that was so, it was predictable that Ovid would omit or transpose material which was already preempted for his own treatment of Acontius and Cydippe. He chose rather to emphasize another, equally elegiac, motif, that of the Hellespont as barrier, like the river of *Amores* 3.6, a natural equivalent of the *ianua clausa*, an obstacle to the fruition of the lovers' desires.

A priori it would seem unlikely that in the earliest version of the story Hero was a priestess. This motif complicates the simple and adequate motivation of parental opposition[40] and generates inconsistencies. One may hazard a plausible guess as to the source of the idea. Musaeus was clearly influenced by the Greek novelists, especially Achilles Tatius;[41] the notion of making Hero a priestess may have been suggested to him by the *Aethiopica* of Heliodorus. Elements of Musaeus' treatment of Hero's entry into the action—the elaborate description of her beauty and its effects on the beholders in general and the hero in particular—are standard in the novelists: so Chariton has Chaereas and Callirhoe fall instantly for each other at a chance meeting at a great festival of Aphrodite. Heliodorus' handling of this theme, however, is in one respect atypical, indeed unique: his Chariclea is a priestess of Artemis.[42]

---

[38] See below, pp. 68-69.

[39] Cf. Nicander's appropriation of Acontius' ruse with the apple for his story of Hermochares and Ctesylla (Ant. Lib. 1.1). In that case Musaeus' treatment would have been influenced by Callimachus both directly and indirectly—a common scenario.

[40] Reflected still in Hero's arguments against marriage to Leander; she dwells on the unsuitability of this particular proposal rather than the impossibility of marriage in principle (Kost [above, n. 3] 369).

[41] Kost (above, n. 3) 29-30, Gelzer (above, n. 22) 308-9.

[42] Chariton 1.1.1-6, Callirhoe's first appearance in public, chaperoned (like Cydippe) by her mother (§5); Heliod. 2.33.4-5 ἐπανατείνεται ἐκθειάζουσα μὲν

VI

One can only guess why Musaeus chose to complicate and embellish the plot in this way. The line of attack adopted by Leander in his wooing—"like a bold sharpe Sophister", as Marlowe describes it—suggests that the poet perceived an opportunity to create and develop a paradox which lent itself to rhetorical exploitation: a priestess of the goddess of love perversely wedded to celibacy. A virgin priestess of Aphrodite appearing as a veritable incarnation of the goddess, Κύπριδος ἀρήτειρα, νέη διεφαίνετο Κύπρις (68), is a contradiction in terms,[43] a point seized on and exploited by Leander, whose plea indeed turns on it: Κύπριδος ὡς ἱέρεια μετέρχεο Κύπριδος ἔργα (141). This is the will of the goddess herself, whose example Hero should follow if she really loves her (146-7, 157):

σὺ δ', εἰ φιλέεις Κυθέρειαν,
θελξινόων ἀγάπαζε μελίφρονα θεσμὸν Ἐρώτων...
πείθεο καὶ σύ, φίλη, μὴ Κύπριδι μῆνιν ἐγείρῃς.

The chance to indulge in this sort of thing, for a poet of Musaeus' propensities, was amply sufficient to outweigh (assuming that he gave the matter a moment's thought) the inconsistencies of plot and characterization entailed. Whatever the poet's motives, it remains overwhelmingly probable that in the original Hero was not a priestess and that this was not a feature of the story that Ovid had to edit out.[44]

παρθενίαν καὶ ἐγγὺς ἀθανάτων ἀποφαίνουσα, ἄχραντον καὶ ἀκήρατον καὶ ἀδιάφθορον ὀνομάζουσα, Ἔρωτα δὲ καὶ Ἀφροδίτην καὶ πάντα γαμήλιον θίασον ἀποσκορακίζουσα—protestations more plausible on the lips of a priestess of Artemis than in Hero's case. It is at a festival of Artemis that Xenophoon's hero and heroine fall for each other (Xen. *Ephes.* 1.2-3). Cf. also Men. *Citharistes* 92-7.

[43] That virgin priestesses of Aphrodite had actually existed in the pagan world seems to be attested only by Pausanias of Sicyon (2.10.4); cf. L. Malten, "Motivgeschichtliche Untersuchungen zur Sagenforschung III. Hero und Leander," *RhM* 93 (1949) 76, n. 43. Whether Musaeus, who was probably a Christian (Gelzer [above, n. 22] 299-302), would have known or cared about that is another matter. In laying the scene of the lovers' meeting in the temple itself he reveals himself "als Kind des 6. Jhs." (Kost [above, n. 3] 228). Ovid's *in aede Dianae* (*Her.* 20.5) must refer to the precinct rather than the building.

[44] *Pace* Kost (above, n. 3) 181.

VII

That the storm which forms the climax of Musaeus' narrative was a prominent feature of our postulated original is demonstrated, as has been noted, by what might be called a stemmatic argument, the agreement of Musaeus and Virgil as independent witnesses[45] to the reading of the archetype. As has also been noted, Ovid was precluded from referring to the storm or to the deaths of the lovers except by way of ironic allusion.[46] However, the correspondences, which are too close and too striking to be ascribed to coincidence, between Musaeus and Ovid's story of Ceyx and Alcyone in *Metamorphoses* XI strongly suggest that this part of the original source was not allowed to go to waste. The similarities are not confined to details but embrace situation and plot: the denouement of Ovid's narrative, in which Alcyone throws herself from a breakwater (a surrogate for the tower supplied *ad hoc*) on to her husband's stranded corpse, *moritura super crudeli funere*, exactly mirrors that of Musaeus' poem (and, one must suppose, of the original).[47] The case for this hypothesis has been convincingly presented by Pohlenz and reinforced by Schott, and need not be rehearsed here.[48] Ovid's Ceyx and Alcyone ends on a note of idyllic peace with the transformation of the pair into seabirds. By making Leander fancy that during his moonlight crossing he heard an echo of the halcyons' song—*Alcyones solae, memores Ceycos amatae, | nescioquid uisae sunt mihi dulce queri* (*Her.* 18.81-2)—Ovid both directs attention to his own technical virtuosity and also strikes a note of tragic irony. It is thus unlikely to be an accident that in the next couplet (*Her.* 18.83-4)[49] we encounter a coincidence with Musaeus (325-6) which to the alert reader signals a foreshadowing of Leander's death in the storm:

---

[45] Musaeus is hardly more likely to have read Virgil than Ovid.

[46] Cf. Kost (above, n. 3) 537.

[47] Ov. *Met.* 11.710-31.

[48] M. Pohlenz, "Die Abfassungszeit von Ovids Metamorphosen," *Hermes* 48 (1913) 7-8, G. Schott, *Hero und Leander bei Musaios und Ovid* (Diss. Cologne 1957) 105-8; on Propertius cf. Papanghelis (above, n. 7) 104-10.

[49] Kenney (above, n. 17) *ad loc.*

iamque fatigatis umero sub utroque lacertis
fortiter in summas erigor altus aquas

τυπτόμενος πεφόρητο, ποδῶν δέ οἱ ὤκλασεν ὁρμή,
καὶ σθένος ἦν ἀνόνητον ἀκοιμήτων παλαμάων.

VIII

Since 1982 all previous speculation about our putative Greek original has been overtaken and largely antiquated by the publication of a Hermopolis papyrus now at Berlin, in which the remnants of fifty hexameter verses appear to offer, in disconnected snippets, most of the ingredients of the Hero and Leander story: love, the sea and its terrors, night, a dauntless heart, lovers' dalliance (ὀαριστύν), toils, somebody gasping (?), destruction, beauty, admiration, and, most strikingly (of this word every letter is clear), a tower.[50] The only essential element of the story of which the fragments appear to make no mention is the lamp. If these are not remains of the Greek poem read by Virgil, Ovid, and Musaeus, the long arm of coincidence has been stretched with a vengeance. Even on the most sceptical view, there can now be no doubt as to an awareness of the story by Hellenistic poets.[51] There is however still no secure way of dating the composition of the original. The poet may have been a contemporary of Virgil's; equally his poem may have been blushing unseen on the shelves of the Alexandrian or some other library since the time of Callimachus. Malten, in an influential article, dated it to the third century B.C., associating it with the establishment of two lighthouses on the shores of the Hellespont called after the two

[50] H. Maehler, "Ein Fragment eines hellenistischen Epos," *Mus. Phil. Lond.* 7 (1982) 109-18, *SH* 901A (cf. the crisp editorial summary, p. 402). The papyrus is dated to the fourth or fifth century A.D., roughly contemporary with Musaeus. Cf. Kenney (above, n. 17) 11, n. 40. On an earlier candidate, *SH* 951, dating to the first century A.D., see Kost (above, n. 3) 20-1 and the edd. *ad loc.* The two fragments cannot be part of the same poem, as apparently assumed by A. Cameron, *Callimachus and his Critics* (Princeton 1995) 49; see Kenney, *loc. cit.*

[51] C. M. Hintermeier, *Die Briefpaare in Ovids Heroides*, Palingenesia (Stuttgart 1993) 59. See now J. Bingen *et al.*, edd., *Mons Claudianus: Ostraca graeca et latina*, II (Cairo 1997) 254-6 for fragments of another poem on Hero and Leander. (I thank Prof. Maehler for kindly drawing my attention to this publication.)

lovers.[52] Gelzer's comment on Malten's dating is perhaps unwarrantably sceptical;[53] nevertheless on balance it seems probable that the poem was written, rather than rediscovered, not long before Virgil first ushered the story into the literary tradition.

IX

It is a fair guess[54] that in the story as Virgil, Ovid and Musaeus found it the spotlight was on Leander and that Hero will have taken centre stage only at the very end. From correspondences between the complaints of Hero in both Musaeus and Ovid about her forlorn and restricted way of life, together with an epigram of Agathias (sixth century A.D.), it was argued by Knaack that she figured in the original as a type of the housebound female familiar to us from Horace's Neobule.[55] Kost's sceptical treatment of this theory may well be justified, but it does an injustice to Ovid when he ascribes to Musaeus the first successful application of this widespread motif to Hero's situation.[56] This credit belongs to Ovid, and very gracefully and delicately did he contrive the adaptation. Lacking the capacity of an Ovid or an Apuleius to vary and

---

[52] Malten (above, n. 43) 76-7.

[53] Gelzer (above, n. 22) 302, n. *a*. His theory is assailable on other grounds if it is true that the use of *pharoi* as lighthouses (as distinct from daytime sailing marks) cannot be attested before the first century A.D. (Ebert, *RE* 12 (1925) 2150-2). That seems inherently unlikely; and Fraser evidently takes it for granted, as Malten did, that the original Pharos was from the first also a beacon (P.M. Fraser, *Ptolemaic Alexandria* [Oxford 1972] I 188 and nn. 101, 102).

[54] This may understate the case; but it must be allowed (a) that arguments *ex silentio*, always hazardous, cannot be pressed in interpreting Virgil's highly compressed allusion; and (b) that we have no sure way of estimating the length of the Greek original. The analogies of Moschus' *Europa*, with 166 verses, and ps.–Moschus' *Megara* (125), go some way to support Maehler's estimate of 120-150 verses; but even that would have allowed the poet some scope for expatiating on Hero's feelings as well as Leander's and indeed for a description of their first meeting. (Cf. on the speaking role possibly to be ascribed to Hero in *SH* 951 the edd. *ad loc.*) If Ovid did indeed invest an originally inert character with life, he was doing with Hero precisely what he did with Cydippe in the following pair of epistles (Kenney [above, n. 17] 17-18). Helen he had had to reinvent rather than create from scratch (E.J. Kenney, "'Dear Helen…': the *pithanotate prophasis*," *PLLS* 8 (1995) 198-9.

[55] Musaeus 186-93, Ov. *Her.* 19.5-16, Agath. *A.P.* 5.297; Knaack (above, n. 36) 65.

[56] Kost (above, n. 3) 387.

give fresh emphasis to an earlier utterance, I may perhaps be allowed to express the conclusion to which this paper, in perhaps an unduly discursive fashion, has been tending by repeating words of my own which are already in print:

> The folk-tale motif of the girl who lives, alone and inaccessible, in a tower, he domesticates. We hear nothing of Hero's parents or family, but her situation is familiar, that of the housebound female ... with nothing to do but sit indoors and spin while she envies men their freedom to enjoy what the world outside has to offer. The beginning of her letter—"Euripidean in its simplicity and sentiment"[57]—is beautifully managed so as to bring out the contrast between the protagonists. Leander has all the action ... Hero can only love and suffer. Her long introspective soliloquy, though occasional coincidences with Musaeus may point to this or that detail in the original, is substantially Ovid's own creation. This is a case-study in egotism versus self-abnegation in love. Hero does indeed love Leander more than he can possibly love her in return (19.17-18). What at the beginning of a love-letter may sound like a routine protestation turns out to be all too tragically accurate. Centrally placed in the triptych [of *Her.* 16-21], Hero is the most vulnerable of Ovid's three heroines, and the most appealing.[58]

It seems to have been Musaeus who was responsible for the fact that the story has come down to us and is generally familiar under the title of "Hero and Leander" rather than "Leander".[59] Ovid, however, had

---

[57] Palmer (above, n. 22) 468.

[58] Kenney (above, n. 17) 14-15. On the theme that it is the one who stays behind rather than the one who goes away who is the greater sufferer see the passage of Byron's letter to Teresa Guiccioli, cit. ibid. 14, n. 37. Cf. W. Kraus, "Die Briefpaare in Ovids Heroiden," *WS* 65 (1950-1) 72: "Für Heros Gefühl findet Ovid so innige Töne, wie er sie selten hat".

[59] Kost (above, n. 3) 117-18.

been there first. It was he who brought Hero to life and made a real woman of her—and gave her the last word.[60]

[60] "... besonders in den letzten beiden Briefpaaren ist an die Stelle des erotischen Bescheidwissens eine selbstlosere Betrachtung des Weibes als eines eigenständigen Wesens getreten" (Kraus [above, n. 58] 77).

ARIADNE ON THE ROCKS: INFLUENCES ON OVID, *HER.* 10

PETER E. KNOX

The abandonment of Ariadne by Theseus on Naxos was celebrated in literature and art throughout antiquity. As the unknown author of the *Aetna* (*c.* A.D. 50?) remarks, *quis non periurae doluit mendacia puppis,* | *desertam uacuo Minoida litore questus?* For most modern readers of Latin poetry the first name that comes to mind upon hearing this rhetorical question is Catullus.[1] So too for Ovid. In the third book of the *Fasti* he describes Ariadne after she has been abandoned again, this time by Dionysus (3.469-75):

> flebat amans coniunx, spatiataque litore curuo
> dedidit incultis talia uerba comis:
> 'en iterum, fluctus, similis audite querellas!
> en iterum lacrimas accipe, harena, meas!
> dicebam, memini, "*periure et perfide Theseu!*";
> ille abiit, eadem crimina Bacchus habet.
> nunc quoque "*nulla uiro*" clamabo "*femina credat.*"'

Ovid's Ariadne quotes from two passages in her celebrated monologue in Catullus 64 (130-35 and 143-44):

> atque haec extremis maestam dixisse querellis,
> frigidulos udo singultus ore cientem:
> 'sicine me patriis auectam, perfide, ab aris,
> *perfide*, deserto liquisti in litore, *Theseu*?
> sicine discedens neglecto numine diuum,
> immemor, a, deuota domum periuria portas?'

> 'nunc iam *nulla uiro* iuranti *femina credat*,
> nulla uiri speret sermones esse fideles.'

---

[1] As it was for Lygdamus in 6.39-42 (= [Tib.] 3.6.39-42): *Cnosia, Theseae quondam periuria linguae* | *fleuisti ignoto sola relicta mari:* | *sic cecinit pro te doctus, Minoi, Catullus* | *ingrati referens impia facta uiri.* For a convenient summary of the distribution of the myth, cf. T.B.L. Webster, "The Myth of Ariadne from Homer to Catullus," *G&R* 13 (1966) 22-31, as well as *LIMC* s.v. "Ariadne".

Ovid's imitation was noted by G.B. Conte as a paradigm for the process of imitation in Roman poetry.[2] The allusion is signalled by *memini*: as Conte puts it, "Ovid's Ariadne has 'lived' her experience as a poetic self, in Catullus' poem, and she remembers the tears she wept there—the tears and lamentation of Catullus' Ariadne." But would an ancient reader of Latin poetry have thought in such exclusive terms of Catullus alone? Snd when reading Ovid's Ariadne, in the *Fasti* and elsewhere, can we recover traces of another important model? In none of Ovid's works is the importance of imitation as an organizing principle as great as it is in his *Heroides*. The characters in these poems are self-consciously grounded in their literary experiences of the past, an aspect of the epistles which was highlighted in an important article by Duncan Kennedy.[3] My purpose here is to introduce a note of caution into our analyses of the processes of imitation in ancient poetry when we have good reason to suspect that not all of the relevant texts have survived. The focus for this investigation is Ovid's portrayal of Ariadne in the tenth Epistle of the *Heroides*.

The extent of Hellenistic influence on Catullus 64 is an old subject of scholarly inquiry. Isaac Voss thought the poem a translation of an epithalamium by Sappho, and other hypotheses regarding the identity of the lost "original" have frequently been floated.[4] An early favorite was a lost work of Callimachus, as was argued by A. Riese in 1866.[5] Riese began by attempting to explain apparent borrowings by Catullus from extant Hellenistic works as common imitation of the lost poem of

[2] G.B. Conte, *The Rhetoric of Imitation. Genre and Poetic Memory in Virgil and Other Latin Poets* (Ithaca 1986) 60-63 (= *Memoria dei poeti e sistema letterario* [Turin 1974] 38-40).

[3] D. Kennedy, "The Epistolary Mode and the First of Ovid's *Heroides*," *CQ* 34 (1984) 413-22. Further remarks in A. Barchiesi, *P. Ovidi Nasonis Epistulae Heroidum 1-3* (Florence 1992) 15-41, P.E. Knox, *Ovid: Heroides. Select Epistles* (Cambridge 1995) 18-25.

[4] *Cajus Valerius Catullus et in eum Isaaci Vossii observationes* (Leiden 1684) 189. R. Reitzenstein, "Die Hochzeit des Peleus und Thetis," *Hermes* 35 (1900) 73-105 takes a similar line, arguing that the Hellenistic poem was a wedding poem composed for a Hellenistic monarch, a theory embraced by Kroll in the first edition of his commentary (1923), but rejected in the "Nachträge" to the second edition (1929).

[5] "Catulls 64stes Gedicht aus Kallimachus übersetzt," *RhM* 21 (1866) 498-509.

Callimachus,[6] but he focused his attention in particular upon a single verse preserved in a quotation by Cicero (*Att.* 8.5.1) without attribution and imitated by Catullus: πολλὰ μάτην κεράεσσιν ἐς ἠέρα θυμήναντα ~ 64. 111 *nequiquam uanis iactantem cornua uentis*. M. Haupt first assigned this hexameter to Callimachus' *Hecale*,[7] and several factors, including other evidence of Cicero's familiarity with this poem, support the attribution.[8] Dissent has been based primarily on the unsupported assertion that Catullus' poem is a translation, with the corollary that since the *Hecale* could not be the original, this fragment *must* belong to a different poem, Catullus' actual source.[9] This approach was rightly termed "perverse" by Wilamowitz,[10] and the attribution to the *Hecale* remains the most plausible, with implications for the interpretation of Catullus that have not been fully appreciated. If this verse describes the Marathonian bull, then Catullus has transferred it from its original context in order to describe Theseus' other adventure with the Minotaur, and a slight discrepancy gives the imitation away. The Marathonian bull may well toss its horns to the empty winds (μάτην ... ἐς ἠέρα, *nequiquam uanis ... uentis*), but the Minotaur cannot: it is indoors.[11] Catullus' imitation might thus be viewed as a creative adaptation of Callimachus to a new context, and as such it is wholly consistent with his practice in the use of Greek

---

[6] Thus, Cat. 64.96 *quaeque regis Golgos quaeque Idalium frondosum*, which echoes Theocr. 15.100 δέσποιν', ἃ Γολγώς τε καὶ Ἰδάλιον ἐφίλησας, and Cat. 64.30 *Oceanusque mari totum qui amplectitur orbem*, an imitation of Euphorion, fr. 122 P ὠκεανός, τῷ πᾶσα περίρρυτος ἐνδέδεται χθών are explained (p. 505) as ultimately Callimachean: Callimachus, Riese claims, quoted the Theocritean line in his lost poem, and Callimachus was in turn quoted by Euphorion, but Ovid knew only Callimachus.

[7] *Opuscula*, Vol. II (Leipzig 1875) 81. Pfeiffer includes it in his edition as fr. 732 inc. auct. with the remark, "hexameter hic Callimacho vere dignus est".

[8] See Hollis on *Hecale*, fr. 165 inc. auct.

[9] Such objections (e.g. K.P. Schulze, *Jahrbücher für Philologie* (1882) 208 ff.) led Riese to abandon his hypothesis in his commentary, *Die Gedichte des Catullus* (Leipzig 1884) 154. This view was also adopted by, e.g., Reitzenstein (above, n. 4) 101, n. 4; Kroll on Cat. 64.111.

[10] *Hellenistische Dichtung*, Vol. II (Berlin 1924) 300, n. 2, echoed by Pfeiffer on fr. 732 inc. auct.

[11] This has been noticed, so far as I am aware, only by G. Friedrich, *Catulli Veronensis Liber* (Leipzig & Berlin 1908) ad loc., who goes on to offer a futile attempt at an alternative explanation of *uentis*.

models. But while this is inconsistent with the theory of a lost original translated by Catullus, but it would be entirely consistent with the less extreme hypothesis that Catullus knew and alluded to a poem on the subject of Ariadne's abandonment by Theseus.

A different approach to the question was indicated by E. Maass in a brief note on echoes of Hellenistic poetry in Nonnus' *Dionysiaca*.[12] Maass compared three passages in Nonnus' account of Ariadne in Book 47 of the *Dionysiaca* with parallel passages in Catullus, and suggested that the coincidences he noticed were due to imitation of a common source by the two poets.[13] This plausible scenario has met with resistance on two grounds, the first that these passages represent commonplace motifs in Hellenistic narratives of abandoned women and thus do not constitute evidence of imitation.[14] But it is unreasonable on this basis alone to draw the conclusion that Nonnus and Catullus incorporated these same themes independently into Ariadne's lament, especially if other evidence suggests that they are incorporating familiar elements of her story. The other argument against the existence of a Hellenistic source known to Catullus and Nonnus is that Nonnus was imitating Ovid, not a Hellenistic poem. This hypothesis was advanced by G. D'Ippolito,[15] in an attempt to show that the similarities between Catullus and Nonnus are to be explained by Ovid's imitation of Catullus 64, elements of which were transmitted through him to Nonnus. A further consideration, quite separate from the historically implausible scenario postulated by this theory that Nonnus read Roman poets and used them as sources for Greek myths,[16] is the

[12] "Alexandrinische Fragmente," *Hermes* 24 (1889) 520-29; cf. 527-29 for his remarks on Cat. 64.

[13] The passages in question are Cat. 64.158-63 ~ *Dion.* 47.390-96, Cat. 64.139-40 ~ *Dion.* 47.368-69, Cat. 64.177-82 ~ *Dion.* 47.377-81. Maass' arguments are accepted, without acknowledgement, by Kroll on 64.158-63 and Fordyce on 64.139 and 160. See also the fuller development of this hypothesis by L. Castiglioni, *Studi Alessandrini I. Arianna e Teseo* (Pisa 1907) 31-52.

[14] G. Perrotta, "Il carme 64 di Catullo e i suoi pretesi originali ellenistici," *Athenaeum* 9 (1931) 371 ff. Perrotta's main conclusions concerning the influence of Apollonius on Catullus, made elsewhere in the article, remain valuable.

[15] *Studi Nonniani* (Palermo 1961) 115-30.

[16] See my "Phaethon in Ovid and Nonnus," *CQ* 38 (1988) 536-51, with reference to earlier literature; cf. also N. Hopkinson's introduction to *Studies in the*

extraordinarily unlikely literary background it assumes. It supposes that Nonnus conflated two different versions of Ariadne in Ovid—*Ars* 1.527-64 and *Her.* 10—to construct his narrative, rather than seeking a source in his native literature. That Nonnus was intimately familiar with Hellenistic poetry has long been known and is amply documented.[17] In particular, it should be noted that Book 47 of the *Dionysiaca* constitutes something of a tour-de-force exhibition of his learning in this field. The book opens with an extended narrative (1-264) of the story of Erigone, daughter of Icarios, who introduced wine to Attica. For this myth Nonnus drew on one of the most celebrated narrative poems of Hellenistic Alexandria, the *Erigone* of Eratosthenes—διὰ πάντων ἀμώμητον ποιημάτιον, as Longinus characterized it (*De Subl.* 33.5).[18] The concluding section of the book (470-741) is devoted to Perseus' resistance to Dionysus in Argos. The story is sparsely attested, but it was related in some form by Euphorion, probably in his *Dionysos* (fr. 18 P).[19] Nonnus' familiarity with Euphorion has been demonstrated in other parts of the *Dionysiaca*,[20] and it is likely that he draws on him in this part as well. A. Barigazzi noticed Nonnus' rare metaphorical use of the verb πρηνίζω in *Dion.* 47.668 καί νύ κεν Ἄργος ἔπερσε καὶ ἐπρήνιξε Μυκήνας and its ap-

---

Dionysiaca of Nonnus, N. Hopkinson, ed., Cambridge Philological Society, Supplementary Volume 17 (Cambridge 1994) 3.

[17] E.g., I. Cazzaniga, "Temi Poetici Alessandrini in Nonno Panopolitano," in *Miscellanea di Studi Alessandrini in memoria di A. Rostagni* (Turin 1963) 626-46; A.S. Hollis, "Some Allusions to Earlier Hellenistic Poetry in Nonnus," *CQ* 26 (1976) 142-50; and, most recently, Hollis, "Nonnus and Hellenistic Poetry," in *Studies in the Dionysiaca of Nonnus*, N. Hopkinson, ed., Cambridge Philological Society, Supplementary Volume 17 (Cambridge 1994) 43-62.

[18] On the *Erigone* (fr. 22-7 P), see E. Maass, *Analecta Eratosthenica*, Philologische Untersuchungen 6 (Berlin 1883) 60–138; G.A. Keller, *Eratosthenes und die alexandrinische Sterndichtung* (Diss. Zürich 1946) 53-94; F. Solmsen, "Eratosthenes' Erigone: A Reconstruction," *TAPA* 78 (1947) 252-75; R. Merkelbach, "Die Erigone des Eratosthenes," in *Miscellanea di Studi Alessandrini in memoria di Augusto Rostagni* (Turin 1963) 469-526; and P.M. Fraser, *Ptolemaic Alexandria,* Vol. II (Oxford 1972) 903, n. 202.

[19] As was argued by A. Barigazzi, "Il Dionysos di Euforione," in *Miscellanea di Studi Alessandrini in memoria di Augusto Rostagni* (Turin 1963) 416-54.

[20] Cf. Hollis (above, n. 19) 146-8, and note the parallel drawn there between Euphorion, *SH* 429.7 and *Dion.* 47.711.

pearance in Euphorion, *SH* 418.41, apparently also with Dionysus as subject and a city again as object:[21]

ἄϲτυ] γὰρ ε[ὐθὺ] δεκάϲϲι γυνα[ικείαιϲι
κρ]εξ[άμ]εν[οϲ πρήνιξε δορυϲϲόο[υ Εὐρυμέδοντοϲ

This text is paraphrased in the entry under πρηνής in the *Et. Magn.* (= fr. 18 P), where Euphorion is faulted on lexical grounds for his metaphorical use of this verb: ὁ δὲ Εὐφορίων οὐκ ὀρθῶς λέγει περὶ τοῦ Διονύσου, ὅτι ταῖς γυναικείαις τάξεσιν ἐγκελευσάμενος ἐπρήνιξε τὴν Εὐρυμέδοντος πόλιν, τουτέστι τὸ Ἄργος. Εὐρυμέδων δὲ ὁ Περσεὺς ἐκαλεῖτο.[22] The connection between Nonnus and Euphorion is strengthened by the fact that it is precisely this metaphorical use that Nonnus applies in Book 47 in his narrative of the same myth. Euphorion need not have been Nonnus' sole, or even principal, source for the story,[23] but the most likely explanation is that Nonnus has carefully chosen to close this book as he began it, with a narrative of distinctly Hellenistic coloring.

Against this literary background it is a plausible inference that in the central panel of this book, devoted to Ariadne (265-469), Nonnus is also looking back to a narrative poem of the earlier Hellenistic period, even though no trace of such a work survives.[24] But D'Ippolito's flawed argument for Nonnus' dependence on Ovid offers new possibilities for addressing this question. Earlier discussions do not refer to Ovid's various Ariadnes, because they assume that Catullus 64 is the only text to which he refers.[25] That can hardly be the case: if Catullus knew an earlier narrative of Ariadne, then Ovid surely knew it too.

---

[21] Barigazzi (above, n. 19) 424-25.

[22] Barigazzi was misled by Powell's abbreviated citation, as given above, into thinking that *Et. Magn.* is rejecting Euphorion's version of the myth. Cf. A. Meineke, *Analecta Alexandrina* (Berlin 1843) 50-52.

[23] As F. Bömer puts it, *P. Ovidius Naso: Metamorphosen, Buch IV-V* (Heidelberg 1976) 186, the story is at any rate "im wesentlichen Zügen erst ein Kunstprodukt des hellenistischen Zeit."

[24] The effort by Barigazzi (above, n. 19) 445-54 to identify the source of this part of the book also as Euphorion's *Dionysos* and thus, too, the source of Catullus' Ariadne, is overly optimistic; cf. also his "Sull' *Ecale* di Callimaco," *Hermes* 82 (1954) 316, n. 2.

[25] A notable exception is Castiglioni (above, n. 13).

And it is likely that Ovid referred to it in his works in such a way that his reader would have noted the allusion.[26]

The first parallel adduced by D'Ippolito is indeed suggestive.[27] Ovid describes Ariadne mad with grief on the seashore, *Cnosis in ignotis amens errabat harenis* (*Ars* 1.527), and calling out to Theseus (1.531-32):

Thesea crudelem surdas clamabat ad undas,
indigno teneras imbre rigante genas.

Nonnus' description focuses on the same details (*Dion.* 47.298-301):

ἀλλὰ σὺν ἀλκυόνεσσι Κυδωνίας ἔστενε νύμφη
ἠιόνας μεθέπουσα, βαρύβρομον ἔδνον ἐρώτων·
ἠίθεον δ' ὀνόμηνεν· ἐμαίνετο δ' ἐγγύθι πόντου
ὁλκάδα διζομένη

And the same scene is related by Ariadne in *Her.* 10.17-22, a parallel not noted by D'Ippolito:

...specto, si quid nisi litora cernam.
quod uideant oculi, nil nisi litus habent.
nunc huc, nunc illuc, et utroque sine ordine curro;
alta puellares tardat harena pedes.
interea toto clamanti litore 'Theseu!'
reddebant nomen concaua saxa tuum.

In the *Heroides* Ovid includes the detail present also in Nonnus but not in the *Ars* that Ariadne scanned the sea for signs of Theseus' ship. To find Ovid's—and Nonnus'—point of reference for this moment, we must look through Catullus 64 (124-27):

saepe illam perhibent ardenti corde furentem
clarisonas imo fudisse e pectore uoces,
ac tum praeruptos tristem conscendere montes,
unde aciem <in> pelagi uastos protenderet aestus.

[26] This style of allusion to an earlier source by a poet working mainly from a more proximate text is called "multiple reference" by R.F. Thomas, "Virgil's *Georgics* and the Art of Reference," *HSCP* 90 (1986) 193.

[27] D'Ippolito (above, n. 15) 126.

Catullus' report of this scene is couched in a familiar formula, referring the details to earlier authorities: *perhibent*.[28] To the question of whether he had a model, Catullus himself provides the answer; in their descriptions of this moment, it is likely that Ovid and Nonnus—and their contemporary readers—had recourse to that text.

Two other parallels between Ovid's *Ars* and Nonnus are equally instructive,[29] and since they correspond to nothing in either *Her.* 10 or Catullus, these passages must have been independently formulated. First is the conceit that her grief only makes Ariadne more beautiful, found both in the *Ars* (1.533-34) and the *Dionysiaca* (47.314-15):

> clamabat, flebatque simul, sed utrumque decebat;
> non facta est lacrimis turpior illa suis

> ἦν δὲ φαινοτέρη καὶ ἐν ἄλγεσι, καί μιν ἀνίη
> ἀχνυμένην κόσμησε

And Dionysus' promise of marriage and catasterism is likewise found only in the *Ars* (1.556-8) and the *Dionysiaca* (47.428-29, 446-47, 451-52):

> 'pone metum: Bacchi, Cnosias, uxor eris.
> munus habe caelum: caelo spectabere sidus;
> saepe reges dubiam Cressa corona ratem.'

> παρθένε, τί στενάχεις ἀπατήλιον ἀστὸν Ἀθήνης;
> μνῆστιν ἔα Θησῆος· ἔχεις Διόνυσον ἀκοίτην.

> τί πλέον ἤθελες εὖχος ὑπέρτερον; ἀμφοτέρον γὰρ
> οὐρανὸν οἶκον ἔχεις, ἑκυρὸς δέ σοί ἐστι Κρονίων

> ἀλλὰ σοι ἀστερόεν τελέσω στέφος, ὥς κεν ἀκούσῃς
> εὐνέτις αἰγλήεσσα φιλοστεφάνου Διονύσου

These parallels suggest a common background other than Catullus, and there is at least one indication that this could have been a Greek poem. At *Ars* 1.541 Ovid refers to the Bacchants accompanying Dionysus as *Mimallonides*. This is a unique variant form of Μιμαλλών, a Macedonian term for Bacchant, first attested in Callimachus, fr. 503 Pf. and

[28] For this style of reference in Roman poetry, cf. Fordyce on Cat. 64.1 and Norden on *Aen*. 6.14.

[29] Both noted by D'Ippolito (above, n. 15) 126-27.

Euphorion, *SH* 418.36, a fragment which may come from the *Dionysos*. It is attested only once thereafter in verse in Lycophron, and there are two occurrences in Latin poetry (Persius and Statius), but it is common in Nonnus.[30] Nonnus did not learn the rare word from Ovid; more likely, they both found it in the Hellenistic poetry that informed their work.

Ovid's role in assessing Catullus' and Nonnus' sources has been discounted by earlier critics largely because he was taken to be little more than a slavish imitator of Catullus, especially in *Her.* 10. But if we adopt as a working hypothesis that Catullus alludes to an earlier Hellenistic narrative poem and this poem was also known to Ovid, some interesting consequences for our reading of *Her.* 10 emerge, even as we acknowledge that the poem's primary focus remains Catullus.

In *Her.* 89-92 Ariadne concludes a catalog of her fears of what dangers await her on this deserted island. Lions, tigers and seals are one area of concern, but she also fears falling into the hands of pirates:

> tantum ne religer dura captiua catena,
> neue traham serua grandia pensa manu,
> cui pater est Minos, cui mater filia Phoebi,
> quodque magis memini, quae tibi pacta fui.

Besides being killed by bandits, Ariadne might be made a captive and sold into slavery, a fate she fears even more than death. And one that she regards as particularly disgraceful since she had once been betrothed to Theseus, as she recalls (*magis memini*) from her earlier experience in literature. Ovid here inverts an important point in Ariadne's lament in Catullus, where she declares her willingness to have served Theseus as a slave in his household (64.158-63):

> si tibi non cordi fuerant conubia nostra,
> saeua quod horrebas prisci praecepta parentis,
> attamen in uestras potuisti ducere sedes,
> quae tibi iucundo famularer serua labore,
> candida permulcens liquidis uestigia lymphis,
> purpureaue tuum consternens ueste cubile.

---

[30] For the details, see Pfeiffer's note on Call., fr. 503.

Catullus' Ariadne is not only able to endure the thought that she might not marry Theseus, she considers the possibility of being a slave in his home as attractive.[31] Ovid's reader is thus a bit surprised to hear that Ariadne abhors slavery. This sense of surprise might be deepened if Ovid is playing off not only Ariadne's declaration in Catullus 64, but also an early work in which this same motif figured. The possibility that this is the case is raised by a similar passage in Book 47 of the *Dionysiaca* (47.390-96):

> δέξο με σῶν λεχέων θαλαμηπόλον, ἢν ἐθελήσηις·
> καὶ στορέσω σέο λέκτρα * * * * * * * * * * *
> * * * * * * * * * * * * * * μετὰ Κρήτην Ἀριάδνη
> οἷα τε ληισθεῖσα· καὶ ὀλβίστηι σέο νύμφηι
> τλήσομαι, ὡς θεράπαινα, πολύκροτον ἱστὸν ὑφαίνειν
> καὶ φθονεροῖς ὤμοισιν ἀήθεα κάλπιν ἀείρειν,
> καὶ γλυκερῶι Θησῆι φέρειν ἐπιδόρπιον ὕδωρ·
> μοῦνον ἴδω Θησῆα.

If the close resemblance of this passage to Catullus 64 is attributable to imitation by Nonnus of a common model, then the point of the allusion in Ovid's epistle is sharpened and the contrast with the earlier tradition is the more pronounced.

The second instance reverses our perspective on the process of allusion at work in these texts. In this case, Catullus' Ariadne reproaches the absent Theseus with breaking his promise to her (Cat. 64.139-42):

> at non haec quondam blanda promissa dedisti
> uoce mihi, non haec miserae sperare iubebas,
> sed conubia laeta, sed optatos hymenaeos,
> quae cuncta aerii discerpunt irrita uenti.

Theseus' breaking of faith is of course a defining feature of his portrayal in Ariadne's lament. It is not surprising therefore that Nonnus raises the question in his representation of Ariadne; it is more than a

---

[31] That the motif is relatively common in ancient literature was adduced by Perrotta (above, n. 14) as evidence against Catullus' use of a specific Hellenistic source, but that is hardly persuasive. Indeed, the declaration by Ovid's Ariadne is all the more striking against this background.

little suggestive, however, that he frames the motif in a manner resembling Catullus (*Dion.* 47.368-69):

> οὐ τάδε μοι κατέλεξεν ἐμὸν μίτον εἰσέτι πάλλων·
> οὐ τάδε μοι κατέλεξε παρ' ἡμετέρωι λαβυρίνθωι.

Assume for a moment the existence of a Hellenistic narrative of Theseus and Ariadne. It is difficult to believe that Nonnus could have been ignorant of such a poem, or that knowing of it he ignored it. With this background in mind, let us now consider how Ovid's Ariadne represents Theseus' promise to marry her (*Her.* 10.71-74):

> cum tibi, ne uictor tecto morerere recuruo,
> quae regerent passus, pro duce fila dedi,
> tum mihi dicebas, 'per ego ipsa pericula iuro,
> te fore, dum nostrum uiuet uterque, meam.'

The moment to which Catullus and Nonnus only allude is reported in detail by Ovid's Ariadne, who quotes Theseus' words on that occasion. What is Ovid's source for the quotation? Perhaps he had none. On the other hand, perhaps he did. The manner in which Theseus' words are formulated resembles quite closely Ovid's method of referring to an earlier text.[32] If that is indeed the case here, the reader's sense of Ariadne's bitterness and the depth of Theseus' betrayal would be sharpened by the recollection of the moment in literature when he made his empty promise.

It is unfortunate that Nonnus' worth as a poet perhaps does not encourage many readers to look deeper into the *Dionysiaca* as a source not only of recondite phrases borrowed from Hellenistic predecessors but also for hints of the contents of the poems he knew. And he knew a great many, for his learning was as deep as his inspiration was shallow. Still, it was a clever stroke, when he let Ariadne speak one last time, to remind his readers of her long history in literature. In the final book of the *Dionysiaca*, Ariadne, now deceased, appears to Dionysus in a dream. He now longs for yet another bride and Ariadne laments his faithlessness (48.543-44):

---

[32] Conte (above, n. 1); J. F. Miller, "Ovidian Allusion and the Vocabulary of Memory," *MD* 30 (1993) 153-64.

ἀλλὰ πολυσπερέων γαμίων ἐπιβήτορα λέκτρων,
νυμφίον ὀρκαπάτην, μετὰ Θησέα καὶ σὲ καλέσσω·

When did Ariadne call Theseus "a perjured bridgroom"? Of course, Nonnus had himself told her story in the previous book. But like Ovid's Ariadne in the *Fasti*, Nonnus' heroine may be recalling her life in an earlier text. And it is not likely that she is thinking of Catullus.

LATIN POETS AND THE *CERTAMEN CAPITOLINUM*

PETER WHITE

For W. V. C., *tenellorum nutriculae philologorum*

In A.D. 86, the emperor Domitian established in Rome a quadrennial competition that offered Latin poets the opportunity to display their talents before a large cosmopolitan audience. Among competitors in other categories, the new festival achieved instant prestige. Inscriptions show that the Capitoline Games ranked alongside the four great festivals of Greece and drew athletes and musicians from all around the Mediterranean.[1]

For poets, at least for Latin poets, the Games provided a showcase that was not just grand but unique. Although the Greek world boasted scores of festivals at which poets could vie for crowns or prizes, poetry competitions are rarely documented in the Latin West, and there was nothing that matched the event in the capital.[2] And unlike a similar festival which Nero had attempted to establish two decades earlier, Domitian's festival was destined to enjoy a long run. The Capitoline

---

[1] Study of the Capitoline Games has been facilitated by the recent monograph of M.L. Caldelli, *L'Agon Capitolinus: Storia e protagonisti dall' istituzione domizianea al IV* secolo. Studi Pubblicati dall' Istituto Italiano per la Storia Antica 54 (Rome 1993): it contains an excellent bibliography, a history of the festival, and the most up-to-date prosopography of participants. But with new inscriptions continually appearing, no prosopography can remain up-to-date for very long: for a recent addendum, see S. Şahin, "Inschriften aus Seleukia am Kalydnos," *EA* 17 (1991) 145. For orientation to the nature of agonistic festivals in general, see J. Frei, *De certaminibus thymelicis* (Diss. Basel 1901), L. Robert, "Discours d'ouverture," *Proceedings of the VIIIth International Congress of Greek and Latin Epigraphy (Athens, 3-9 October 1982)* (Athens 1984) 1: 35-45 = *Opera Minora Selecta*, vol. 6 (Amsterdam 1989) 709-19, and P. Herz, "Die musische Agonistik und der Kunstbetrieb der Kaiserzeit" in *Theater und Gesellschaft im Imperium Romanum* Mainzer Forschungen zu Drama und Theater, ed. J. Blänsdorf (Tubingen 1990) 175-95. Other titles will be cited as they become relevant to the following exposition.

[2] For a list of music festivals in the Latin-speaking part of the empire, see H. Leppin, "Musische Agone im Westen," in *Histrionen: Untersuchungen zur sozialen Stellung von Bühnenkünstlern im Westen des Römischen Reiches zur Zeit der Republik und des Principats*, Antiquitas 1, Abhandlungen zur alten Geschichte 41 (Bonn 1992) 169-76.

Games went on being celebrated for at least the next two hundred years. They invited participation from the widest imaginable pool of talent and brought the victors to the attention of a large public. Not the least important spectators were the emperors themselves. The festival took place under their presidency and served their ideological objective of asserting Rome's cultural parity with Greece, as the planting of public libraries had done beginning in the reign of Augustus.

For all these reasons, one might have expected the Capitoline Festival to play a defining role in Roman literary culture. Yet the poetry contest attracted scant comment from Latin writers, and winners of it have left almost no trace in the historical record. Specifically, we know of one certain winner, L. Valerius Pudens, who was crowned in 106, and perhaps two others, Scaevus Memor at the end of the first century and Attius Tiro Delphidius in the middle of the fourth.[3] As for other information, Martial, whose career as an epigrammatist overlapped with the first four celebrations of the Games and who treats them as a matter of some topical interest, only once alludes to the poetic part of the pro-

[3] Pudens is known from *ILS* 5178. Scaevus Memor is identified as a Capitoline victor in Mart. *Epigr.* 11.9, but he may have competed in a tragedy competition rather than in the poetry competition (I will return to this point below). Delphidius' status as a victor depends on the interpretation of a passage in Ausonius' *Professores* (5.4-7, p. 45 Green), which says that as a boy poet he celebrated Jupiter, "wearing the wreath of an Olympian crown." If this line implies a a competitive venue, and if it can be assumed from Delphidius' Gallic background that his prize poem would have been in Latin rather than Greek, then the Capitoline Contest in Rome is the only known event which would have accommodated it. All other candidates for the Capitoline honor roll in Latin poetry are supposititious. In various prosopographical sources one will find the Collinus of Mart. *Epigr.* 4.54 registered as one of the champions. Martial identifies him as a Capitoline victor, but says nothing to indicate that he was a poet, let alone a Latin poet (*pace* Caldelli [above, n. 1] 124-25), and Martial is at least as likely to have celebrated a victory by a popular athlete or musician as a victory by a poet. C. Concordius Syriacus of Beneventum can be added to the list of Capitoline victors only if one adopts Mommsen's expedient of inserting a mention of Rome into *CIL* 9.1663 = *ILS* 5179. Nemesianus' credentials as a contest winner similarly rest on emendation of a passage in the *Historia Augusta* (*Vita Car.* 11.2), an emendation rightly contested by Leppin, *Histrionen* (above, n. 2) 317 and Caldelli (above, n. 1) 162-63. I am less concerned here with the contest in Greek poetry, but let it be noted that we appear to know of only one victory in that, by the veteran contestant M. Aurelius Ptolemaeus (*Fouilles de Delphes* 3.1 [Paris 1929] 53, no. 89).

gram.[4] Martial's contemporary Statius has no more to say about it than that he once competed unsuccessfully himself, probably in the year 90 (*Silvae* 3.5.31-33 and 5.3.231-33). Of involvement by any of his various poet friends there is no hint. In the decade after Martial and Statius retire, Pliny continues to gossip about literary actitivities in the capital, and even insists at one point that "there is no one with literary interests who is not my friend" (*Epist.* 1.13.5). He is a tireless auditor and chronicler of recitations. But he never mentions anyone who competed in the Capitoline poetry contest.[5] Nor does the subject come up later in the reminiscences of Gellius or the correspondence of Fronto. The only author to glamorize the event is Florus, who makes it the turning point of his life in the dialog fragment *Vergilius Orator An Poeta.* But his picture of the contest is in every sense a view from the margins: Florus is a North African, marooned for the time being at Tarraco in northern Spain, while his interlocutor is a provincial from the south of Spain. Otherwise, so far from bolstering Roman national pride, the Capitoline crown in Latin poetry seems to rank among the most uncoveted and obscure prizes in European literary history.

The apparent insignificance of the competition may reflect in part the fact that it was only one event on a program that stretched over many days.[6] The Capitoline Festival comprised races and athletic competitions

[4] *Epigr.* 4.1.6, 4.54.1-2, 9.3.8, 9.23.5, 9.35.10, 9.40.1, 9.101.22, and 11.9; half of these references are concentrated in a book that appeared in a contest year (94). Martial's only certain reference to participation by a poet is in 11.9. Although the Diodorus of of 9.40 is often identified in prosopographical sources as a poet, Martial says nothing that supports that inference (*pace* Caldelli [above, n. 1] 127, and the tone of the poem suggests that Diodorus is in any case fictitious. Similarly, there is no good reason to think that the Carus of 9.23 is a poet; Leppin (above, n. 2), 224, concluded that he competed "in einer unbekannten Disziplin." If he is the same man as the recipient of 7.74, he should be a rhetor.

[5] It would be simpler to be able to say that Pliny never refers to the Capitoline Games at all. He does, however, make one (disparaging) reference to a competition by pantomimes which J. Beaujeu has argued must have taken place in the context of the Capitoline festival ("Les jeux sacerdotaux du haut empire (à propos de Plin. *Epist.* VII.24.6)," *BICS* 22 [1975] 109-24). For dissent, see Leppin (above, n. 2) 175, n. 38.

[6] We do not yet know the date or duration of the *certamen Capitolinum.* The problem has become tangled up in discussion of another unknown, the date of the assassination of the emperors Pupienus and Maximus, who Herodian (8.8.3) says were slain during the Capitoline Games in A.D. 238. For arguments focusing on this latter episode, see M. Peachin, "Once More A.D. 238," *Athenaeum* 67 (1989) 594-

as well as musical events, and even the music portion of the program included many events in addition to the two contests in Greek and Latin poetry.

Furthermore, the kind of poetry that was required of contestants may have discouraged some from participating. Louis Robert observed on more than one occasion that the Capitoline Games imitated the program of a Greek *agon* down to the smallest details.[7] This insight can profitably be applied to the competition in poetry. We know that as originally established, the Capitoline program included contests in Greek and Latin oratory as well as the two contests in poetry.[8] The pairing of oratory and poetry should mean that this part of the program conformed to the model of Greek festivals in which orators first and then poets vied in uttering panegyrics on the god to whom the festival was dedicated or on the imperial house.[9] As a matter of fact, Quintilian informs us that the "praise of Capitoline Jupiter" served as "the standing theme of the sacred contest" in oratory (*Inst.* 3.7.3). From the only sample that survives of a performance at the festival, we know that in A.D. 94 a contestant for the crown in Greek poetry also took Zeus as the theme of his piece (*IGRR* 1: 116-17, no. 350). Finally, Statius lets us know that when he competed for the crown in Latin poetry in 90, he too offered verse in praise

604; M. Sartre, "Le Dies Imperii de Gordien III: Une inscription inédite," *Syria* 61 (1984) 49-61; X. Loriot, "Les premières années de la grande crise du IIIe siècle: De l'avènement de Maximin le Thrace (235) à la mort de Gordien III (244)," *ANRW* 2.2 (Berlin, 1975) 721-22; and J. R. Rea, "O. Leid. 144 and the Chronology of A.D. 238," *ZPE* 9 (1972) 1-19. A more promising clue to the date of the Games may be a fragment of the Fasti Ostienses for A.D. 146 (A. Degrassi, *Inscriptiones Italiae* 13.1 [Rome, 1947] 207, frag. 28 = L. Vidman, *Fasti Ostienses* [Prague, 1982] 50), which seems to show an *agon quinquennalis* being celebrated in May and June; but of the crucial line only traces remain. Whenever the Capitoline Games fell, they will have occupied at least a couple of weeks of the calendar. A recently published inscription shows that the program of a lesser *agon* in the minor town of Oenoanda took up 24 days: see M. Wörrle, *Stadt und Fest im kaiserzeitlichen Kleinasien: Studien zu einer agonistischen Stiftung aus Oinoanda,* Vestigia 39 (Munich 1988), who discusses the duration of agonistic festivals on pp. 245-48.

[7] For example, L. Robert, "Deux concours grecs à Rome," *CRAI* (1970) 7 = *Opera Minora Selecta,* vol. 5 (Amsterdam 1989) 648 and "Une vision de Perpétue," *CRAI* (1982) 209 = *Opera Minora Selecta,* vol. 5, p. 822.

[8] We know about the contests in oratory from Suetonius, who indicates that they had been dropped from the program by the time he was writing the *Caesares* (*Dom.* 4.4, cf. 13.1).

[9] See Wörrle, (above, n. 6) 229-34 and 248-50 and Frei (above, n. 1) 34-41.

of Jupiter. At *Silvae* 3.5.31-33 he writes that when he was denied the crown, his loyal wife reproached Jupiter as being "cruel and ungrateful." "Ungrateful" only has point here if the poem that Statius presented was a eulogy of Jupiter.[10] That this is what he means is confirmed by a close parallel elsewhere in his work. *Silvae* 2.7 contains a typically Statian set-piece in which the Muse Calliope hails the birth of Lucan by forecasting his poetic achievement. Midway through a catalog of his future poems, she prophesies: "to fond theaters you will present ungrateful Nero" (lines 58-59). As other sources make clear, Calliope is looking ahead to a performance by Lucan in the Neronian Festival of A.D. 60, when he delivered a panegyric in honor of Nero.[11] The emperor is called "ungrateful" because, in spite of having been made the subject of Lucan's panegyric on that occasion, he later put him to death.

This mosaic of evidence makes it as clear as could be desired that Latin poets who competed in the Capitoline Festival presented poems in praise of Jupiter. To borrow a distinction from a modern *agon*, they competed in a short compulsory program rather than in a free program. This narrow focus must surely have narrowed the appeal of the poetry contest. There will have been many poets—and many connoisseurs of poetry, for that matter—who did not consider panegyric the most compelling proof of art, and some whose genius was wholly incompatible with panegyric. Imagine Juvenal having to produce a serious short piece extolling Jupiter.

One other factor having to do with the conditions of performance may also have inhibited participation in the contest. An inscription that preserves the Greek hexameters recited by a contestant in A.D. 94 records that his verses were composed ex tempore (*IGRR* 1: 116-17, no. 350, lines 6 and 10, and no. 351, line 7). If this reflects normal procedure at the Capitoline Festival, then of course the other contestants

---

[10] F. Vollmer correctly interpreted the line in his commentary, but the implication has sometimes been overlooked by subsequent scholars. One should perhaps add that although Latin poets often equate Jupiter with the reigning emperor, the context makes it difficult to suppose that that is what Statius intends with his reference to a "cruel and ungrateful" Jupiter here. Just three lines earlier, he refers to a victory he had won in Domitian's Alban Games, and there Domitian appears in his own human guise as the emperor: *me ... sancto ... indutum Caesaris auro.*

[11] See the Suetonian life of Lucan, p. 332.1-2 Hosius, and the Vaccan life, p. 335.21-22.

must have had to improvise too, and presumably the same rule applied to entries in Latin poetry as in Greek.[12] It is difficult to know how far one can generalize from an indication provided by one text. So far as I am aware, the inscriptions which constitute the bulk of our information about agonistic festivals do not disclose whether poets and orators ordinarily recited prepared texts or composed on the spot.[13] Perhaps, indeed, the practice varied from place to place. But there is one other clue which suggests that at the contest in Rome, poets improvised. During Nero's festival, which was the clear forerunner of Domitian's Capitoline Games, Lucan seems to have taken two prizes. In addition to the panegyric I mentioned earlier, he presented a poem on Orpheus which is described as "written ex tempore against several poets."[14]

Lucan is said to have possessed great facility as an impromptu composer, and it was a knack he shared with Martial and Statius among others. But it was hardly a universal attribute or a universally admired one. The images that typically convey the ideology of Roman poets are of labor by lamplight, fine-spun verses, working with both ends of the stylus, and use of the file. The ability to churn out verse *stans pede in uno* implies values that are just the opposite. Serious poets could not help but have harbored reservations about the Capitoline contest if success in it depended on facility at improvisation.

But if the terms of competition discouraged participation by limiting the sort of poetry that could be presented, Latin poets must have felt still greater misgivings about the prospect of competing publicly at all. That the poetry contest created a problem of status maintenance for would-be competitors is suggested by what we know about actual competitors. The winner in A.D. 106 was a thirteen-year-old (*ILS* 5178). Florus describes himself as a boy (*puer*) when he competed a few years earlier, and from other things he says about himself, it is clear that he cannot

[12] As has been most recently pointed out by A. Hardie, *Statius and the Silvae: Poets, Patrons and Epideixis in the Graeco-Roman World* (Liverpool 1983) 83.

[13] Hardie (above, n. 12) 74-85, supplies helpful evidence about the role of improvisation in epideixis, but not evidence which directly answers this question.

[14] The Vaccan life, p. 335.21-25 Hosius; Statius *Silvae* 2.7.58-59 confirms that the "Orpheus" was a competition piece.

have been over 20.[15] Delphidius also delivered his performance as a *puer* (if that performance took place in the context of the Capitoline Games).[16] One of the participants in the Greek poetry contest in the year 94 was only 11 years old.[17] An event which pitted young against old in a public arena was wholly out of keeping with the Roman sense of hierarchy.

Nor are age levels the only noteworthy statistic about the contestants. The eleven-year old prodigy just mentioned was apparently the child of freedman parents.[18] Scaevus Memor, who won the crown for either poetry or tragedy in the 90s, had a brother of freedman origin.[19] Florus describes himself as a youth without resources or prospects, obliged ultimately to earn his living by setting up as a schoolteacher in Spain.[20] Valerius Pudens, the Latin champion of A.D. 106, logged his only subsequent public distinction more than thirty years later, when he became the *curator* of a town in central Italy.[21] Among the four or five competitors in Latin poetry who are known to us, not one can be identi-

[15] *Vergilius Orator An Poeta* 1.4 Jal; Jal estimates Florus' age at about 16: *Florus: Oeuvres* (Paris 1967) 2:103-4.

[16] Ausonius *Prof.* 5.7 p. 45 Green.

[17] *ILS* 5177. To these examples from the Capitoline Games can be added some from other poetry contests. Statius' father was not yet an adult when he debuted at the Sebasta in Naples (*Silvae* 5.3.133-37); Lucan was only 20 when he was crowned at the Neronia (above, n. 11); and Augustine was about the same age when he won a contest in Carthage (*Conf.* 4.2.3 and 4.3.5). But despite the prominence of young competitors, no evidence has yet emerged to suggest that poetry contests were divided into a boys' category and a men's category, as athletic events and some other musical events were. Statius' words about his father, *ad patrii raperis certamina lustri | vix implenda viris, laudum festinus et audax | ingenii* (*Silvae* 5.3.134-36) imply that the young Papinius was competing against grown-ups.

[18] See T. Mommsen, *Römisches Staatsrecht*, vol. 3, 3rd ed. (Leipzig, 1887) 789, n. 6.

[19] Scaevus is identified as the *frater* of the satirist Turnus in Mart. *Epigr.* 11.10 and in Valla's Proban scholia on Juv. *Sat.* 1.20, and Turnus is identified as *libertini generis* in the latter passage.

[20] Florus' circumstances are illuminated throughout the *Vergilius Orator An Poeta*, but see especially 3.1-2 Jal.

[21] To the mention of Pudens' victory at the age of 13, *ILS* 5178 adds only the detail *curat. rei p. Aeserninor. dato ab imp. optimo Antonino Aug. Pio.* Compare the assessment of F. Jacques, *Les curateurs des cités dans l'occident romain: Études prosopographiques* (Paris 1983) 272: "Nous ne savons pas si Pudens était chevalier; en tout cas, il ne fît pas de carrière procuratorienne."

fied as either a senator or an *eques*.[22] This too is a curious pattern. So far as the data permit us to form an impression of Roman literary society otherwise, senators and knights were preponderant among Latin poets by the end of the first century.[23] The Capitoline competition, by contrast, seems to have attracted poets who were disproportionately young or undistinguished.

The obscurity of the participants is all the more interesting because it is a point on which Domitian's festival departed strikingly from its Neronian model. Nero produced his first cultural extravaganza when he held the Games of Youth (*Ludi Iuvenales*) in A.D. 59. These were evidently closed theatrical performances which took place on a private or quasi-private stage and featured local talent: the sources say that Nero compelled the equestrian and senatorial orders to participate.[24] Then in the following year the experiment was greatly expanded. Nero proclaimed a sacred festival on the Greek model, with a full program of artistic and athletic contests and races. The artistic competitions were held in Pompey's theater and although participation by senators and knights was evidently no longer compulsory (perhaps because the games were now thrown open to a much broader range of contestants), it was vigorously promoted.[25] The young Lucan competed and Nero himself claimed the starring role.

From the standpoint of those authors who report on Nero's festivals, the spectacle of senators and knights vying before the public was an enormity second only to Nero's own performances. Tacitus speaks of Roman notables being led by the excuse of poems and orations to "pollute themselves" upon the stage (*Ann.* 14.20.4). This harsh phrase expresses an attitude toward theatrical performance that went well beyond mere reprobation. The publication of the *senatusconsultum* from

---

[22] Note also that Palfurius Sura, a victor in Latin oratory during Domitian's reign, competed as a defrocked senator (Suet. *Dom.* 13.1).

[23] I have collected the evidence in *Promised Verse* (Cambridge, Mass. 1993) 8-13 and 211-22.

[24] Tac. *Ann.* 14.15.1, 16.21.1, *Hist.* 3.62.2, Suet. *Nero* 11.1, and Cass. Dio 61.19.

[25] Tac. *Ann.* 14.20.4 and 21.3, Suet. *Nero* 12.3 (*honestissimus quisque contenderat*), Valla's Probus on Juv. *Sat.* 4.53, and the sources for the life of Lucan in n. 11.

Larinum has recently drawn fresh attention to efforts to legislate against stage appearances by members of the upper orders.[26] The sanctions were progressively increased, but one consequence which regularly followed was that when senators or knights took to the stage, they forfeited the privileges of their rank. That the intent of these enactments was not merely to discourage full-blown acting careers, but to prohibit any sort of public performance by members of the elite is suggested by the jurists' broad definition of a stage. To quote the *Digest*, "a stage, according to Labeo's definition, is a structure set up for giving shows in any place where anyone may stand and move so as to display himself, whether the place be public, private, or in the street, provided only that the general public is admitted to the performance."[27]

Whatever ambition a poet might have cherished to be crowned by the emperor before thousands of spectators, then, the official strictures on performance posed a serious disincentive for those poets who were Roman senators or knights. It may be that they did not actually run the risk of incurring *ignominia* by participating in the Capitoline competition. Contestants in this as in all sacred festivals competed for a crown and not for a cash prize. That meant that they could not be described as acting *quaestus causa*, which mitigated the offense of performing in public. On these grounds the jurists had ruled that athletes in sacred contests were not *famosi*, and the same consideration evidently applied to other participants as well.[28] But the legal impairment only

---

[26] B. Levick, "The *Senatus Consultum* from Larinum," *JRS* 73 (1983) 97-115 and W. D. Lebek, "Standeswürde und Berufsverbot unter Tiberius: Das SC der Tabula Larinas," *ZPE* 81 (1990) 37-96. Also helpful are M. Ducos, "La condition des acteurs à Rome: Données juridiques et sociales" in *Theater und Gesellschaft* (above, n. 1) 19-30 (though Ducos takes no account of the new inscription) and W. J. Slater, "Pantomime Riots," *CA* 13 (1994) 120-44.

[27] *Scaena est, ut Labeo definit, quae ludorum faciendorum causa quolibet loco, ubi quis consistat moveaturque spectaculum sui praebiturus, posita sit in publico privatove vel in vico, quo tamen loco passim homines spectaculi causa admittantur, Dig*. 3.2.2.5. The translation is from Watson's edition (*The Digest of Justinian*, ed. T. Mommsen, P. Krueger, and A. Watson [Philadelphia, 1985]), slightly modified. Note also the breadth of the statement in the praetor's edict: *infamia notatur ... qui artis ludicrae pronuntiandive causa in scaenam prodierit, Dig*. 3.2.1.

[28] *Athletas autem Sabinus et Cassius responderunt omnino artem ludicram non facere: virtutis enim gratia hoc facere. et generaliter ita omnes opinantur et utile videtur ut neque thymelici neque xystici neque agitatores nec qui aquam equis spargunt ceterave eorum ministeria qui certaminibus sacris deserviunt, ignominiosi*

translated an underlying social standard.[29] To the extent that a poet was at the same time a prominent citizen with an important role to play in capital society, performing in public exposed him to criticism no matter what the legal consequences were, as narrative accounts of Nero's music festival plainly show.[30]

Hence when Domitian launched his own version of a sacred festival, he took pains to decouple it from the area of enterprise of the Roman elite. Rather than name it after himself, as Nero had done, he placed it under the patronage of Capitoline Jupiter. Rather than exploit the occasion to showcase his own poetical talent, he contented himself with the relatively self-effacing role of festival president.[31] Most important and again unlike Nero, so far as the indications we have permit us to judge, he refrained from pressing senators and knights to compete. By safeguarding their respectability, he gained acceptance of the Capitoline Festival as a national institution and so so gave it a greater chance of success than Nero's festival had had. But the result was to be that the poetry contest would lack input from the poets in Rome who were most active and most creative. The only significant case of an established and important poet known to have competed in it is Statius,[32]

habentur, *Dig*. 3.2.4.1. Note, however, that this general principle seems not to have protected Palfurius Sura, who was stripped of senatorial rank after having taken part in the Neronia. But Sura had offended Roman sensibilities in a particularly egregious way by entering a wrestling contest with a Laconian maiden (schol. Juv. *Sat*. 4.53, cf. Suet. *Dom*. 13.1).

[29] A point well made by Ducos (above, n. 26) 20: "Il ne s'agit pas d'une *ignominia*, liée aux décisions des censeurs ou à une sanction judiciaire, mais d'une infamie, d'une réprobation générale de la société qui a des conséquences diverses en matière de droit. En un mot, les règles juridiques ne font que traduire cette indignité par des exclusions et des incapacités dans divers domaines."

[30] To the texts cited above in notes 24 and 25, add also Tac. *Ann*. 15.65 (on the opprobrium that the conspirator Piso had incurred by earlier performances as a *tragoedus*) and Tacitus' own judgment on Piso at *Ann*. 15.48.3. On Roman attitudes toward performance generally, see E. Aubrion, "L'historien Tacite face à l'évolution des jeux et des autres spectacles" in *Theater und Gesellschaft* (above, n. 1) 197-211.

[31] One should keep in mind that Domitian too fancied himself a poet, and it is clear from Quint. *Inst*. 10.1.91-92 that he would have been crowned had he chosen to compete.

[32] I have hedged this statement slightly in order to bypass the uncertainties surrounding Scaevus Memor, whom Martial identifies as a Capitoline victor in *Epigr*. 11.9. He is almost totally unknown: the sum of our information about him consists of two couplets of Martial (*Epigr*. 11.9 and 10), a scholium on Juv. *Sat*. 1.20, and

and Statius is unrepresentative for two reasons. First, it is doubtful that he can be classified as belonging to the equestrian elite.[33] But much more important, Statius' origins were Greek and his attitude toward public competition was therefore shaped by Greek rather than Roman values. Statius' father was a veteran contestant on the Greek festival circuit in both prose and poetry competitions, and Statius himself made his poetic debut by competing in the Sebasta of his native Naples.[34]

I have argued that the Capitoline poetry contest proved irrelevant to Roman literary culture in large part because public competition in the Greek style was abhorrent to the city's elite. On this point Roman feeling remained faithful to the past. So far as I know, we are still without evidence that poetry competitions had ever had a place in Roman literary life before the imperial festivals—for example, in the mysterious gatherings of the *collegium poetarum*.

Yet everyone knows that Latin poets did stake out an arena for performance. This was the *recitatio*, a practice pioneered by Asinius Pollio in the latter half of the first century B.C. (Sen. *Cont.* 4 *praef.* 2). For all the barbs it elicited from satirists, the formal recitation was a lively and successful institution, and some of the reasons for its success can be appreciated if we now compare it with the *certamen Capitolinum*. There are other differences between the two performance contexts (regarding

two snippets of verse. As indicated in n. 19, he may not have belonged to the equestrian order, in which case he would have faced no obstacle to competing publicly. But the larger uncertainty concerns the event in which he competed. Every datum we have about him identifies him as a tragic poet, which is also the context in which Martial seems to place his Capitoline victory (*Clarus fronde Iovis, Romani fama cothurni, | spirat Apellea redditus arte Memor*). It is known that the Capitoline Games featured a competition for tragic actors, and it is possible that in the more elaborate festival program of the early years there was also a competition for tragic playwrights (as there was in some Greek music festivals). It was no doubt some such reasoning as this that led Leppin, *Histrionen*, p. 259 (above, n. 2) to conclude that Memor "gewann unter Domitian den Sieg in der tragischen Dichtung beim Kapitolinischen Agon." If Memor competed in the contest for tragic playwrights rather than in the poetry contest, he would not have actually performed on stage.

[33] There is no positive indication that Statius possessed equestrian status; and on the other side there are suggestive though not decisive counter-indications in his father's life-long occupation as a *grammaticus* (*Silvae* 5.3) and in references to the penury of both Statius and his father (Juv. *Sat.* 7.79-92 and Stat. *Silvae* 5.3.117-18).

[34] The primary evidence is all in Stat. *Silvae* 5.3, but it is most helpfully interpreted by Hardie (above, n. 13), in chapters 1, 2, 5, and 6.

the kind of material that could be presented, for example), but I want to emphasize three considerations which are social rather than literary. First, recitations seem to have taken place predominantly in private houses. The term "a stage" hardly ever arises in this context (any more than it does in connection with the actitivity of public speakers), and so recitation was tacitly dissociated from the kind of performance that incurred obloquy. Second, poets did not compete with one another at recitations. Among countless allusions to the practice, there is not one which shows that two or more poets even appeared together on the same program. There could be no mistake about who enjoyed precedence. And finally, in the recitation setting poets communicated with a public through the mediation of socialites who sponsored them. Literary activity was thus brought under the hegemony of the *principes civitatis* even when they were not themselves literary producers. The success of the home-grown institution goes some way to expose the reasons for the sterility of Domitian's Greek transplant.

II

Virgil

VIRGIL'S PINDAR?

RICHARD F. THOMAS

A recent work on the components of Virgil's epic states the following: "Some sources are simply neglected, for instance Herodotus for the third book of the *Aeneid*, or Pindar whose works are overlooked and minimized as texts closely studied by Virgil."[1] The view that Pindaric poetry is not relevant for intertextual studies, not only of Virgil, but generally of non-Horatian Roman poetry, is a deeply entrenched one, and is, for instance, reflected in Douglas Gerber's 1969 bibliography on Pindar, which recorded only a single unremarkable item under the category of "influence on Virgil", namely C. Nardi's 1956 article.[2] There has been more activity in the years since Gerber's book, chiefly on the subject of katabasis, on the connection between the Pindaric and the Virgilian Isles of the Blest, and on the proem to the third *Georgic* (and Setaioli summarizes this scholarship in his article on Pindar in the *Enciclopedia Virgiliana*), but there has been no general examination of the subject since A. Seidel's 1925 unpublished dissertation, which I have been unable to examine.[3]

Bacchylides

Before turning to Pindar, we might look briefly to the topic of Pindar's fellow epinician poet, Bacchylides, whose possible influence on Virgil will concern us a little later; in general, I find little support for the view, mainly intuitive and based on the likelihood that Horace knew Bacchylides, that suggests Virgil had any special familiarity with this poet.[4] It is true that Servius on three occasions invokes Bacchylides

---

[1] N. Horsfall, *Virgilio: l'epopea in alambicco* (Naples 1991) 45.

[2] C. Nardi, "Reminiscenze pindariche in Vergilio," in Ἀντίδωρον *Hugoni Paoli oblatum* (Genoa 1956) 242-9.

[3] A. Seidel, *De Vergilii studiis Pindaricis* (diss. Breslau 1925); its thesis is summarized by A. Setaioli, *Enciclopedia Virgiliana*, s.v. "Pindaro".

[4] Setaioli (above, n. 3) refers to the coincidence between *Aen.* 6.309-10 and Bacchylides 5.65-67; for various views on the coincidences, cf. H. Lloyd-Jones, "Heracles at Eleusis: P. Oxy. 2622 and P.S.I. 1391," *Maia* 19 (1967) 206-29; R.J.

while talking of Virgilian details, but the specifics of his discussion do not provide confidence of a Virgilian engagement with Bacchylides: on *Aen.* 11.93 Servius refers to the Greek poet for the topos of inverted weapons as a feature of military funerals (fr. 8 Bacch.); on *Aen.* 2.201 he reports that Bacchylides treated Laocoon and his wife (Virgil mentions no wife, Bacchylides apparently mentioned no children), and he also treated the arrival of the snake from the Calydnian islands, while Virgil has them come from Tenedos and nowhere mentions the Calydnians (fr. 9 Bacch.); and lastly Servius tells us at *Aen.* 6.21 (where Virgil gives seven as the number of Athenians sent to the labyrinth) that various authors (Sappho, Bacchylides, Euripides and Plato) record the number as fourteen (Bacch.*Dith.* 17.2). To sum up, Servius or, better, his source, seems to have had access to Bacchylides, but on details which differ from those in Virgil himself. This does not remove the possibility of engagement, but there is nothing persuasive of Bacchilydean intertextuality in what Servius says.[5]

Epinician/Pindaric

The way might then seem to be clear for discussing Virgil and Pindar. As I have elsewhere argued, however, the terms "epinician" and "Pindaric" should not be considered interchangeable, particularly in discussions of intertextuality in Roman poetry.[6] The point was made all the more clear by the appearance in 1977 of a papyrus of the Callimachean epinician, the *Victoria Berenices*, but even before that there was already evidence for the representation of the genre in this Hellenistic poet: the *Victoria Sosibii*, the epinician to Polycles of Aegina that is the basis of *Iambus* 8, of which only one line survives, and the treatment of Euthycles the Locrian, victor at the Olympic games (*Aet.* 3.84-5 Pf.), which while not strictly an epinician, shows its debt in the first line, which survives detached by the diegesis (Call. *Aet.* 3.84 Pf): ἦλθες ὅτ' ἐκ Πίσης, Εὐθύκλεες, ἄνδρας ἐλέγξας. And the subject has now re-

Clark, "Two Virgilian Similes and the 'ΗΡΑΚΛΕΟΥΣ ΚΑΤΑΒΑΣΙΣ," *Phoenix* 24 (1970) 244-55; also E. Paratore, "Bacchilide e Virgilio," *WS* 69 (1956) 289-296.

[5] On this subject, cf. B. Gentili, *Enciclopedia Virgiliana*, s.v. "Bacchilide".

[6] Cf. R. F. Thomas, "Callimachus, the Victoria Berenices, and Roman Poetry," *CQ* 33 (1983) 95; *Virgil, Georgics* , Vol. 2 (Cambridge 1988) 38-39.

ceived a full and good treatment by Therese Fuhrer in her 1992 monograph on the relationship of Callimachean epinician to choral lyric.[7] The editorial and scholarly interests in Pindar, apparent from the *Pinakes* (Tables of published works), and, for instance, from the bare knowledge we have that Callimachus wrote a περὶ ἀγώνων (fr. 403) are fully matched by an interest in Pindar as a poetic model and a source on a variety of matters.[8]

Let me say at the outset that this paper will not simply become another Virgil and Callimachus offering, and Pindar will remain in the spotlight throughout; he will not, however, be there alone. It will, I think, be possible to make some general and fundamental observations about the precise way in which Virgil used Pindar; the procedure I postulate is paralleled in other areas of Virgilian compositional technique, where archaic material is likewise filtered through the Hellenistic literary experience (most obviously Hesiod and Aratus for didactic, Homer and Apollonius for epic).

Genre and Form

One of the reasons that Pindar did not have an effect on Roman poetry consonant with his reputation in antiquity and with his current popularity perhaps has to do with genre: the performative aspects of epinicians (and of paeans, dithyrambs, prosodia, hyporchemata, threnoi and encomia) are so strongly felt that such poetry perhaps intrinsically resists becoming a model for a poetry, such as Latin poetry, whose function is no longer really performative. One might expect even more in Latin lyric, but that genre, which tends towards abbreviation, particularly abbreviation of myth (a poem such as Hor. *Odes* 3.27 is unusual in length and in its exploration of mythic narrative), more easily involves itself in performative fictions that are outside the surviving Pin-

---

[7] T. Fuhrer, *Die Auseinandersetzung mit der Chorlyrikern in den Epinikien des Kallimachos*, Schweiz. Beitr. zur Altertumswsissenschaft 25 (Basel/Kassel 1992) *passim.*

[8] See Fuhrer (above, n. 7) 15, n. 19 for bibliography on the connection between Pindar and Hellenistic (including Callimachean) poetry; see also L. Lehnus, *Bibliografia Callimachea 1489-1988* (Genoa 1989) 339-45.

daric types—symposium, hymn. These are also types which have greater cross-cultural plausibility.

To the extent that Pindar *is* a model for a poet such as Virgil, Callimachus again plays a role in this regard; the Hellenistic poet had himself already minimized the performative or cultural aspects of epinician. Although the *Victoria Berenices* functions as an epinician to the poet's queen, it is accommodated to a genre and a poem of which praise is not the chief function; and the actual epinician portions even of this relatively small part of the *Aetia* are quite limited, possibly occupying only 10 lines before the transition to the story of Herakles and Molorchus and the aetion on the Nemean Games. Likewise *Iambus* 8 is part of a larger non-epinician collection: epinician iambic is a generic hybrid. And from the account in the *Diegesis* the poem, although classified as an epinician (Ἐπίνικος Πολυκλεῖ Αἰγινήτῃ νικήσαντι διαύλῳ Ἀμφορίτῃ ἐν τῇ πατρίδι), seems to have centred not so much on the victory of Polycles as on the details of the festival *Hydrophoria* and its aetiological basis in the Argonaut's competing to fetch water on Aegina. Epinician in the service of aetiology, or the other way around? Or both? And the *Victoria Sosibii* narrates three separate victories won over a number of years; to this extent the performative immediacy that is a key feature of epinician is absent from the Callimachean genre, which may be seen as an epinician catalog.

The conversion of epinician from strophic lyric to elegiac couplets in two of the Callimachean instances and (probably) to stichic iambic trimeters in the other, further contributes to the disruption of the original genre and to its transformation to a more malleable form, capable of crossing further generic boundaries in Rome. Virgil could doubtless have performed this transformation himself, but the fact that Callimachus did so before him is worth noting; the fact that we have fragmentary, generically transformed epinician in Callimachus, a poet Virgil demonstrably read extensively, should at the very least be in our consciousness as we ask whether features of his poetry are indebted to Pindar, a poet who is not generally considered one of Virgil's chief sources. What I will be suggesting in much of this paper, then, is that Pindaric presence may be mediated by the presence of something else, by a poetry which had already developed an intertextual relationship with the Pindaric model.

Let us start with a matter relevant to this general issue, one already treated in my commentary on the *Georgics*, where one reviewer accused me of taking a "phrase translated from Pindar" and "father[ing it] on Callimachus."[9] The phrase comes from the proem to the third *Georgic*, on which L. P. Wilkinson had claimed, in 1970, that Pindaric influence was "seminal and pervasive", though "little recognized by scholars so far."[10] Since there seems to me even now to be a classicizing resistance in some quarters to acknowledging the depth of Hellenistic intertextuality in Virgil, and since there is also additional evidence in this particular matter, it is worth briefly reconsidering the issue (with apologies for some repetition from the earlier, more restricted context):[11]

> Te quoque, magna Pales, et te memorande canemus
> pastor ab Amphryso, uos, silvae amnesque Lycaei.
> cetera, quae vacuas tenuissent carmine mentes,
> omnia iam vulgata: quis aut Eurysthea durum
> aut inlaudati nescit Busiridis aras?
> cui non dictus Hylas puer et Latonia Delos
> Hippodameque umeroque Pelops insignis eburno,
> acer equis? temptanda via est, qua me quoque possim
> tollere humo victorque virum volitare per ora.
> primus ego in patriam mecum, modo vita supersit,
> Aonio rediens deducam vertice Musas;
> primus Idumaeas referam tibi, Mantua, palmas,
> et viridi in campo templum de marmore ponam
> propter aquam, tardis ingens ubi flexibus errat

---

9 R.E. Fantham, rev. of Thomas (1988), *CP* 86 (1991) 166.

10 Wilkinson, "Pindar and the Proem to the third Georgic," *Forsch. zur römischen Literatur, Festschr. zum 60. Geburtstag von K. Büchner* (Wiesbaden, 1970) 286-91.

11 I have not here repeated from my commentary all the arguments, nor rehearsed all the evidence, for muting the purely Pindaric voice in these lines. Setaioli (above, n. 3) is the most recent resource to mention Pindar, but not Callimachus, in the context of the proem. While this volume was in production there appeared J. Henderson, "Pump up the Volume: Juvenal, *Satires* 1.1-21," *PCPS* 41 (1995) 101-37, which represents my position as exclusive of Pindar, a position I did not intend. I shall be more careful in a future discussion of these lines.

Mincius et tenera praetexit harundine ripas.
in medio mihi Caesar erit templumque tenebit:
illi victor ego et Tyrio conspectus in ostro
centum quadriiugos agitabo ad flumina currus.
cuncta mihi Alpheum linquens lucosque Molorchi
cursibus et crudo decernet Graecia caestu.
ipse caput tonsae foliis ornatus olivae
dona feram. iam nunc sollemnis ducere pompas
ad delubra iuvat caesosque videre iuvencos,
purpurea intexti tollant aulaea Britanni.
in foribus pugnam ex auro solidoque elephanto
Gangaridum faciam victorisque arma Quirini,
atque hic undantem bello magnumque fluentem
Nilum ac navali surgentis aere columnas.
addam urbes Asiae domitas pulsumque Niphaten
fidentemque fuga Parthum versisque sagittis;
et duo rapta manu diverso ex hoste tropaea
bisque triumphatas utroque ab litore gentis.
stabunt et Parii lapides, spirantia signa,
Assaraci proles demissaeque ab Iove gentis
nomina, Trosque parens et Troiae Cynthius auctor.
Invidia infelix Furias amnemque severum
Cocyti metuet tortosque Ixionis anguis
immanemque rotam et non exsuperabile saxum.
interea Dryadum silvas saltusque sequamur
intactos, tua, Maecenas, haud mollia iussa:
te sine nil altum mens incohat. en age segnis
rumpe moras; vocat ingenti clamore Cithaeron
Taygetique canes domitrixque Epidaurus equorum,
et vox adsensu nemorum ingeminata remugit.
mox tamen ardentis accingar dicere pugnas
Caesaris et nomen fama tot ferre per annos,
Tithoni prima quot abest ab origine Caesar.

Wilkinson found six features of the proem to be Pindaric, and claimed that in these Virgil looked directly to Pindar, against the view of U. Fleischer, who had postulated a Hellenistic intermediary:[12]

W1 The architectural metaphor of *Geo.* 3.13ff. is developed from Pindar *Ol.* 6.1-5

W2 The general epinician flavor suggests Pindar.

W3 *Geo.* 3.7 (*umeroque Pelops insignis eburno*) seems to be a verbal reminiscence of *Ol.* 1.27 ἐλέφαντι φαίδιμον ὦμον κεκάδμενον

W4 At 3.17 *illi victor ego*, the confidence of Virgil in including himself along with Octavian is a mark of Pindaric epinician.

W5 On the appearance of *Invidia* (37) Phthonos occurs at least 8 times in Pindar—"naturally enough in a writer of epinicians."

W6 Outside the proem of *Geo* 3, in the final lines of the preceding book, we find the following (*Geo.* 2.541-2):

sed nos immensum spatiis confecimus aequor,
et iam tempus equum spumantia solvere colla.

The metaphor bears a resemblance to Pindar, *Ol.* 6.22 ὦ Φίντις, ἀλλὰ ζεῦξον ἤδη μοι σθένος ἡμιόνων ("but now Phintis yoke for me the sturdy mules')—of poetics but where the victor has won in the mule chariot race.

I am not denying that Pindar may be present in these lines, but he is there with others. In 1976, S. Lundström (unaware of Wilkinson's article) traced some of these other possibilities, the Greek tragedians, Callimachus, Theocritus, Ennius, Accius, Lucretius, and Varius Rufus.[13]

But things changed in 1977, a fact for which Wilkinson and Lundström cannot be faulted, but also one that others seemed unable to

---

[12] "Musentempel und Octavianehrung des Vergil im Proemium zum dritten Buche der Georgica," *Hermes* 88 (1960) 280-331; cf. 281-2

[13] "Der Eingang des Proömiums zum dritten Buche der Georgica," *Hermes* 104 (1976) 163-91.

grasp.[14] In short, our corpus of relevant material was expanded: we now had, at the beginning of the third book of Callimachus' *Aetia*, an epinician, just as we had always had epinician elements at the beginning of the third book of the *Georgics*. This fact at least qualifies **W2**, the claim that Virgil and Pindar share an epinician tone. On **W1**, concerning architectural metaphors, it may be noted that Callimachus also, somewhere in the *Aetia*, seems to have produced such a metaphor (*Aet.* fr. 118), and a temple may have figured in the epinician for Berenice. As for **W4**, Virgil's positioning himself along with Octavian (*illi victor ego*), this feature is not simply Pindaric, and the third line of the *Victoria Berenices* has the same linking of poet and ruler: ἡμ[ε]τερο̣[. . . . . .]̣ εῶν ἐπινίκιον ἵππω̣[ν Whether we read ἡμέτερον παιᾶνα / δώρημα / δ' ἄνθημα / ἑῶν in the lacuna, we clearly have a parallel with Virgil's *victor* and Callimachus' ἐπινίκιον, as well as with the intertwining of poet and ruler—ἡμετέρον and *ego*; and whether ̣εῶν represents τεῶν, or ἑῶν (= σεαυτοῦ), we have mention of the addressee in close proximity (*illi* = *Caesari*). As for **W5**, Phthonos figured not only in Callimachus' programmatic passages, but also in the epinician *Victoria Sosibii*, possibly in the *Victoria Berenices* too; and while the Virgilian attitude of overcoming *Invidia* finds no parallel in Pindar, it registers distinctly with the generally applicable Callimachean dictum ὁ δ' ᾔεισεν κρέσσονα βασκανίης (*Epigr.* 21.4). On **W6**, Virgil's *et iam tempus equum spumantia solvere colla* may find its metaphorical impulse in Pindar, but Pindar nowhere mentions the unyoking of chariots as a marker of the end of song. For this wording at any rate, I think Virgil may have gone to Callimachus, *Hymn* 5.9-10 where Athena looses her horses' necks after the Gigantomachy ὑφ' ἅρματος αὐχένας ἵππων | λυσαμένα, where Bulloch shows that αὐχένας ἵππων (cf. Virgil's *equum ... colla*) is a Callimachean innovation on the simple Homeric ἵππους in similar expressions.[15] If so, that leaves us with **W3**, the one possible dictional coincidence where Pindar and Virgil both refer to the ivory shoulder of Pelops. I would first of all simply state that in the

---

[14] Latinists have often been slow in responding and adjusting to the papyrus finds of relevant Greek material, while some might think they respond excessively to the few Latin finds.

[15] Cf. A. W. Bulloch, *Callimachus, The Fifth Hymn* (Cambridge 1985) ad 9-10.

process of referring to Callimachean epinician, it would be in Virgil's manner to refer through him to the ultimate epinician master model—that is, to make a double reference. This would be true even if Callimachus nowhere treated the infamous Pindaric *zetema* concerning the way Pelops lost his shoulder. My strong suspicion is still that Callimachus, author of a treatise on the games in general (περὶ ἀγῶνων, fr. 403 Pf.) did treat that matter, and it is strengthened by the fact that Callimachus elsewhere seems to have adapted a line of the same first *Olympian*: in that poem Pindar had referred to the superiority of the Olympic games (*Ol.* 1.7 μηδ' Ὀλυμπίας ἀγῶνα φέρτερον αὐδάσομεν). At *Iambus* 4.57-9, this same argument for the superiority of the Olympic games was adopted and polemicized by the olive-tree in its agon with the laurel: κῶς ἄεθλον οὐκ ἐγὼ κρέσσων | σεῦ; καὶ γὰρ ὠγὼν οὑν Ὀλυμπίῃ μέζων | ἢ 'ν τοῖσι Δελφοῖς; On a larger scale, J. K. Newman has recently argued that the entire structure of the Callimachean *Hymn to Zeus* is modelled on that of the *First Olympian*.[16] Virgil's reference to Pelops, then, given these facts, given the context of the opening of *Georgics* 3, and given the Callimachean material throughout, is likely to include a reference to Callimachus as well as to Pindar.

But, of course, it is Molorchus who now clinches it. Before 1977, the first appearance of this figure was at *Georgics* 3.19-20:

> cuncta mihi Alpheum linquens lucosque Molorchi
> cursibus et crudo decernet Graecia caestu.

Now that the first appearance of Molorchus occurs precisely at the beginning of *Aetia* 3, we see that Virgil can only be referring to Callimachus. All Greece, Virgil says, will leave the Alpheus and the groves of Molorchus and cede to the new Mantuan setting. This simply underscores the utterances of *Geo.* 3.3-10: Virgil's own fame is to be constructed on themes other than those now done to death by Hellenistic (chiefly Callimachean) treatments and their Roman neoteric followers. For how can you say such themes are *omnia iam vulgata* if you are thinking only of Pindar, rather than an intermediary treatment of Pindar and others? Other than a single quote in a letter of Cicero to Atticus (13.38.2 = 341 ShB), and three other general references, there is no se-

[16] Cf. J.K. Newman, "Pindar and Callimachus," *ICS* 10 (1985) 185-88

cure evidence of knowledge of Pindar in Rome before the writing of the *Georgics*—his works can hardly be *vulgata*, then.[17] In short, any reference to Pindar at the opening of the third *Georgic* manifests itself through the filter of Hellenistic, and specifically Callimachean, epinician and poetics.

Poetics

"No Greek poet says so much as Pindar about his art." So begins Bowra's first chapter on "The Theory of Poetry", the first in his book on Pindar.[18] Saying a lot about their art is also of course one of the things that the Roman poets do, some more than Virgil, but also Virgil—if in a more embedded way than Horace, Propertius and Ovid. But again, with the exception of Horace, Pindar is not a poet who figures much in discussions of programmatics and metapoetics in Rome. Again it is Callimachean poetic metaphor, and Callimachean *recusatio*, that receive the focus in studies of the Roman poets.

Does this critical state reflect reality, or simply a "privileging" of Callimachus? J. K. Newman has stated: "What the Romans took from Greek Alexandria ... requires far more careful definition than has been customary." This is, I think, a fair comment, and yet I think it is also the case that the Augustan poets did look chiefly to Callimachus and other Hellenistic poets in this matter.[19] The subject is too large to treat satisfactorily here, but we may look at some exemplary material which shows that Callimachus systematized Pindaric metapoetics, "packaged" it in ways that made it accessible as a coherent programme, from which the Roman poets could then draw. Outside the epinicians, in *Dithyramb* 2 (= fr. 70b, p. 75 Snell-Maehler) we find the intensely metapoetic lines:[20]

---

[17] Catullus' lyric seems unaware of Pindar (not surprisingly), and Horace's *Pindarum quisquis studet aemulari* (*Odes* 4.2) suggests an avant-garde literary phenomenon.

[18] C. M. Bowra, *Pindar* (Oxford 1964) 1.

[19] I have studied this issue in "Callimachus Back in Rome," *Hellenistica Groningana* 1 (1993) 197-215.

[20] Newman (above, n. 16) 181-2 uses these instances to show Callimachus' debt to Pindar.

Πρὶν μὲν ἕρπε σχοινοτένειά τ' ἀοιδὰ
 διθυράμβων
καὶ τὸ σὰν κίβδηλον ἀνθρώποισιν ἀπὸ στομάτων...

"In the old days there flowed from the lips of men the drawn out song of dithyrambs, with the *san* discarded." The reference is both to length[21] and to the earlier practice of writing asigmatic odes (as was done, e.g. by Lasus of Hermione). Another Pindaric example, much discussed recently, is found at *Paean* 7b 11-14. I give the text of di Benedetto, also tentatively accepted by Ian Rutherford in his forthcoming study of the *Paeans*:

 Ὁμήρου [πολύτρι]π̣τὸν κατ' ἀμαξιτόν
 ἰόντες, ἀ̣[λλ' οὐκ ἀλ]λοτρίαις ἀν' ἵπποις
ἐπεὶ αὐ̣[τοὶ τὸ πο]τανὸν ἅρμα
Μ̣οισα[ῖον ἐλαύνο]μεν.

The sense then would be, "going along the much-worn track of Homer, but on someone else's mares, since we ourselves are driving the winged chariot of the Muses." What Pindar seems to be claiming, then, is originality and innovation, within a larger Homeric tradition.

Each of these two separate utterances of course finds a new life in the *Aetia* Preface, the first at 17-18:

ἔλλετε Βασκανίης ὀλοὸν γένος· αὖθι δὲ τέχνῃ
κρίνετε,] μὴ σχοίνῳ Περσίδι τὴν σοφίην

As in the first of the Pindaric fragments, the injunction is against excessive length, σχοίνῳ looks to σχοινοτένεια, and the ingredient of τέχνη is added. The second Pindaric passage is modified at lines 25-8:

προς δέ σε] καὶ τόδ' ἄνωγα, τὰ μὴ πατέουσιν ἅμαξαι
 τὰ στείβειν, ἑτέρων ἴχνια μὴ καθ' ὁμά
δίφρον ἐλ]ᾶν μηδ' οἶμον ἀνὰ πλατύν, ἀλλὰ κελεύθους
 ἀτρίπτο]υς, εἰ καὶ στεινοτέρην ἐλάσεις.

Where Pindar seems to have allowed following in the path of Homer, provided the vehicle was individual, Callimachus, with apparent refer-

---

[21] Newman (above, n.16) 181, n. 32 collects other Pindaric rejections of length.

ence to Pindar, insists that the path be independent, that it both be original, and at the same time narrow—that is generically restrained.[22]

These distinctions are reflected in Roman programmatics, where it is the Callimachean versions that seem to find favour. *Ars*, Roman *techne*, becomes the motto for all Augustan, and much other Roman, poetry, and in Lucretius (4.1 *loca nullius ante | trita solo*), Virgil (*Georgics* 3.292-3 *qua nulla priorum ... devertitur orbita*), and Propertius (3.1.18 *intacta ... via*) the integrity of the path is what matters, as well, in the case of Propertius, as its narrowness (3.1.14 *non datur ad Musas currere lata via*).

I believe that the Roman poets of the Augustan period looked to Callimachus chiefly for two reasons: first, his program coincided largely with theirs, and secondly it is precisely that—a sustained, accessible program. This sustained, metapoetic voice, which unlike that of Pindar speaks not just for the poet but for the scholar-poet, who talks openly and systematically about his own work and about the work of other poets in his tradition, made Callimachus the inevitable standard against which the Roman poets judged themselves and talked about themselves, whether they embraced the Hellenistic poet, or as in the case of Virgil, came to modify their relationship with him. Nor I am denying the influence of Pindar—and more work needs to be done on this subject—but in these and other instances the primary reference is to Callimachus.

Intertextual Paradigms

One area, perhaps the most important, in which Virgil is held to have used Pindar as a model is in the treatment of myth. Given how much we have lost—of archaic lyric (including Pindar), of Greek tragedy, and of Hellenistic poetry (these being three of the chief genres which are vehicles for the expression of myth)—we are on very uncertain ground when we try to recover Virgil's sources. At the same time I do think Virgil had some access to Pindar, but as in what I have said so

[22] This is hardly the place to respond to A. Cameron, *Callimachus and his Critics* (Princeton 1995), which appeared while this volume was in production. Cf. my paper, "Virgil after Cameron," delivered at CAMWS (Boulder, Spring, 1997) and subsequently to be published in *HSCP*.

far, that access is mediated through the incorporation of other models, as well as through his own reworking and his own originality—a quality too often denied to Virgil, particularly by nineteenth century Hellenic scholarship. Here are three three areas of potential Pindaric influence.[23]

Not much need be said here on one of the most complex mythological issues in Virgil, the katabasis of Aeneas. Norden is clearly correct in his view that Virgil drew not only from the Homeric katabasis of Odysseus, but also from katabases of Orpheus and Heracles. Which poets he drew from we will never know, and Norden's and Lloyd-Jones' suggestion that he worked with a handbook (for instance Apollodorus treated the katabasis of Heracles at 2.5.12) cannot be discounted. I will mention two aspects: Pindar treated the katabasis of Heracles at *Dith.* 2, as did Callimachus (fr. 515, 628Pf.), although neither Norden in his commentary on *Aen.* 6, nor Ganschinietz in his long article on katabasis in Pauly-Wissowa, seems aware of this. Such was the state of the world before Pfeiffer. Pindar and Callimachus, then, with a Virgilian treatment following—a pattern that becomes familiar.

Birds and leaves

Then there is the issue of the similes at *Aen.* 6.309-12, on which a great deal has been written:[24]

> quam multa in silvis autumni frigore primo
> lapsa cadunt folia, aut ad terram gurgite ab alto
> quam multae glomerantur aves, ubi frigidus annus
> trans pontum fugat et terris immittit apricis.

Virgil here likens the souls first to falling leaves, and then to birds fleeing to the shelter of land. In *Odyssey* 11 there is no simile, but Norden pointed to Bacchylides 5.63-7, which contains the leaf simile but no bird simile. He used this and other evidence to claim that Virgil had access to a Heracles katabasis now lost (Bacch. *Ep.* 5.63-67):

> ἔνθα δυστάνων βροτῶν
> ψυχὰς ἐδάη παρὰ Κωκυτοῦ ῥεέθροις,

[23] Cf. Setaioli (above, n. 3).

[24] Cf. also the forthcoming article of A. Giesecke in *Hermes*.

οἷά τε φύλλ' ἄνεμος
Ἴδας ἀνὰ μηλοβότους
πρῶνας ἀργηστὰς δονεῖ

In 1967, Lloyd-Jones suggested that Virgil's source was Pindar; the evidence was to be found in P. Oxy 2622a: 12-15, which is probably by Pindar (P. Oxy 2622 a: 12-15):[25]

αὐ]τίκα νιν φθιμένων [
]τρέφεται καὶ ὅσ' ἐν πόντῳ [
]μενοσ[ ]
]ạ μ[έγα]ν Διὸς υἱὸν

The subject is clearly similar to that of Bacchylides, and on the basis of the mention of the sea (πόντῳ), Lloyd-Jones proposed that Pindar also referred to the land—although no trace of the reference survived. If so, we would have had leaves on the land and birds at sea, and Virgil now has a source for his double simile. This proposal, all of these proposals, which presuppose that Virgil needed a precise source, containing both similes, need examination from the point of view of Virgilian composition—about which we can know a great deal. Scholars have long recognized that Bacchylides' and Virgil's leaf simile ultimately go back to Glaucus' comparison of the generation of men to leaves in *Iliad* 6.146-9:

οἵη περ φύλλων γενεή, τοίη δὲ καὶ ἀνδρῶν.
φύλλα τὰ μέν τ' ἄνεμος χαμάδις χέει, ἄλλα δέ θ' ὕλη
τηλεθόωσα φύει, ἔαρος δ' ἐπιγίγνεται ὥρη.
ὣς ἀνδρῶν γενεὴ ἡ μὲν φύει, ἡ δ' ἀπολήγει.

And the bird simile at the start of *Iliad* 3 provides a model for Virgil's bird simile (2-7):

Τρῶες μὲν κλαγγῇ τ' ἐνοπῇ τ' ἴσαν ὄρνιθες ὥς,
ἠΰτε περ κλαγγὴ γεράνων πέλει οὐρανόθι πρό,
αἵ τ' ἐπεὶ οὖν χειμῶνα φύγον καὶ ἀθέσφατον ὄμβρον,
κλαγγῇ ταί τε πέτονται ἐπ' Ὠκεανοῖο ῥοάων,
ἀνδράσι Πυγμαίοισι φόνον καὶ κῆρα φέρουσαι·
ἠέριαι δ' ἄρα ταί γε κακὴν ἔριδα προφέρονται.

---

[25] H. Lloyd-Jones, "Herakles at Eleusis: P.Oxy. 2622 and P.S.I. 1391," *Maia* 19 (1967) 206-29.

Now the first of these similes is already in a context which lends itself readily to reapplication to the souls of the dead, whether that reapplication is performed by Bacchylides and Virgil independently, or by the latter with the help of the former, or of some other writer on katabasis. I will return to this last option in a moment, but for now, would just note that the Homeric bird simile is elsewhere, on three separate occasions, used with modification by Virgil; in one of these the context is close to the Homeric one (*Aen.* 10. 262-6):[26]

clamorem ad sidera tollunt
Dardanidae e muris, spes addita suscitat iras,
tela manu iaciunt, quales sub nubibus atris
Strymoniae dant signa grues atque aethera tranant
cum sonitu, fugiuntque Notos clamore secundo.

Back to katabasis. Clark refers to *Georgics* 4.471-80 as another instance of katabasis with similes, but he does not seem to have realized what emerges from these lines:

umbrae ibant tenues simulacraque luce carentum,
quam multa in foliis auium se milia condunt,
Vesper ubi aut hibernus agit de montibus imber.

Here in the earlier version Virgil has the bird simile (not in Bacchylides), but no leaf simile (in Bacchylides). Leaves, however, do figure; the birds are roosting in them. Here, again is the double simile of *Aeneid* 6.309-12:

quam multa in silvis autumni frigore primo
lapsa cadunt folia, aut ad terram gurgite ab alto
quam multae glomerantur aves, ubi frigidus annus
trans pontum fugat et terris immittit apricis.

When he came to write *Aeneid* 6 Virgil re-employed his earlier single (bird) simile, but added a new simile, available from Homer (leaves). He footnotes the expansion with the removal of leaves from the bird simile: *quam multa in foliis* becomes *quam multa in silvis*, with *foliis* become the subject (*folia*) of the Aeneid simile. Did Virgil need Bacchylides to do this? Or was Homer and his own invention sufficient? I

---

[26] Cf. 7.698ff; *Geo.* 1.374ff.

suspect the latter. And I suspect we are meant to notice the development from *Georgics* 4 to *Aeneid* 6, notice it and see it as a Virgilian development. Thaniel, who responds to Clark with healthy scepticism for the notion of an archetypal katabasis of Heracles, does not even mention the simile from *Georgics* 4.[27] Incidentally Clark and Thaniel also point to two similes at Apollonius *Argonautica* 4.214-219 and 238-40 which may also be influential for Virgil. If so, again we would be dealing with what has emerged as a typical Virgilian procedure: reference to archaic Greek poetry through the filter of Hellenistic rewriting of that earlier tradition. The process may be set in terms of intertextual stemmatics:

Homer birds*Il.* 3.2-7) Homer leaves(*Il.* 6.146-9)
\ /
Virgil birds (*Aen.* 10-262-6) Virgil leaves (*Geo.* 4.472-4)
\ /
Virgil birds and leaves (*Aen.* 6.309-12)

Jove's Way

The second perceived area of connection between Pindar and Virgil is in the two poets' treatments of the Isles of the Blest in *Olympian* 2 and Elysium in *Aeneid* 6.[28] There is little dictional connection, and in the complexity of the metempsychosis of *Aeneid* 6 Pindar would only be one source along with a large number of others. I would like to turn from this familiar connection to one that has not been noticed. *Olympian* 2 may also be a model for a different Virgilian construct, to be found in the *Georgics*. Here is the relevant Pindaric material: the souls of the lawless, who have performed evil in the realm of Zeus, are punished; different is the treatment of the good (Pind., *Ol.* 2.61-71):

ἴσαις δὲ νύκτεσσιν αἰεί,
ἴσαις δ᾽ ἁμέραις ἅλιον ἔχοντες, ἀπονέστερον
ἐσλοὶ δέκονται βίοτον, οὐ χθόνα τα-
ράσσοντες ἐν χερὸς ἀκμᾷ
οὐδὲ πόντιον ὕδωρ

---

[27] G. Thaniel, "Vergil's Leaf- and Bird-Similes of Ghosts," Phoenix 25 (1971) 237-45.

[28] Setaioli (above, n. 3) 108-9

κεινὰν παρὰ δίαιταν, ἀλλὰ παρὰ μεεν τιμίοις
θεῶν οἵτινες ἔχαιρον εὐορκίαις
  ἄδακρυν νέμονται
αἰῶνα, τοὶ δ' ἀπροσόρατον ὀκχέοντι πόνον.
ὅσοι δ' ἐτόλμασαν ἐστρίς
ἑκατέρωθι μείναντες ἀπὸ πάμπαν ἀδίκων ἔχειν
ψυχάν, ἔτειλαν Διὸς ὁδὸν παρὰ Κρό-
  νου τύρσιν· ἔνθα μακάρων
νᾶσον ὠκεανίδες...

This is a fascinating passage, surely one of the most striking in any literature; for the present purposes, however, I would view it just from the focalization of the poet of the *Georgics*—though I might say that it would be unsurprising to find that Virgil had been attracted to it. There are some dictional parallels, unnoted by Mynors or myself in our commentaries. The phrase οὐ χθόνα ταράσσοντες ἐν χερὸς ἀκμᾷ οὐδὲ πόντιον ὕδωρ characterizes the ease of life in the world of Kronos; Virgil uses the same language, but from the other perspective, to describe the work that must go into attaining success in the agricultural world (*Geo.* 2.418-19):

sollicitanda tamen tellus pulvisque movendus
et iam maturis metuendus Iuppiter uvis

Pindar's use of of ταράσσω for ploughing is a striking neologism, which did not take hold, and is unparalleled in Greek; likewise *sollicitare* in the same sense otherwise occurs only in Tibullus and Ovid, who in their usage look to Virgil.[29] And a little later in *Georgics* 2, Virgil criticizes the aggressive seafaring that characterizes the age of Jupiter (503-4):

sollicitant alii remis freta caeca, ruuntque
in ferrum, penetrant aulas et limina regum.

Where Pindar's ταράσσοντες has two objects, χθόνα and πόντιον ὕδωρ, Virgil once has *sollicitanda* with *tellus* as subject, once *sollicitant* with *freta caeca* as object.

---

[29] ταράσσω in this sense (=ploughing) is not paralleled by LSJ. *Sollicitare* first in this sense in Virgil; other instances look to the same metaphor (Tib. 1.7.30; Ovid, *Fast.* 4.396).

For Pindar the blessed state is attained by passing along the road of Zeus and coming to the tower of Kronos/Saturn, ἔτειλαν Διὸς ὁδὸν. For Virgil, in the so-called theodicy of *Georgics* 1, the road leads in the other direction, towards toil, as Jupiter ends the ease of Saturnian existence (121-22): *pater ipse colendi* I *haud facilem esse viam voluit.* Finally, for Pindar, the good have a life that is free from toil (*Ol.* 2 62-3 ἀπονέστερον I ἐσλοὶ δέκονται βίοτον), while those who do not pass on to the Blessed Isles endure "toil that can not be looked upon" (ἀπροσόρατον ὀκχέοντι πόνον). These utterances likewise resonate in the Virgilian theodicy; there too the time of Saturn had been free from toil (*ipsaque tellus* I *omnia liberius nullo poscente ferebat*, 127-8). In fact, could these instances of πόνος not be the source of Virgil's notorious *labor improbus*, the condition of the age of Jupiter (*labor omnia uicit* I *improbus et duris urgens in rebus egestas*, 145-46)? The epithet *improbus* is a fairly good representation of Pindar's ἀπροσόρατον. Leonard Woodbury talks of Pindar's uncharacteristic eschatological interests in *Olympian* 2—"ordinarily his reverence is given to the Olympian gods."[30] Perhaps Virgil read the poem and was similarly struck by it, and saw the potential it contained for constructing an anti-Saturnian world, but in life, not the afterlife—a potential he realized throughout the poem. The *Georgics* functions as a sustained metaphor in which the age of toil, the age of modern agriculture, is presented as existing in the realm of Jupiter, which early in the poem replaces the pre-agricultural, spontaneously beneficent age of Saturn.

Aetna's Giants

I will conclude with the best known Pindaric/Virgilian coincidence, recognized and discussed since antiquity.[31] Aulus Gellius 17.10 documents Favorinus' views of Virgil's reworking, at *Aeneid* 3.570 ff., of Pindar's description of the eruption of Aetna in the *First Pythian* (*Aen.* 3.570-87; *Pyth.* 1.13-28[32]):

---

[30] L. Woodbury, "Equinox at Acragas: Pindar, *Ol.* 2.61-62," *TAPA* (1966) 598.

[31] On this see Horsfall (above, n. 1) 122-3; F. P. Rizzo, *EV* s.v. *Etna*.

[32] See also Macrobius, Sat. 5.17.7-14.

portus ab accessu ventorum immotus et ingens
ipse: sed horrificis iuxta tonat Aetna ruinis,
interdumque atram prorumpit ad aethera nubem
turbine fumantem piceo et candente favilla,
attollitque globos flammarum et sidera lambit;
interdum scopulos avulsaque viscera montis
erigit eructans, liquefactaque saxa sub auras
cum gemitu glomerat fundoque exaestuat imo.
fama est Enceladi semustum fulmine corpus
urgeri mole hac, ingentemque insuper Aetnam
impositam ruptis flammam exspirare caminis,
et fessum quotiens mutet latus, intremere omnem
murmure Trinacriam et caelum subtexere fumo.
noctem illam tecti silvis immania monstra
perferimus, nec quae sonitum det causa videmus.
nam neque erant astrorum ignes nec lucidus aethra
siderea polus, obscuro sed nubila caelo,
et lunam in nimbo nox intempesta tenebat.

ὅσσα δὲ μὴ πεφίληκε Ζεύς, ἀτύζονται βοάν
Πιερίδων ἀΐοντα, γᾶν τε καὶ πόν-
    τον κατ' ἀμαιμάκετον,
ὅς τ' ἐν αἰνᾷ Ταρτάρῳ κεῖται, θεῶν πολέμιος,
Τυφὼς ἑκατοντακάρανος· τόν ποτε
Κιλίκιον θρέψεν πολυώνυμον ἄντρον· νῦν γε μάν
ταί θ' ὑπὲρ Κύμας ἁλιερκέες ὄχθαι
Σικελία τ' αὐτοῦ πιέζει
    στέρνα λαχνάεντα· κίων δ' οὐρανία συνέχει,
νιφόεσσ' Αἴτνα, πάνετες χιόνος ὀξείας τιθήνα·

τᾶς ἐρεύγονται μὲν ἀπλάτου πυρὸς ἁγνόταται
ἐκ μυχῶν παγαί· ποταμοὶ δ' ἁμέραισιν
    μὲν προχέοντι ῥόον καπνοῦ
αἴθων'· ἀλλ' ἐν ὄρφναισιν πέτρας
φοίνισσα κυλινδομένα φλὸξ ἐς βαθεῖ-
    αν φέρει πόντου πλάκα σὺν πατάγῳ.
κεῖνο δ' Ἁφαίστοιο κρουνοὺς ἑρπετόν

δεινοτάτους ἀναπέμπει· τέρας μὲν
  θαυμάσιον προσιδέσθαι,
  θαῦμα δὲ καὶ παρεόντων ἀκοῦσαι,
οἷον Αἴτνας ἐν μελαμφύλλοις δέδεται κορυφαῖς
καὶ πέδῳ, στρωμνὰ δὲ χαράσσοισ᾽ ἅπαν νῶ-
  τον ποτικεκλιμένον κεντεῖ.

I am not here primarily concerned to rebut Favorinus' low estimation of the Virgilian version, though I may perhaps be able to explain in passing why such an impression could be formed. Clearly this is a direct reworking of Pindar, but a few observations and qualifications are worth making: first, although Favorinus did not notice this, Virgil's lines occur in a Homeric context and from the perspective of their place in the narrative they recall the *Odyssey* more than any other work. The line before the passage begins makes this quite clear (568 *ignarique viae Cyclopum adlabimur oris*). Moreover, the frame of the passage is Homeric: the opening line (*portus ab accessu uentorum immotus et ingens*) is in fact a reference to *Od.* 9.136 which describes Odysseus' arrival at the land of the Cyclopes (ἐν δὲ λιμὴν εὔορμος); similarly the closing frame of Virgil's passage (583-7 *noctem illam tecti silvis immania monstra | perferimus, nec quae sonitum det causa videmus. | nam neque erant astrorum ignes nec lucidus aethra | siderea polus, obscuro sed nubila caelo, | et lunam in nimbo nox intempesta tenebat*) refers to the gloom of the Homeric account (*Od.* 9.142-48—the coincidences between Homer and Virgil are underlined here for Homer, and above for Virgil):

ἔνθα κατεπλέομεν, καί τις θεὸς ἡγεμόνευε
<u>νύκτα</u> δι᾽ ὀρφναίην, <u>οὐδὲ</u> προὐφαίνετ᾽ <u>ἰδέσθαι</u>·
ἀὴρ γὰρ περὶ νηυσὶ βαθεῖ᾽ ἦν, <u>οὐδὲ</u> <u>σελήνη</u>
<u>οὐρανόθεν</u> <u>προὔφαινε</u>, <u>κατείχετο</u> δὲ <u>νεφέεσσιν</u>.
ἔνθ᾽ οὔ τις τὴν νῆσον ἐσέδρακεν ὀφθαλμοῖσιν·
<u>οὐδ</u>᾽ οὖν κύματα μακρὰ κυλινδόμενα προτὶ χέρσον
<u>ἐσίδομεν</u>, πρὶν νῆας ἐϋσσέλμους ἐπικέλσαι.

So the Pindaric description is embedded in the middle of an overall adaptation of Homer, which incidentally, Knauer lists in his index of Homeric parallels—though he does not mention Pindar of course. Pindar, like Homer, had also referred to night-time and darkness (as did Virgil at 583-7), at *Pyth.* 1.22-24:

ποταμοὶ δ' ἁμέραισιν
μὲν προχέοντι ῥόον καπνοῦ
αἴθων'· ἀλλ' ἐν ὄρφναισιν πέτρας
φοίνισσα κυλινδομένα φλὸξ ἐς βαθεῖ-
αν φέρει πόντου πλάκα σὺν πατάγῳ.

In Gellius' account, Favorinus in fact criticised Virgil's treatment of day and night (as opposed to that of Pindar, to which he felt it did not respond correctly). But Virgil is here following Homer, not Pindar, in his treatment of gloom associated with the area. Physical blindness in Homer—a failure to see the island of the Cyclopes (οὐδὲ ... ἰδέσθαι ... οὐδ' ... ἐσίδομεν) becomes for Virgil a more figurative blindness, a failure to see the cause of the sound of Aetna (*nec quae sonitum det causa videmus*). Both are caused by darkness, which in Homer is provided by the gods, in Virgil caused by the eruption. Virgil is possibly even correcting the Homeric explanation. We may criticize Virgil's treatment on these and other grounds, but we should not criticize them as translation of Pindar when they are in fact adaptation and integration of something else into the central Pindaric description. And at the same time it is worth noting that Virgil is possibly alluding to versions in Aeschylus (*Prom.* 351ff.) and Lucretius (1.722ff.; 2.593; 6.639ff., 669, 680ff.), although he avoids Lucretian language.

A further distancing from the Pindaric core of the passage occurs in the opening of the aition: 578 *fama est Enceladi semustum fulmine corpus* "the story goes that Enceladus' body was scorched by Jove's bolt." Horsfall has most recently and most fully studied the way such phrases as *fama est* function as references to the tradition and to previous details of aetia and the like. But this is not how the story goes in Pindar, for whom it is Typhoeus who is buried under Aetna (*Ol.* 4.6-7):

ἀλλὰ Κρόνου παῖ, ὃς Αἴτναν ἔχεις
ἶπον ἀνεμόεσσαν ἑκατογκεφάλα
Τυφῶνος ὀβρίμου.

A scholium on this passage states that Pindar thought Typhoeus was buried under Aetna, while Callimachus thought the figure was Enceladus (Σ Pind., *Ol.* 4.6-7): ὁ μὲν Πίνδαρος τῷ Τυφῶνι ἐπικεῖσθαί φησι τὴν Αἴτνην, ὁ δὲ Καλλίμαχος τῷ Ἐγκελάδῳ. In other words the issue was a *zetema* in Alexandria, and Callimachus pointedly disagrees with Pindar (*Aet.* fr. 1.35-36):

αὖθι τὸ δ' ἐκδύοιμι, τό μοι βάρος ὅσσον ἔπεστι
τριγλώχιν ὀλοῷ νῆσος ἐπ' Ἐγκελάδῳ.

Virgil, then, while translating Pindar into a Homeric frame, takes Callimachus' side against the overall Pindaric model, as he emphatically identifies Enceladus as the buried giant,[33] marking as he does so his own intertextuality (*fama est Enceladi*) Favorinus may even have been right in his literary judgement of Virgil, who was here perhaps trying to do too much.

In conclusion, I would say that there is overall little pure Pindar in Virgil. But Pindar does matter to Virgil, and in ways not touched on here presumably, even if he is almost always used and interpreted against other, chiefly Hellenistic, poets. Books have been written on Pindar's Homer and Callimachus' Hesiod; but with Virgil we need to think of Virgil's Homer / Hesiod / Pindar / Callimachus—always, I think, Callimachus. Then we need only to add all prior Latin poetry to the mix. The allusive scope of his whole corpus is complex and our knowledge of that scope, as of his methods, will remain as fragmentary as the remains of the models, but scrutiny is usually worthwhile.[34]

---

[33] Horsfall (above, n. 1) 122-3 does not discuss the variation in giants; it is discussed at Rizzo (above, n. 28) but without awareness of the polemical aspects, and Callimachus is misquoted, although he does mention later tradition in, e.g., Ovid.

[34] Callimachus, as we saw, set himself against the Homeric (*Il.* 15.187ff.) and Pindaric (*Ol.* 7.54ff.) versions in which lots were cast by Zeus and his brothers for the various realms.

** Versions of this paper were given at the University of Leeds, and at the University of Colorado, Boulder, in both of which places I was fortunate to receive useful criticism and warm hospitality.

IMAGES OF FALLEN TROY IN THE *AENEID*

DAVID O. ROSS

Aeneas' awareness that Troy is lost comes to him in the quiet of night's first sleep, when Hector appears to him in a dream (*Aen.* 2.268-97). This was the first night of peace for the city after ten years of war, a night of triumph and celebration. Only Cassandra knew that it was also their last (2.246-9).

Hector is Troy, in a way that no Achaean hero, even Achilles, stands for the Greek side. He is the son of Priam and Hecuba; he is Troy's first and strongest defense, slayer of men; he is the husband of Andromache, the father of Scamandrius, as he called his son, whom others called Astyanax, "for Hector alone protected Ilion" (*Iliad* 6.402-3). Homer's Hector knows well that, after his death in battle, the day will come when Troy will fall, but the fate neither of its citizens nor of his parents is such a grief to him as that of his wife (*Il.* 6.447-54), but by then the earth will cover him (464-5).

The Hector of Aeneas' dream is not, though, the great warrior (2.274-6):

> ei mihi, qualis erat, quantum mutatus ab illo
> Hectore qui redit exuuias indutus Achilli
> uel Danaum Phrygios iaculatus puppibus ignis!

Nor is this simply a figure in defeat, returning from death to ask some simple favor (as does Patroclus when he appears in a dream to Achilles in *Iliad* 23.62-107). Had Virgil meant simply to have this Hector rouse Aeneas, to tell him that Troy has fallen and must be abandoned, and to entrust to him the city's gods for the quest for a new Troy, he would have done so. What remains so vivid to Aeneas as he tells the story seven years later (this is the first time that he enters his narrative to Dido—*mihi ... ei mihi*, 271, 274) is the savage brutality that Hector has suffered. He comes to announce the death of his city, but he appears in all the horror of his mutilation as well (2.270-3):

> in somnis, ecce, ante oculos maestissimus Hector
> uisus adesse mihi largosque effundere fletus,

raptatus bigis ut quondam, aterque cruento
puluere perque pedes traiectus lora tumentis.

His hair and beard are matted with dried blood; he shows the wounds suffered at the hands of Achilles as he was dragged around his city's walls (2.277-9).[1]

Aeneas' response (2.281-6) is in character: incomprehension.[2]

'o lux Dardaniae, spes o fidissima Teucrum,
quae tantae tenuere morae? quibus Hector ab oris
exspectate uenis? ut te post multa tuorum
funera, post uarios hominumque urbisque labores
defessi aspicimus! quae causa indigna serenos
foedauit uultus? aut cur haec uulnera cerno?'

What he cannot understand is (first) the fact of death (*'quibus Hector ab oris expectate uenis?'*) and (then) the signs of savage brutality. Virgil, I think, assumes here a distinction, frequent elsewhere, between an idealized world of heroes and heroic action, and the stark reality of violence and death. Aeneas' first words are those of one hero addressing another, "O life of our country, our best hope...": Hector has returned, after so much, at the moment of final victory (so it seems). Aeneas, still very much the hero, at first reacts to the vision as if it must be a figure of glory, come to share in glory; thus it is that he cannot understand this silent and mutilated apparition, so changed from that Hector who once returned, glorious and proud, in the spoils of Achilles.

Aeneas will meet a similar figure in the underworld. Of the three significant ghosts of his past whom he sees there, only Dido might have

---

[1] R.G.Austin in his comment on line 278 (*Aeneidos Liber Secundus* [Oxford 1964]) is undoubtedly right to understand *uulnera illa* as those received from the dragging: the context does not allow for a proud displaying of wounds received fighting for his country (so Servius auctus, "*uelut insignia praeferens et ostentans, quae a diuersis pugnans pro patria susceperat*"). As in 1.483 (pointedly), Virgil has in mind the post-Homeric tradition, according to which Hector's body was dragged around the walls of Troy, not around Patroclus' tomb (*Il.* 24.14-17).

[2] Austin (*ad loc.*) finds narrative realism in these lines: "Virgil gets the atmosphere of a dream with much subtlety. Aeneas has forgotten Hector's death, and at first he does not take in the significance of his ghastly appearance; he thinks only that Hector has been away, and asks him, so naturally, why he has been so long in coming."

been expected. The first and the third are thoroughly Virgilian. Palinurus, the Trojan helmsman and guide, whose death is both cruel and inexplicable, who is constrained to the wrong side of the Styx and expected to be content with an aetiological consolation (*duri solacia casus*, 6.377), is an important part of Virgil's conception of Troy: the steersman from Troy cannot come to Italy, but his personal suffering must be buried and forgotten beneath his lasting (heroic) monument. Deiphobus, though, is the figure that concerns us here.

Immediately after leaving Dido, Aeneas comes to the gathering of the heroic dead—Argive, Trojan, and Greek (6.477-93). Eight of the long rank of Trojans are named. We might suppose Hector to be one of the throng gathered around Aeneas, but he is not named, or even hinted at; a less restrained poet might have brought him forward. Among Virgil's ghosts, though, he is hardly to be noticed, until Aeneas sees Deiphobus (494): then we might remember that he is Hector's brother and that it was his appearance that Athena had assumed when she urged Hector to stand against Achilles for the last time (*Il.* 22.226-47). For Virgil, though, Hector is made present only through the mutilation Deiphobus has suffered, and only because we must recall the mutilated Hector of Aeneas' dream (6.494-502):

> Atque hic Priamiden laniatum corpore toto
> Deiphobum uidet et lacerum crudeliter ora,
> ora manusque ambas, populataque tempora raptis
> auribus et truncas inhonesto uulnere naris.
> uix adeo agnovit pauitantem ac dira tegentem
> supplicia, et notis compellat uocibus ultro:
> 'Deiphobe armipotens, genus alto a sanguine Teucri,
> quis tam crudelis optauit sumere poenas?
> cui tantum de te licuit?'

Just as with Hector, Aeneas speaks first,[3] and again with an address of epic hyperbole, "Deiphobus, powerful in arms, descended from the lofty line of Teucer..."; again questions follow, "What are these dire wounds?"

---

[3] *ultro compellare* occurs in both passages, which may not mean anything. It could suggest that Virgil was working on both passages at the same time.

Deiphobus' answer takes us back to the deception of Troy's last night (*supremam falsa inter gaudia noctem*, 513), to that deep sleep like death itself (*dulcis et alta quies placidaeque simillima morti*, 522). We are among well-known Virgilian themes: the heroic world reduced at the end to an obscure destiny and human passions (*sed me fata mea et scelus exitiale Lacaenae | his mersere malis*, 511-2). The result is the mutilation Aeneas sees so vividly among these ghosts of the heroic past: we see again the final reality of heroic glory.

These images would not be out of place in another gallery, among the pictures of war that Aeneas and Achates see while waiting for Dido (1.441-97). Virgil does not tell us what these pictures are (we assume that they are wall paintings), but he is very explicit as to where they are: this is the temple of Juno (the Trojans' greatest enemy), in a grove at the very center of the city, on the spot there the Tyrians first found the head of a horse, the omen of their future supremacy in war (1.441-5). The tension between what Aeneas knows and what the reader knows is obvious: Aeneas, at this time of his greatest need and weakness, has been brought (the reader knows) to the very center of hostile power. The tension continues in what follows (1.450-2):

> hoc primum in luco noua res oblata timorem
> leniit, hic primum Aeneas sperare salutem
> ausus et adflictis melius confidere rebus.

The reader, who knows the future of Carthage, cannot share in the hope and confidence offered by this "strange fortuity" (*noua res*); Aeneas' words of encouragement to Achates seem therefore too full of promise (1.459-63): "What land, what region is not full of our struggle, Achates? Look, there's Priam. Even here heroic fame has its reward, here is sympathy for suffering and death. Don't fear; this renown of ours promises us safety."

> constitit et lacrimans 'quis iam locus,' inquit, 'Achate,
> quae regio in terris nostri non plena laboris?
> en Priamus. sunt hic etiam sua praemia laudi,
> sunt lacrimae rerum et mentem mortalia tangunt.
> solue metus; feret haec aliquam tibi fama salutem.'

Aeneas' hope is expressed in the familiar terms of heroism. Heroic glory and renown (*laus*, *fama*) have their rewards, and the recognition of heroic struggle and death will assure sympathy and safety even here in this strange city.[4] If we listen to Aeneas, if we put ourselves in the positon of the faithful Achates, then we will be prepared to see scenes of heroic valor, of epic struggle and death, and we too will feel encouraged and hopeful. But we know the future of Carthage, and we can see things in these pictures that Aeneas does not.

Perhaps, though, Aeneas does see as clearly as we do. It has not been that long since he addressed his men on the beach—again, words of encouragement, borrowed in part from Odysseus at a similar moment: "Be brave, don't fear, we will go on" (1.198-207). Encouraging words, but he himself felt no such confidence (1.208-9):

> Talia uoce refert curisque ingentibus aeger
> spem uultu simulat, premit altum corde dolorem.

Here at the center of Carthage there is a similar gulf between what he says and what he feels. He weeps as he begins to speak (*constitit et lacrimans...*, 459—surely not tears of joy), and he concludes with a fuller flood of emotion (1.464-5),

> sic ait atque animum pictura pascit inani
> multa gemens, largoque umectat flumine uultum.

As Aeneas looks at the pictures, he groans and weeps, and what he sees is empty, without substance (*pictura inani*). *Sunt lacrimae rerum et mentem mortalia tangunt*: for Achates these words were intended to convey simple encouragement—"here we will find sympathy"—, but they may also be Virgil's way of directing us more deeply to Aeneas' tears as he views his heroic past (its values now without substance), and sees its

---

[4] R.D.Williams' comment (*The Aeneid of Virgil: Books 1-6* [London 1972] *ad loc.*) is worth quoting: "Line 462 is often detached from its context and quoted to summarise the note of pathos in the *Aeneid*; there is no harm in this provided that it is understood that the meaning is 'people are sympathetic', not 'the world is full of sorrows, is a vale of tears'." Aeneas says simply "People are sympathetic," but Virgil surely wants his reader to consider further.

cruelty and senselessness as well. What are these images of his city at war?[5]

First, as an introduction, the back and forth flow of battle (1.466-8): Greeks in flight, Trojans in pursuit; then Trojans, and "helmeted Achilles pressing them with his chariot."

Next Aeneas sees the camp of Rhesus (1.469-73), raided during its first night by Diomedes and Odysseus (so *Iliad* 10):

nec procul hinc Rhesi niueis tentoria uelis
agnoscit lacrimans, primo quae prodita somno
Tydides multa uastabat caede cruentus,
ardentisque auertit equos in castra prius quam
pabula gustassent Troiae Xanthumque bibissent.

We might expect to see Diomedes "bloody with great slaughter", but it is not likely that the artist could have shown what Virgil brings out here, that the camp was "betrayed in its first sleep." We think inevitably ahead to Troy's last night and that first sleep (*prima quies*, 2.268) when bloody Hector appeared to Aeneas; Aeneas weeps again as he recognizes the scene. Virgil goes beyond Homer (and Euripides too), and again beyond pictorial art, by referring to an oracle (known to the later tradition) according to which Troy would not fall if Rhesus' horses ate the grass of the Trojan plain and drank from the Xanthus: what might have been.

Next is Troilus, "unhappy boy," out of place in such combat (1.474-8):

parte alia fugiens amissis Troilus armis,
infelix puer atque impar congressus Achilli,
fertur equis curruque haeret resupinus inani,
lora tenens tamen; huic ceruixque comaeque trahuntur
per terram, et uersa puluis inscribitur hasta.

---

[5] W.V. Clausen comments on this passage as one of the three major ecphrases of the poem (*Virgil's Aeneid and the Tradition of Hellenistic Poetry* [Berkeley 1987] 16-18): illuminating and (as always) understated. Virgil (Clausen sees) has added the spectator (Aeneas) to the traditional ecphrasis, in order to focus on Aeneas' emotions: "Aeneas is present as spectator, completely absorbed, the various scenes viewed as through his eyes, colored by his passion. Standing here, on alien ground, confronting his ghosts, Aeneas is all but overwhelmed by a sense of displacement and loss. Nowhere else is his grief so powerfully expressed" (p.17); "Aeneas is, by now, emotionally exhausted" (p.18); on the language of Aeneas' tears, see 130 n. 7.

Virgil again recalls a later story, with a prophecy that had Troilus, a son of Priam, reached the age of twenty, Troy would not have fallen; again what might have been is brought to mind. The scene is purposely one of horror: the youth is being dragged by his horses, flung from his chariot on his back, still grasping the reins, "his neck and hair are dragged along the ground, the dust is furrowed by his turned spear." Inevitably, we think again of Hector as he appears to Aeneas. This too is a son of Priam: Hector, Deiphobus, Troilus. Virgil reminds us as well of another figure out of place in the heroic world, Euryalus, boy and lover (*ora puer prima signans intonsa iuuenta*, 9.181), whose night raid, with disastrous results, is also patterned on Homer's raid on Rhesus' camp. Troilus is no better suited than Euryalus to heroic combat (*impar congressus Achilli*). The "turned spear" is another motif. It is not Achilles' spear (*versa* would be meaningless if it were),[6] but rather is Troilus' own, with which he had been driving the horses; Camilla bears a similar token (*et pastoralem praefixa cuspide myrtum*, 7.817—the last line of Book VII).

"Meanwhile" (*interea*) the Trojan women in sorrow and supplication are bringing the peplos to the temple of Athena, "unjust" (*ad templum non aequae Palladis*, 479).[7] The goddess, turned away, fixes her gaze on the ground (*diua solo fixos oculos auersa tenebat*, 482). Here is another aspect of Troy's fall, a revelation of the women's suffering and of divine rejection. *Non aequae* is Virgil's comment; the goddess's refusal is also Homer's (*Il.* 6.311).

The last great scene follows with no connective, as if Aeneas' gaze had suddenly, without warning, fallen upon it—and in fact Aeneas' reaction becomes an important and vivid part of the panel (1.483-7):

> ter circum Iliacos raptauerat Hectora muros
> exanimumque auro corpus uendebat Achilles.
> tum uero ingentem gemitum dat pectore ab imo

---

[6] So, rightly, Williams, "The Pictures on Dido's Temple," *CQ* 10 (1960) 145-6 (and his commentary ad loc.), who refers to Remulus' taunt, *uersaque iuuencum* | *terga fatigamus hasta*, 9.609-10.

[7] The Homeric model is, of course, the reason for Hector's return from battle in *Iliad* 6, which will lead to his meeting with Andromache. Theano's prayer (to destroy Diomedes) is answered by the goddess with a simple refusal (6.297-311): what Virgil has brought to the scene is clear.

ut spolia, ut currus, utque ipsum corpus amici
tendentemque manus Priamum conspexit inermis.

Here is Hector, as he will appear to Aeneas in Book II, having been dragged in the dust just as Troilus is seen, and here is Priam, their father, another suppliant, a king "unarmed", begging for the corpse of the city's greatest protector. This is no scene of "ransom," though: Virgil has made something quite different from that heroic convention, and something that this Carthaginian artist could not have been able to represent. Achilles is "selling Hector's lifeless body for gold." This is the third time the Greek hero has been shown: he was first named in the initial ebb and flow of battle (468), then as the warrior whom the boy Troilus unhappily met (475), and now is revealed as a figure of ruthless greed, not just savage (*saeuum*, his epithet at 458), but simply a mercenary.[8] Virgil forces us again to recognize what heroism had become, and it is this, we may conclude, that draws the groan from Aeneas.

There is a final scene of general fighting (488-93), intended to balance the opening scene, in which Aeneas recognizes himself, along with Memnon and (finally) Penthesilea, leader of the Amazons, raging in the fire of battle (1.491-3),

... furens mediisque in milibus ardet,
aurea subnectens exsertae cingula mammae
bellatrix, audetque uiris concurrere uirgo.

Here too is a suggestion of greed: gold will play an important part in Camilla's death in battle. Here too is a figure (reserved for last, just as is Camilla in the Italian catalogue in Book VII) from a world out of joint (we remember Troilus), the virgin warrior not just "daring to do battle with men," but swept away by the raging fire of war "in the midst of thousands"—as wild a phrase as Virgil ever wrote, and his own touch, not the artist's.

---

[8] Clausen has observed (above, n. 5) 131, n. 8, "The figure of Achilles dominates these scenes...The intended effect? The man of sensibility contemplates the man of violence." And earlier (p. 17), "Then, with a convulsive outburst of grief...he sees Achilles selling Hector's mangled body for gold; the chariot, his friend's body, Priam's outstretched helpless hands. A brutal, ignoble scene, deliberately unHomeric."

Troy was Rome's heroic past, just as King Arthur's court and Camelot could be said to be our heroic past; as such, Troy was the object of easy idealization, both in the Roman mind and often (intentionally, of course) in Virgil's poem—Homeric heroism without Homer's realism. But there are many Troys in the *Aeneid*, just as there are many images of heroism, and often there is tension between the nobility of heroic deeds and sacrifice, and savage brutality. In these images of fallen Troy, Virgil shows us the savagery that was Troy's conclusion.

THE STATUE OF ATHENA AT TROY AND CARTHAGE

ALESSANDRO BARCHIESI

For Virgil, epic has a serious responsibility for the destiny of nations. Homeric words being prophecies, the *Aeneid* must be careful in its own words, and hints, about the fate of towns and empires. The language of the gods in the poem is particularly loaded with historical significance. When Delian Apollo foretells the rule of the house of Aeneas—*hic domus Aeneae cunctis dominabitur oris* (*Aen.* 3.97)—he is not only updating and correcting an Iliadic prophecy,[1] but also aligning his own divine nature with the mission of Aeneas and Augustus: he had been born in Delos to rule over mankind (*h. Hom. Ap.* 29): ἔνθεν ἀπορνύμενος πᾶσι θνητοῖσιν ἀνάσσεις. It was perhaps a momentary expression of anger, when Hera said to Zeus, "Yes, take my favourite towns and destroy them, Argos, Sparte, and Mycenae..." (*Il.* 4.51-54), provided only she can satisfy her wrath against Troy, yet Virgil is scrupulously exact in showing that these words are not without consequence: they will come true when Jupiter picks (capriciously?) two of these towns (*Aen.* 1.284-85; 6.838) and adds one more, Juno's favourite (1.15-16), though one in North Africa not Greece.

In the temple of Juno at Carthage, Aeneas sees an episode of Trojan supplication to Athena (*Aen.* 1.482): *illa solo fixos oculos aversa tenebat.* It is quite easy, in retrospect, to unpack the message of doom that Homer had conveyed in a brief hint (*Il.* 6.311): ἀνένευε δὲ Παλλὰς Ἀθήνη. Aeneas rereads, through the pictures of the Epic Cycle at Carthage, a scene from the Iliad. Its importance is clearly signalled by its setting as a centerpiece. The behavior of the statue has a clear bearing on Troy's fate, but the exact meaning of the image is not easy to pin down. Many layers of representation are involved:[2] the author's voice; the reaction of Aeneas, the Virgilian focalizer; the effect of the images, and the visual strategy employed by the artists; the text of the Iliadic model; as-

[1] See my comments in "Immovable Delos," *CQ* 44 (1994) 441-43.

[2] On the narratology of descriptions see D.P. Fowler, "Narrate and Describe: The Problem of Ecphrasis," *JRS* 81 (1991) 25-35.

sumptions about the "original" aspect of Athena's statue in the Trojan temple. *Oculos auersa tenebat* implies that the gaze of the statue reflects Athena's refusal, but it would be difficult to decide whether the statue's description suggests an interpretation by the beholder(s) or some kind of miraculous event. This difficulty is compounded by *tenebat*, denoting immobility, which is contrasted with *aversa*, which *could* imply a supernatural movement, but is also the normal expression for divine hostility.

This indecision is made only too clear if we consider the problems of the Homeric model: Aristarchus athetized line 6.311. Our scholia assume that the poet is describing an actual gesture of the statue and reject the paradoxon as laughable: ἀθετεῖται ... γελοία δὲ καὶ ἡ ἀνανεύουσα Ἀθηνᾶ (schol. A *Il.* 6.311, p. 185 Erbse). But it would be quite legitimate to take ἀνανεύω not as a description of a gesture, a sign oriented towards the supplicating Trojans, but as an aside of the omniscient narrator referring to the Olympian reception of a prayer (cf. *Il.* 16.250). The name "Pallas Athene" of course can denote both "the ritual image" (e.g. 6.303) and "the goddess herself"; many ancient interpretive communities would have seen this ambiguity as purposeful—including Virgil as a Homeric reader of his age.[3] It is also conceivable that Virgil wanted to answer the Homeric *zetema* with an insoluble knot, *aversa tenebat*—a challenge to future commentators.

Since, however, ἀνανεύω is a common gesture of refusal in Homer, we might ask what the implications of the Virgilian substitute are: looking down and away instead of throwing the head back. On one level, as we saw, this is an apt gesture if Virgil intended to keep a balance between a miraculous event and a subjective (even retrospective) interpretation of a normal sculptural pattern. More can be inferred if we combine the message, the context, and some kind of historical perspective.[4] What is at stake in the Trojan supplication? *Iliad* VI describes a

[3] Similarly, art historians argue that it is meaningless to ask whether Athena as represented in versions of the Ajax/Cassandra myth in archaic Greek vases is the goddess or the statue: K. Schefold, "Statuen auf Vasenbildern," *JdI* 52 (1937) 45 (see also G. Hedreen, "Image, Text, and Story in the Recovery of Helen," *CA* 15 [1996] 180 n. 125); J.B. Connelly, "Narrative and Image in Attic Vase Painting: Ajax and Kassandra at the Trojan Palladion," in P.J. Holliday, ed., *Narrative and Event in Ancient Art* (Cambridge 1993) 101.

[4] I have suggested a similar sense of catastrophic irony for the Delian prophecy in Aen. 3.90-98, (above, n. 1) 440-42.

failed attempt to conjure Athena's help against her favourite warrior Diomedes; Virgil is staging this Homeric episode, but out of context and in retrospect: the relevance of Athena's hostility is now enhanced by a sense of catastrophe. Athena will do the Trojans much more harm than by protecting Diomedes: the next image mentioned by Virgil (1.483) is the corpse of Hector, the most famous victim of Pallas Athena. The town will fall as a result of the Trojans' inability to preserve divine favour through their communal cult—a message of central importance for the figurative program of this temple of Juno, patroness of the new town Carthage. It is natural to read this episode—which Virgil sets as a centerpiece in Aeneas' rehearsal of the pictures—as a pivotal prophecy of Troy's divinely sanctioned annihilation. Aeneas' reception of the scene and the Carthaginians' concern with it can be seen to converge. I have argued elsewhere that the future destruction of Carthage by the Romans is a powerful focus of this episode,[5] with suggestions of a sort of historical vertigo: the rising buildings of the new town can be seen as an epitome of future destruction. Could the statue of Athena play a role in this strategy?

The gaze of the statue avoids contact with her former protégés. Athena's role in the destruction of Troy will be decisive. But the first speech of the *Aeneid* (1.39-45) informs us that she has changed again. The disaster of the Greek *nostos* begins in Athena's temple at Troy. This was the site of a famous miracle: the statue *raises* its eyes to avoid the sight of what Ajax Oileus did to Cassandra. Most of our sources for this episode focus on an *upwards* movement of the eyes, an omen of destruction for the conquerors of Troy as Athena glances to the ceiling, not to the floor (schol. AD N 13.66 = Call. fr. 35 Pf): ὥcτε τὴν θεὸν τοὺς ὀφθαλμοὺς τοῦ ξοάνου εἰς τὴν ὀροφὴν τρέψαι (στρέψαι Schwartz). The same detail is found in Lycophron, *Alex.* 361 γλήνας ἄνω στρέψασα χώσεται στρατῷ and Quintus of Smyrna 13.426-7 βλοcυρὰς ἔτρεψεν ὀπωπὰς | νηὸν ἐς ὑψόροφον. Interestingly, Triphiodorus (649) describes the prodigy with the same verb, ἀνανεύω, which Homer had used for Athena's reaction to Trojan supplication.

---

[5] See my discussion in "Unità e diffrazione: la rappresentazione del dolore nell' *Eneide*," *A & A* 40 (1994) 115-24.

Virgil's new interest in the eyes of the statue, unparalleled in Homer, shows that the Greeks' punishment is already inscribed at the very center of the narrative of Trojan fall. This prophecy of reversal is by no means a solitary hint: the pathos of the Virgilian episode links the rising temple of Carthage with the Scipionic destruction, a thousand years later, and the tears shed by a Trojan/Roman leader for Troy at the site of Carthage anticipate by a millennium Scipio quoting the Homeric prophecy of Trojan doom while his men sack the temples and level Carthage to the ground. Scipio's tears, it will be remembered, were concerned with Trojan past but also with Roman future as seen in the mirror of the sack of Carthage. And the sack of Carthage had been coterminous with the final revenge of 'Troy' over the Achaeans in the sack of Corinth.[6]

The poetic lineage of Athena's miracle could be older than Callimachus. Alcaeus' narrative (fr. 298.8-27 V) of Ajax' sin and punishment (along with much other lost poetry) could be relevant to this tradition. Just like Virgil (Aen. 1.41 *unius ob noxam...*), Alcaeus had emphasized that the collective disaster irradiates from the fatal guilt of one individual.[7] It would be good to know, for instance, the meaning of ὑπ' ὄφρυσιν πελιδνωθεῖσαν (fr. 298. 25 V = *POxy* 2303.11-12). Athena's reaction to the violence is being described: "under the eyebrows" is most naturally taken in reference to the eyes, and the verb normally means "to become livid."[8] I wonder if we could start from the standard epithet *glaukopis*: this would give point to an alteration of the eyes, provoked by anger. The corresponding epithet in Quintus of Smyrna (13.426), βλοσυράς, is problematic already in Alexandrian lexicography, but could point to a

---

[6] On the historical perception of the two events, see N. Purcell in D. Innes, H. Hine, and C. Pelling, edd., *Ethics and Rhetoric. Classical Essays for D. Russell* (Oxford 1995) 133-48. Denis Feeney reminds me that Callimachus had combined Athena's revenge, Palladion cult, and dangerous visions in the *Bath of Pallas*.

[7] More precisely, S. West, "Notes on the Text of Lycophron," *CQ* 33 (1983) 132 has noticed the analogy between *unius ob noxam* and Lyc. *Alex*. 365 ἑνὸς δὲ λώβης ἀντί. The "basic conception" of the Alcaic poem was, according to A. M. van Erp Taalman Kip, "Aias and Kassandra," in S. Slings et al., edd., *Some Recently Found Greek poems* (Leiden 1987) 113, that "the punishment of one man may be necessary to prevent a whole community from suffering."

[8] Cf. van Erp Taalman Kip (above, n. 7) 120-21; Pfeiffer on Call. fr. 374.

similar change.[9] Note also that when Virgil describes Cassandra she is being dragged outside the temple and raising her eyes to the sky (2.403-6):

> ecce trahebatur passis Priameia virgo
> crinibus a templo Cassandra adytisque Mineruae
> ad caelum tendens ardentia *lumina* frustra,
> *lumina*, nam teneras arcebant uincula palmas.

The gesture suggests a desperate cry for divine intervention, and possibly implies analogy with Athena's famous reaction,[10] omitted in the selective narration of the Virgilian Ilioupersis. In this learned epic omission, of course, does not mean absence of implication.

In the heat of battle, Aeneas could not foresee the future: in the line before Cassandra appears, he mentions *invitis ... diuis*, and clearly means that divine protection helps the Greek invaders. Yet his choice of words—*uirgo, passis crinibus, adytis*—focuses on the sacrilege that will overturn divine benevolence. The image of helplessness and impotence—the girl in chains raising *ardentia lumina* to the sky like a big-eyed martyr—is in fact the hinge of this narrative of destruction: a promise of reversal out of a Trojan nadir.

Ovid responds to the Virgilian iconography with a more standard image: Cassandra supplicates with a gesture, not a gaze (*Met.* 13.410-11):

> tractisque comis[11] antistita Phoebi
> non profecturas tendebat ad aethera palmas

The imitator thereby constructs Virgil's version as a strained, manneristic deviation: the new epic has the true version, the mainstream image

---

[9] On the evolution of the epic adjective see M. Leumann, *Homerische Wörter* (Basel 1950) 141-48 and Livrea on A.R. 4.1437.

[10] So, independently from each other, West (above, n. 7) 135 and, with some hesitation, A. Pardini, "La colpa di Aiace e la poesia augustea," *MD* 22 (1989) 201-6.

[11] The text is problematic: perhaps *tractis* is the best solution because it responds to *trahebatur passis* in Virgil. I suppose that Ovid is interested in a close verbal recall of Virgil precisely because he changes the iconography of the reference. In both cases, *traho* reflects the standard use of ἕλκω to describe the sin of Aias (see van Erp Taalman Kip [above, n. 7] 117).

reflected by the visual arts, which often show Cassandra stretching her arm(s), and dragged by the hair.[12] Ovid displaces the Greeks as a "model" for Virgil. (It seems to be an article of faith that epic poets ignore the rape of Cassandra, but since the tradition is so explicit and persistent,[13] it is much more in Virgil's and Ovid's style to acknowledge this alternative tradition obliquely)

A link between the omen in *Aeneid* I and the Cassandra myth has been seen by Frances Muecke,[14] but her fine paper has a different agenda—the gesture of looking down in the Aeneid—and she goes too far when she brings in as a parallel the witness of Sinon about the Palladion (2.170-75):

> ... aversa deae mens
> nec dubiis ea signa dedit Tritonia monstris.
> vix positum castris simulacrum: arsere coruscae
> luminibus flammae arrectis, salsusque per artus
> sudor iit, terque ipsa solo (mirabile dictu)
> emicuit parmamque ferens hastamque trementem.

In fact the Trojan epic tradition offers three narrative situations where a "Palladion"—i.e. an image of Athena—plays a role: the supplication in *Iliad* VI, the theft of the Palladion, and the rape of Cassandra. The first and the last are normally thought of as compatible: a cult image of Athena Ilias in the citadel of Troy can be both the object of a supplication (involving the dedication of a *peplos*) and the silent but fatal witness of Ajax' sacrilege. Conversely, the second episode clearly presupposes a different kind of "Palladion": an image small enough to be carried away in a night theft, and endowed with magical, talismanic power. Understandably, very few ancient sources try to merge this object with the

---

[12] On the different iconographies see Connelly (above, n. 3). The use of *antistita Phoebi* suggests Roman tragedy (Acc. *trag.* 167 R^2) and unties the image from Virgilian monopoly: *antistita* is an Ovidian hapax, never used in Virgil.

[13] E.g. Ov. *am.* 1.7.17-18; cf. McKeown ad loc.

[14] "Turning Away and Looking Down: Some Gestures in the *Aeneid*," *BICS* 31 (1984) 107: "the episode may be regarded as a source for the portent at 1, 482"; see also the thorough discussion by Licinia Ricottilli, "'Tum breviter voltum demissa profatur' (*Aen.* 1.561): individuazione di un 'cogitantis gestus' e delle sue funzioni e modalità di rappresentazione nell' *Eneide*," *MD* 28 (1992) 179-227.

'official' cultic image of Athena offended by Ajax' sin in the *Iliupersis* and supplicated by Hecuba in the *Iliad*.[15]

In the world of the *Aeneid*, the distinction is confirmed when the snakes take their place under the statue of Athena in the sanctuary (2.226-27): this cannot be the same image which Sinon had just mentioned as having been stolen by the Greeks.[16] If the Trojans knew more about the epic tradition, they would easily realize that Sinon is a sham: this cannot possibly be true epic. This prodigy of a sweating statue that gives "clear indications"—flashing red eyes like a toy vampire, stomping thrice on the ground in full panoply like a cartoon—breaks the dignified rules of epic supernatural,[17] and is exposed as outré fiction by the self-conscious narrator himself (with his provocative aside *mirabile dictu*). *Aversa deae mens*, of course, implies the epic truth that Minerva is *aversa* to the Trojans, not to the Greeks.

Yet this "rigmarole" (Austin) is inextricably enmeshed with further truths and unintentional ironies. The raising of the eyes—exaggerated, in Sinon's expressionistic style, in *luminibus ... arrectis*—anticipates the disaster of the Greeks. The alleged anger of Athena against the Greek heroes who *uirgineas ausi diuae contingere uittas* anticipates—by a few hours—the much worse offense inflicted by Aiax on the *uirgineae uittae* of the dishevelled Cassandra. The prophecy that if the Trojans bring the horse in town they will some day invade Greece and sack the Greek towns is impeccable: the Trojans believe the prophecy, bring the horse in—and after one millennium their descendants will take Argos as a revenge (cf. 6, 840 *ultus avos Troiae templa et temerata Minervae*).[18]

---

[15] For the different traditions I refer briefly to Ziehen, *RE* s.v. "Palladion", 36.2 (1949) 171-89; U. von Wilamowitz Moellendorff, *Die Ilias und Homer* (Berlin 1920) 379-95. The difference between the Homeric cult statue and the stolen Palladion is well illustrated by C.A. Faraone, *Talismans and Trojan horses* (New York-Oxford 1992) 7 (the latter is "not a cult statue, but rather a highly specialized talisman") and 137 ("small, armed wooden statue").

[16] The alternative examined by Conington-Nettleship ad loc., that the statue had been replaced in the meantime by a new one, is unsatisfactory.

[17] Virgil's Trojan narrative is clearly more Homeric than cyclic if one uses as a divide the distinction drawn by J. Griffin, "The epic cycle and the uniqueness of Homer," *JHS* 97 (1977) 39-53.

[18] Cf. Odysseus' lying speech to Polyphemus: "Poseidon wrecked our ship" (*Od.* 9.282-86), an unfortunate, unnecessary invention which turns to a bad omen.

But Virgil, who avoids the Palladion lore, has good reasons to link the cult statue of *Iliad* 6 with the catastrophic *nostos* of the Greeks. This is, after all, the first event mentioned in the first speech of Juno in the poem, and the emphasis on disaster created by one individual sin (*unius ob noxam*) could be reminiscent of Alcaeus' narrative, with its emphasis on individual guilt and collective destiny. The Carthaginians—if one has to guess—would be interested both in the fate of an Eastern town threatened by the Greeks, and in the dangerous excesses which can suddenly stop divine protection. They could not feel the same identification with the Achaeans which is normally assumed for the figurative programme of the Akragas Olympieion, where on the western pediment "eroberten Griechen die Stadt Troias als Sinnbild Karthagos."[19] And after all, there was a belief that the statue, which once fell from the sky, survived that catastrophe and more: ironically, this image of an image, briefly glimpsed in the doomed temple at Carthage, could be the most lasting of all monuments mentioned in the Aeneid—with the Ara Maxima, of course.

The Hellenized people of modern Troy seem to have had an Iliupersis in their city sanctuary of Athena Ilias, "in apparent imitation of the Parthenon in Athens."[20] It is worth questioning the identification: perhaps an Iliupersis was "too painful" a choice for the Ilians.[21] But perhaps not, it can be argued,[22] since those people must have perceived themselves as Greek, and had also a vested interest in the Homeric identity of their city. What interests me most is that at some stage the local interpretive community could have negotiated a *sympathetic* reading of their

Similarly, Sinon starts his lying tale by associating himself with Palamedes; his grievance against the Greeks will be, together with Ajax' crime, the cause of the disastrous *nostos*.

[19] So G. Gruben, *Die Tempel der Griechen* (Munich 1976) 309. Contrast N. Horsfall (below, n. 23), who argues that the images in Juno's temple are a program for Carthaginian imperialism based on identification with the Greek aggression. Even for fifth century Greek art, historians would normally assume that the sack of Troy is represented so as to suggest identification with the victims, with an eye on Persian aggression to Greek towns and sanctuaries; see, e.g., Connelly (above n. 3).

[20] C. Brian Rose, *The 1991 Post-Bronze Age Excavations at Troia*, Studia Troica 2 (1992) 44.

[21] B. Ridgway, *Hellenistic Sculpture. I. The styles of ca. 331-200* (Bristol 1990) 152.

[22] Brian Rose (above, n. 20) 57 n. 11.

Iliupersis, while the prosperity of the town depended on its *syggeneia* to the Romans and then the Julians: a pro-Ilian reading not unlike the one that Aeneas infers from the Carthaginian monument, with the support of Dido and of her informant—Teucer, a Greek settler and exile branded by fellow-Greeks as a half-Trojan.[23] Of course we cannot say much on the intended programme of a Carthaginian temple where a Roman poet inscribes—with bitter irony—the charter myth of the nation that will some day erase Carthaginian culture. But Virgil's readers were probably in a good position to understand that sympathy for the losers is a reasonable reaction to representations of the fall of Troy. The most popular vehicle for this kind of representation, Greek vase painting, was by no means exclusively centered on nationalistic pride, and images of sacrilege and tainted victory—Priam, Cassandra, Troilus—were likely to suggest complex reactions.[24]

Meanwhile the cult statue remained as a powerful focus of (dis)continuity: the same, it was claimed,[25] a different one, as the skeptical Demetrios of Skepsis retorted with serious arguments.[26] After all, Homer seems to imply a sitting Athena, while the extant statue was a standing, armed goddess. Roman students of Homer could have been impressed by Aristarchus' opinion[27] that the Homeric statue too was a standing one: in Virgil, the parallelism of 1.482 *diva solo fixos oculos aversa tenebat* with Dido's simulacrum in the Underworld (6.469-71, *illa solo fixos oculos aversa tenebat ... quam si dura silex aut stet Marpesia cautes*) obliquely suggests that Homer had a standing simulacrum

---

[23] Barchiesi (above, n. 5); the discussion had been started by N. Horsfall, who argues that Teucer hates the Trojans, Dido deceives Aeneas, and Virgil hates the Carthaginians: S.J. Harrison, ed., *Oxford readings in Vergil's Aeneid*, (Oxford 1994) 131-44.

[24] One might compare recent discussions of the Onesimos kylix: e.g., D. Williams, *Onesimos and the Getty Iliupersis*, Greek Vases in the J. Paul Getty Museum 5 (1991) 61; M.J. Anderson, "Onesimos and the interpretation of Iliupersis iconography," *JHS* 115 (1995) 130-35.

[25] Hellanicos, *FGrHist* 4 F 25 b ("intent on pleasing the Ilians", Strab. 13.1.42) defended the continuity of Homeric and contemporary Troia. As it happens, he was right about the location.

[26] Strab. 13.1.41-42: the whole chapter on Troiá is now reedited and annotated by S. Radt-J.W. Drijvers, *Studia Troica* 3 (1993) 201-31.

[27] Schol. Hom. *Il.* 6.92, p. 147 Erbse.

of Athena, not a sitting one. So too does the incidental reference to a *clipeata Minerva* at the end of the Laocoon story (2.227).

Between Alexander and Julius Caesar, with their archaeological manoeuvres, another Roman provided a new example of mutability and continuity out of Trojan ashes. In 86 BCE after a siege of eleven days, Fimbria took Ilion, killed some people,[28] and burned the houses and the sanctuary itself. But the statue, they say, was not affected, protected by the ruins.[29] Excited by this success, Fimbria was able to feel like a new and more efficient Agamemnon.

After this new, recapitulative paradoxon, Julian Ilion kept the recent memory inscribed in its architectural planning, now recovered by the 1993 excavations: interestingly, part of the Fimbrian destruction was deliberately preserved as a memorial in the restored citadel.[30] The reconstructed sanctuary offered not only a memento of cyclic destruction and a "poetica delle rovine" but also a spectacular, geopolitical vista on the Dardanelles.[31] The people of New Ilion were now ready to thank the Roman Greats, Julius Caesar and Augustus and Agrippa, for their support of Athena Ilias:[32] the statue, in the meantime, surfaces in *Aeneid* I

---

[28] App. *Mithr.* 53. Cf. C. Brian Rose, *The 1993 Post-Bronze Age Excavations at Troia*, Studia Troica 4 (1994) 78 "someone was in the building when it collapsed ... he clearly died in the fire which destroyed it".

[29] Liv. fr. 17 (Aug. *Civ.* 3.7); Obs. 56b; App. *Mithr.* 53.

[30] Brian Rose (above, n. 28) 75: "It is now clear that the north building was destroyed in the Fimbrian attack of 85 BC, and that it was left as a ruin during the subsequent Augustan restoration of the sanctuary. The building was apparently intended to stand as a symbol of the wars which had nearly destroyed life in the city"; and 80: "the ruin was showcased rather than hidden".

[31] See Brian Rose (above, n. 20) 45: "a portico bordered on the sanctuary on the east, south, and west sides; the north was left unenclosed so as not to hinder a breathtakingly beautiful view of the Dardanelles".

[32] P. Frisch, *Die Inschriften von Ilion* (Bonn 1975): Athena Ilias is mentioned with emphasis at no. 77 (a document by a Roman magistrate, perhaps Julius Caesar, granting privileges to the town) and 86 (Agrippa's pietas towards Athena and goodwill for the Ilians); cf. 84 (probably acknowledging the Augustan restoration or reconstruction of the temple of Athena, cf. Frisch 183-84; Brian Rose [above, n. 20] 45).

as an anxious token of persistence and mutability[33] involving the empires of Troy, Greece, Carthage, and Rome.[34]

[33] See the highly suggestive analysis of P. Hardie, "Augustan Poets and the Mutability of Rome," in A. Powell, ed., *Roman Poetry and Propaganda in the Age of Augustus*, London 1992, 59-82.

[34] I am grateful to Peter Knox and Christopher Jones for several helpful suggestions. This paper was written during a stay at Harvard in 1995/96 and I hope Wendell Clausen (who was my first host in the United States, in 1992, when he opportunely brought me to Lexington Green) will enjoy it as a mnemosynon of our long conversations about Virgil, Italy, and the Pacific Northwest.

PARENTHETICALLY SPEAKING (IN VIRGIL AND OTHER POETS)

R. J. TARRANT

> **(fatebor enim):** again in *A.* 4.20 'Anna (fatebor enim) miseri post fata Sychaei'; cf. *A.* 12.813 'Iuturnam misero (fateor) succurrere fratri'. The parenthesis is colloquial in tone and qualifies a slightly embarrassed or rueful admission, as by Cicero, *Ad Att.* 12.36.1 'hae meae tibi ineptiae (fateor enim) ferendae sunt', or Ovid, *Trist.* 5.6.5 'sarcina sum (fateor)'; see Tränkle 8.
>
> The use of parenthesis to create an effect of liveliness or spontaneity is a feature of Callimachus' style which V. imitates especially in the *E.*; 2.32-3, 53, 3.9, 29-30, 54, 93, 104, 4.48, 61, 5.70, 89, 6.6-7, 7.22-3, 8.38 (cf. Theocr. 11.25-7), 9.3, 6, 23, 25, 64, 10.38-9, 46. Parenthesis is not found in the pastoral poetry of Theocritus (but cf. [Theocr.] 8.50) and very rarely in his other poetry: 12.13, 15.15-16, 22.64. For parenthesis in Callimachus see F. Lapp, *De Callimachi Cyrenaei tropis et figuris* (diss. inaug., Bonn, 1965), 52-3; for Callimachus and V., G. Williams, *Tradition and Originality in Roman Poetry* (Oxford, 1968), 711, 730-1.

Wendell Clausen's note on *Eclogues* 1.31, like many notes in this admirable commentary, both illuminates an aspect of Virgil's writing and also suggests larger questions to be pursued. In this instance, a reader might be stimulated to learn more about the practice of Hellenistic poets, about Virgil's "neoteric" predecessors, and about Virgil's use of parenthesis in his larger-scale works. The following observations are an attempt to open up some of these lines of inquiry. The results will confirm and extend the assumption underlying Clausen's remarks, that parenthesis is a stylistic device employed with conscious artistry for a variety of effects by Hellenistic poets and their Roman counterparts.[1]

---

[1] Other commentators who offer useful material include Neil Hopkinson on Callimachus' *Hymn to Demeter* (*ad* 7, 22, 70) and R. G. Austin on individual books of the *Aeneid* (see below, nn. 39, 40, 44). By far the fullest discussion of the device is Michael von Albrecht's analysis of parentheses in Ovid's *Metamorphoses* (*Die Paren-*

To begin, however, a problem of method. Parenthesis is a recognized feature of Greek and Latin literature of all periods.[2] But the means by which parenthesis is marked in printed texts, i.e., rounded brackets or double dashes, are elements of punctuation shaped by modern convention rather than by traceable ancient practice.[3] As a result the indication of parentheses, like other aspects of the punctuation of ancient texts, is to a degree determined by editorial convention or by the preferences of individual editors. This factor does not arise for practical purposes in cases where an inserted word or phrase separates closely connected elements of syntax, as in *Ecl.* 3.29-31 *'ego hanc uitulam (ne forte recuses, | bis uenit ad mulctram, binos alit ubere fetus) | depono'* and 4.48 *adgredere o magnos (aderit iam tempus) honores*; or where a parenthesis follows a sense-pause within the sentence, such as *Ecl.* 7.21-23 *'aut mihi carmen, | quale meo Codro, concedite (proxima Phoebi | uersibus ille facit), aut* eqs. In such instances all editors will normally indicate a parenthesis, but many alleged instances do not belong to these clearly defined categories.[4] To take the *Eclogues* as a convenient example, the list of parentheses given in Clausen's note above is a nearly complete account of those marked in Mynors' Oxford text.[5] In the Paravia text of Mario Geymonat, however, fully half of these passages are not formally treated as parentheses, including the one on which I have transcribed

these in Ovids Metamorphosen und ihre dichterische Funktion [Hildesheim 1964]). Marisa Squillante Saccone's article "Parentesi" in *EV* 2.972-74 briefly surveys the Virgilian material; it is most informative on the position of parentheses within the hexameter and on words used to introduce a parenthetic expression.

[2] The most wide-ranging study is E. Schwyzer's "Die Parenthese im engern und weitern Sinn," *APAW phil-hist. Kl.* 1939.6; cf. also H. Boldt, *De liberiore linguae graecae et latinae collocatione verborum* (diss. Göttingen 1884) 159-79.

[3] The use of rounded brackets as a marker of parenthetical words or phrases is a contribution of the Italian humanists: cf. M. B. Parkes, *Pause and Effect* (Berkeley 1993) 212-213 pl. 30 (showing a primitive form of parenthesis in a manuscript annotated by Coluccio Salutati just before 1400). For the present purpose I consider rounded brackets and double dashes as functionally equivalent.

[4] For some examples of borderline phenomena on which editorial practice varies see below, p. 147 and also nn. 26, 37.

[5] Not included are 5.21, 41, 6.18-19, and 8.96; also 3.102 where Clausen *ad loc.* persuasively argues for reading *'hi certe (neque amor causa est) uix ossibus haerent'* with Stephanus. Another conjectural parenthesis has recently been suggested by R. G. M. Nisbet (*PVS* 20 [1991] 6-7 = *Collected Papers on Latin Literature* [Oxford 1995] 330) who proposes punctuating 9.37-38 *'tacitus, Lycida, mecum ipse uoluto | —si ualeam meminisse neque est ignobile—carmen.'*

Clausen's comment.[6] This degree of editorial variation may be atypically large (though it is certainly not unique[7]), but it is enough to show that our sense of an author's usage in this regard can be significantly affected by editorial decisions.[8]

One consequence is that there is little point in compiling numerical tabulations of parentheses in various authors, or in calculating their relative frequency with any precision. My comments on absolute and relative frequency are thus phrased in broad terms and focus on the most striking differences or tendencies, which are not likely to have been obscured by editorial differences.

Among Hellenistic poets Callimachus stands out for the freedom and exuberance with which he uses parenthesis. In the *Iambi* and *Epigrams* its effect is often, as noted by Clausen, to capture the quick turns of informal speech, while the *Aetia* and the *Hymns* feature the explanatory parenthesis, usually containing items of mythical or geographical lore.[9] One of Callimachus' distinctive techniques is the parenthesis in the opening line of a poem, which appears in several poems of various genres; besides suggesting urbane informality, the device shows off the poet's supple control of verse-form.[10] Callimachus also illustrates two other uses of parenthesis that will recur in our investigation, multiple parenthesis and parenthesis as a form of characterization. A virtuoso specimen of the former concludes the *Hymn to Artemis* (260 ff.), a

---

[6] The other points of divergence are 3.9, 93, 104, 4.61, 5.70, 6.69, 9.3, 6, 23, 25, and 10.46 (also 5.21, 41, and 8.96 mentioned in the previous note). In most of these places Geymonat sets off the words in question by commas, which acknowledges their inserted position but produces a somewhat different effect.

[7] There is a comparably wide difference in the treatment of parenthesis by modern editors of Apollonius Rhodius.

[8] My sampling of editions of Virgil suggests that editors who are prone to intervene in the text in other ways (e.g., through conjecture or athetesis) are also more given to marking parentheses than their conservative counterparts: see n. 32 below.

[9] For the latter cf., e.g., *Aet.* 43.70-71, *Hymn.* 1.43, 3.47-48, 172, 244-45, 4.49.

[10] This feature is discussed for the *Iambi* by W. Bühler, "Archilochos und Kallimachos," *Entretiens Hardt* 10 (1963) 245 f.; note also *Hymn.* 3.1, *Epigr.* 4.1, 56.1-2, 61.1-2. Callimachus' fondness for initial parenthesis may be a factor in Ovid's use of the device in the proem of the *Metamorphoses* (1.2-3 *di, coeptis—nam uos mutastis et illa | —adspirate meis*). For a Virgilian parenthesis at the start of a major section within a poem see below, n. 41.

fourfold alternation of negative command ("let no-one dishonor Artemis ... nor let anyone contend with her in archery ... nor let anyone attempt to woo her ... nor avoid the yearly dance in her honor") and parenthetic *exempla* of the punishment attending each violation ("for Oeneus dishonored her ... for Agamemnon boasted to his cost ... for neither Otus nor Orion courted her with success ... for Hippo lamented her refusal to dance"). Of this passage Gregory Hutchinson writes "the static and repetitive form of this extended series falls very strangely in so nervous a writer,"[11] and his comments rightly call attention to the grim urgency of the lines. But there is as well a bravura quality to this conclusion that seems wholly appropriate for Callimachus, along with an interest in exploiting the expressive potential of parenthesis.

That interest appears in another form in the use of parenthesis as a characterizing device. Novelists alert to the way a character's syntax may reflect his or her thoughts have used parenthesis for this purpose; so for example Jane Austen in *Sense and Sensibility* suggests the mercenary thoughts of Mr. John Dashwood as he affects to commiserate with the disinherited Edward Ferrars: "I cannot conceive a situation more deplorable. The interest of two thousand pounds—how can a man live on it?—and when to that is added the recollection, that he might, but for his own folly, within three months have been in the receipt of two thousand, five hundred a year, (for Miss Morton has thirty thousand pounds,) I cannot picture to myself a more wretched condition."[12] Callimachus achieves a comparable effect in the first *Iamb*, where the opening lines of Hipponax' story of Bathycles are punctuated by three asides in rapid succession (32-35, 39, 41-42), the first of which stands out for its abruptness and complexity ("An Arcadian named Bathycles—I won't go on long, don't make faces at me, friend, for even I don't have much time to spare, since I must spin back to the middle of Acheron, alas—was one of the fortunates of old" etc.).[13] The prominently placed parenthesis depicts the speaker's mixture of self-conscious prickliness and peremptory authority.

[11] *Hellenistic Poetry* (Oxford 1988) 70.

[12] Chapter 37.

[13] ἀνὴρ Βαθυκλῆς Ἀρκάς—οὐ μακρὴν ἄξω, | ὦ λῷστε μὴ σίμαινε, καὶ γὰρ οὐδ' αὐτὸς | μέγα σχολάζω· δεῖ με γὰρ μέσον δινεῖν | φεῦ φεῦ Ἀχέροντος—τῶν πάλαι τις εὐδαίμων | ἐγένετο, κτλ.

By comparison Apollonius Rhodius is conspicuously restrained in his use of parenthesis, and perhaps for that reason his practice has received less notice from scholars. Nevertheless, the *Argonautica* contains several noteworthy or innovative examples of the device, some of which clearly struck the imagination of Roman poets.

Unlike Callimachus (and, as we shall see, Virgil), Apollonius makes relatively little use of parenthesis as a means of characterization. All but a few of the clear-cut instances are in the narrative rather than in speeches,[14] and provide background to or added details of the action being described. Of this kind are passages like 2.913 δὴ γὰρ συνανήλυθεν Ἡρακλῆι and 3.1315 δὴ γάρ σφι πάλαι προπεφραδμένον ἦεν. In two places Apollonius combines narrative asides of this sort with a more overtly personal parenthetic remark: 4.984-85 ἵλατε, Μοῦσαι, | οὐκ ἐθέλων ἐνέπω προτέρων ἔπος (followed in 986-990 by an alternative version of the incident in question) and 4.1511 εἴ μοι θέμις ἀμφαδὸν εἰπεῖν (followed in 1514 by a comment on the character's name, καὶ γὰρ τὸ κάλεσκέ, μιν οὔνομα μήτηρ). The doubling of the parenthesis serves to focus particular attention on the narrator's active role in shaping the poem and in reflecting on its content.

A distinctive use of parenthesis in Apollonius is its appearance in a structurally marked position. The longest parenthetic insertion in the poem, the miniature ethnography of the Amazons in 2.987-92, comes between the protasis and apodosis of a conditional sentence ([985 f.] "they would have fought the Amazons, not without bloodshed... [993 f.] if the breezes of Argestes had not come again at Zeus' command"). The extended parenthesis threatens to, but does not, derail the narrative, an effect taken still further by Virgil's description of battling bees at *G.* 4.67 ff., where the thought begun in *sin autem ad pugnam exierint* is interrupted by a vivid scene of combat and is only completed twenty lines later at 86-87 *hi motus animorum atque haec certamina tanta* |

[14] Exceptions include 3.500-501 (Jason), 903-4 (Medea), 4.793-94 (Hera). The relative scarcity of parentheses in speeches is to some extent a product of the comparatively low proportion of speeches in the poem, 29% against 45% or more in the Homeric epics and the *Aeneid* (cf. R. L. Hunter, *The* Argonautica *of Apollonius: Literary Studies* [Cambridge 1993] 138), but cannot be entirely explained in those terms.

pulueris exigui iactu compressa quiescent.[15] The similarity of structure and content makes it likely that Virgil was recalling Apollonius here: the replacement of Amazons by bees adds a further dimension of mock-heroic irony to Virgil's scene, as does the disproportion between the Lilliputian warriors and the enormous narrative aside devoted to them.

Another structure interrupted by parenthesis in Apollonius is the simile. At 1.1243 ff., where Polyphemus son of Eilatus searching for Hylas is compared to a wild animal vainly trying to enter a sheepfold, parenthesis marks a temporal recession within the scene of the simile: "ablaze with hunger he follows but does not come upon the flocks—for the shepherds shut them up in the folds beforehand—but he groans and rages furiously..."[16] Even more remarkable is the simile at 2.541 ff., in which the image of a lone wanderer is no sooner introduced than it is broken into by a generalizing aside: "and as when a man wanders from his homeland—as we suffering mortals wander much—nor is any land too remote" etc.[17] Richard Hunter interprets this intrusion as mimetic of the shifting thoughts of the wanderer,[18] but it also, and perhaps more

---

[15] J. Marouzeau, *Traité de stylistique latine*, ed. 2 (Paris 1946) 242 f. treats 67 *nam*—85 *subegit* as a true parenthesis of mammoth proportions, but *hi motus* in 86 only makes sense with reference to the immediately preceding lines; Mynors and Thomas are therefore right to regard the formal syntax of the condition as abandoned. Marouzeau acutely observes that the complex or extended parenthesis can function as a display of authorial mastery (i.e., the ability to digress and return to the starting-point without losing the thread), even if neither of his two examples (the other is Cic. *Verr.* 4.4-7) may be parenthetic in the strict sense.

[16] λιμῷ δ' αἰθόμενος μετανίσεται, οὐδ' ἐπέκυρσε | ποίμνῃσιν—πρὸ γὰρ αὐτοὶ ἐνὶ σταθμοῖσι νομῆες | ἔλσαν—, ὁ δὲ στενάχων βρέμει ἄσπετον (so Vian's Budé). Fränkel's OCT does not mark πρὸ ... ἔλσαν as a parenthesis.

[17] Ὡς δ' ὅτε τις πάτρηθεν ἀλωμένοςῳοῖα τε πόλλα | πλαζόμεθ' ἄνθρωποι τετληότες —, οὐδέ τις αἶα | τηλουρός, πᾶσαι δὲ κατόψιοί εἰσι κέλευθοι. Fränkel again does not mark a parenthesis, but rather transposes 545b-46a ἄλλοτε ... ἐπιμαίεται to follow τετληότες in 542; cf. *Noten zu den Argonautica des Apollonios* (Munich 1968) 197-201. Vian continues the parenthesis to κέλευθοι at the end of 543.

[18] (Above, n. 12) 138. Hunter's reading is presumably based on Vian's text of the passage, which by including more of the passage in the parenthesis conveys a stronger sense of a broken syntactical structure.

prominently, highlights the narrator's empathetic involvement in the scene.[19]

These may be the only instances in Apollonius of parentheses in the strict sense within a simile, but many of Apollonius' similes are elaborated in a way that makes them syntactically independent of the surrounding narrative; some editors register this aspect of the simile by a dash where the simile rejoins the narrative, thereby placing the simile in a semi-parenthetic relationship to its surroundings.[20] Such expanded similes are not uncommon in Homer, but Homer's paratactic style reduces the disjunctive effect of the elaboration.[21] In at least one case a later poet has reworked an Apollonian simile of this kind into one containing a fully-fledged parenthesis. In 4.1682 ff. Apollonius compares the wounded Talos to a pine tree left half-cut by woodmen at the end of the day: ἀλλ' ὥς τίς τ' ἐν ὄρεσσι πελωρίη ὑψόθι πεύκη, | τήν τε θοοῖς πελέκεσσιν ἔθ' ἡμιπλῆγα λιπόντες | ὑλοτόμοι δρυμοῖο κατήλυθον, ἡ δ' ὑπὸ νυκτὶ | ῥιπῇσιν μὲν πρῶτα τινάσσεται, ὕστερον αὖτε | πρυμνόθεν ἐξεαγεῖσα κατήριπεν, κτλ. This simile is the basis for the description of Theseus' victory over the Minotaur in Catullus 64.105-11:

nam uelut in summo quatientem brachia Tauro
quercum aut conigeram sudanti cortice pinum
indomitus turbo contorquens flamine robur
eruit (illa procul radicitus exturbata
prona cadit, late quaeuis cumque obuia frangens),

---

[19] For the first person in a simile Hunter (138 n. 131) compares Virg. *Aen.* 12.908 ff. *ac uelut in somnis ... nequiquam auidos extendere cursus | uelle uidemur* etc. For a very similar first-person plural aside in Apollonius' narrative proper cf. 4.1081-82 οἷά τε πολλά | ῥέζομεν ἀμπλακίῃσιν, treated as a parenthesis by Fränkel and Livrea.

[20] Some examples: 1.1247 ff., 1269 ff., 4.35 ff., 933 ff. (and compare a similarly expanded time-description, 4.109 ff.) The question of how to punctuate in such cases also arises in Virgil, cf. *Aen.* 7.380-83, 9.61-64, 10.358 f., 407-9, 12.333-36, 717-19 (all marked as parentheses by Ribbeck but not by more recent editors); see also Saccone (above, n. 1) 974. Ovid's *Metamorphoses* seems to contain only one instance, 1.533 ff.

[21] Cf., e.g., *Il.* 11.113 ff., 18.318 ff. On techniques of expansion in Homeric similes, cf. W. C. Scott, *The Oral Nature of the Homeric Simile* (Leiden 1974 [*Meemosyne* Supplement 28]) 140-61, Mark Edwards, *The Iliad: A Commentary* 5 (Cambridge 1991) 26.

sic domito saeuum prostrauit corpore Theseus
nequiquam uanis iactantem cornua uentis.

Catullus' recasting of Apollonian elements is too intricate to be pursued here; for the present purpose the relevant detail is the echo of ἡ δ' (1684) in *illa* (108) at the corresponding syntactical point in the simile. In exploiting the simile's potential for parenthesis Catullus enriches the texture of his adaptation in a manner fully characteristic of his model.[22]

Also on the border between phrasal expansion and parenthesis is the mannered repetition of a proper name, as in Virg. *Ecl.* 6.21-22 *addit se sociam ... Aegle, | Aegle Naiadum pulcherrima.* Repetitions of this kind, elegantly described by Roger Mynors as "eddies in the forward flow of the discourse,"[23] are especially at home in bucolic poetry, but in fact Apollonius offers closer parallels to the Virgilian example just cited than anything adduced from Theocritus; cf. 1.191-192 Λαοκόων τε, | Λαοκόων Οἰνῆος ἀδελφεός, οὐ μὲν ἰῆς γε | μητέρος, ἀλλά ἑ θῆσσα γυνὴ τέκε, 1.202-3 Λέρνου πάις Ὠλενίοιο, | Λέρωου ἐπίκλησιν, γενεήν γε μὲν Ἡφαίστοιο, 3.861-2 Βριμὼ ... | Βριμὼ νυκτιπόλον, χθονίην, ἐνέροισιν ἄνασσαν.[24] These cases fall short of being true parentheses, but Apollonius has at least one example of this pattern in parenthetic form, at 4.827-30: ἠὲ παρὰ Σκύλλης στυγερὸν κευθμῶνα νέεσθαι— | Σκύλλης Αὐσονίης ὀλοόφρονος, ἣν τέκε Φόρκῳ | νυκτιπόλος Ἑκάτη, τήν τε κλείουσι Κράταιιν—, | μή πως κτλ. The arrangement of this passage in turn corresponds closely to Virg. *Ecl.* 2.31-34 *mecum una in siluis imitabere Pana canendo | (Pan primum calamos cera coniungere pluris | instituit, Pan curat ouis ouiumque magistros), | nec te paeniteat* etc.

Catullus resembles Apollonius rather than Callimachus in his restricted use of parenthesis, and is indeed more sparing with it even than Apollonius. It is remarkable, for example, that a poem as overtly Hel-

---

[22] For a similar parenthetically added detail in a simile cf. Virg. *Aen.* 1.502 *Latonae tacitum pertemptant gaudia pectus*, a line of which the Homeric counterpart (*Od.* 6.106) is non-parenthetic. See also below, p. 152.

[23] On *G.* 1.11 *ferte simul Faunique pedem Dryadesque puellae: | munera uestra cano* (which Mynors alone among recent editors treats as a parenthesis).

[24] On epanalepsis of this kind see also Clausen *ad loc.*, Nisbet (above, n. 5) 327.

lenistic in its manner as poem 64 should contain just one example of the device, in the Apollonius-inspired simile discussed shortly before. Catullus' practice in this regard may not have been typically "neoteric," since later poetry influenced by the lost work of his contemporaries shows much greater fondness for parenthesis.[25] At the qualitative level, however, Catullus remains faithful to his Hellenistic models, since nearly all his examples of parenthesis reveal elements of wit and sophistication. Some of these effects may result from direct imitation of Apollonius or Callimachus—such as the *captatio beneuolentiae* of 66.71-74, which probably derives from the corresponding lines of the *Coma Berenices*—but most of them are likely to be original, since they reflect Catullus' distinctive combination of playfulness and powerful emotion.

Among Catullus' whimsical uses of parenthesis is the structural joke at 61.152-53, where one of the ten appearances of the refrain *'io Hymen Hymenaee io, | io Hymen Hymenaee* is so placed as to interrupt the syntax of the surrounding lines (151 *'quae tibi sine seruiat | (io... Hymenaee) | usque dum'* etc.). Disruption of a conventional, and sometimes parenthetic, formula is also found at 66.18, where the everyday expression *ita me di iuuent* yields the dislocated variant *'non, ita me diui, uera gemunt, iuuerint.'*[26] Virgil shows the influence of such interweaving, which flirts with parenthesis but holds it off, in lines like *Ecl.* 8.109 *'parcite, ab urbe uenit, iam parcite carmina, Daphnis'* and 10.77 *ite domum saturae, uenit Hesperus, ite capellae.'*[27] When Catullus does present the "so help me" formula parenthetically, in the speech of the garrulous door in 67.9-10 *'non (ita Caecilio placeam, cui tradita nunc sum) | culpa mea est'* etc., it takes on characterizing force, conveying the

[25] So, for example, the *Ciris* attributed to Virgil, cf. R. O. A. M. Lyne's edition (Cambridge 1978) 30-31.

[26] The phrase appears in its normal shape and as an aside at 61.189-90 *'ita me iuuent | caelites.'* These words could be punctuated as a parenthesis, as in Kroll's text, although the colloquial expression *ita me di ament* as it appears in Comedy is not regarded as parenthetic even when it comes between closely linked syntactical elements, as in *HT* 569 *'ut equidem, ita me di ament, metui!'* or 953 *'non, ita me di ament, auderet.'*

[27] I note, though, that Nisbet (above, n. 5) 337 treats *uenit Hesperus* as a true parenthesis and as one of an accumulation of bucolic features of style clustered in the final lines as a generic sign-off.

speaker's awareness of its servile position and eagerness to remain on its master's good side.

Even more striking and apparently original is Catullus' use of parenthesis to express strong personal feelings. The effect is particularly linked to the mention of Troy, and of the memories of his brother that the name of Troy evokes. In the Protesilaus-Laodamia section of poem 68b Catullus begins to relate the gathering of the Greeks at Troy (87-88 *nam tum Helenae raptu primores Argiuorum | coeperat ad sese Troia ciere uiros*) but is drawn first into an apposition with epanalepsis (89-90 *Troia (nefas!) commune sepulcrum Asiae Europaeque, | Troia uirum et uirtutum omnium acerba cinis*) and then into recalling and mourning his brother's death (91-92 *quaene etiam nostro letum miserabile fratri | attulit* etc.); finally another mention of Troy leads to a resumption of the narrative (101-103 *sed Troia obscena, Troia infelice sepulcrum | detinet extremo terra aliena solo. | ad quam* etc.). Lines 89-100, although functionally an interruption, remain syntactically connected to their context, and so the only true parenthesis is the indignant exclamation *nefas!* in 89, the model for several instances of the word as an authorial comment in the *Aeneid* (7.37, 8.688, 10.673; see also below, p. 152). In poem 65, however, a similar set of thoughts assumes a genuinely parenthetic form: between the opening apology to Hortalus for not being able to compose new poems (1-4) and the offer of a Latin version of the *Coma Berenices* in compensation (15 ff.) comes a lengthy aside that begins with explanation (5-8 *namque...*) but then veers off into self-absorbed lament (10-14 *numquam ego te ... | aspiciam posthac?* etc.). As Gordon Williams writes, "the effect of the device is to make the expression of the poet's grief appear to be an irresistible emotion that has simply broken the connection of ideas and syntax."[28] The length and complexity of the parenthesis recall Callimachean practice, but Catullus invests it with a depth of feeling that is entirely his own. Nevertheless, the artfulness of

---

[28] *Tradition and Originality in Roman Poetry* (Oxford 1968) 708. Williams treats these lines as a case of anacoluthon rather than parenthesis, but the introduction with *namque* and the seamless join of 1-4 (*etsi...*) and 15 ff. (*sed tamen...*) make it legitimate to consider what comes between as parenthetic. In any case his comments remain apt. Cf. also Hutchinson (above, n. 11) 299-300.

this ostensibly spontaneous outburst is highly appropriate in a poem written to accompany a translation from Callimachus.[29]

"In his use of the parenthesis Virgil goes far beyond Catullus; he is the first Roman poet to make it a constant feature of his style."[30] I again cite Gordon Williams, whose acute comments have done much to focus attention on this stylistic device. Williams' primary concern was to note the far higher frequency of parentheses (roughly three to one) in the *Eclogues* as compared to the *Georgics* and *Aeneid*; he attributed the difference to Virgil's wish to produce an impression of artful informality in the *Eclogues*, while in the longer poems "more sustained exposition was needed ... and the movement of thought was no longer ... in short steps that could, to some extent, bear treatment as units."[31] These general observations may be supplemented and refined with regard to the *Aeneid* in particular. There it is striking that more than half of the parentheses appear in speeches (a much higher percentage than in Apollonius, for example), and are thus a means for rendering, in however stylized a way, the rhythms of direct speech. Now the *Eclogues* consist almost entirely of speech, whether dialogue or dramatic monologue, and this fact surely has some influence on the abundance of parentheses they contain. It is also worth emphasizing that parenthesis in Virgil's longer poems (and especially the *Aeneid*), though used relatively less often than in the bucolics, is still noticeably frequent. Ovid, for example, is generally thought of as a writer fond of parenthesis, but the *Aeneid* contains nearly as many parentheses as the *Metamorphoses* in absolute terms, and uses the device fully as often if the greater length of the *Metamorphoses* is taken into account.[32] While Virgil follows Callimachus and

---

[29] Williams (above, n. 28) 712-14 proposes taking 68.141 *ned diuis homines componier aequum est* as a parenthesis to avoid placing a lacuna between 141 and 142, but to do so disrupts the apparent connection between *nec ... aequum est* and 143-44 *nec tamen ... domum.*

[30] Williams (above, n. 28) 730.

[31] *Ibid.*

[32] The absolute figures (based on Mynors' text of Virgil and von Albrecht's list of parentheses in the *Metamorphoses*) are 150 for the *Aeneid* (just under 10,000 lines) and 186 for the *Metamorphoses* (somewhat more than 12,000 lines). Von Albrecht's tally is probably on the high side, since it is based on Rudolf Ehwald's Teubner edition (Leipzig 1915), which contains several examples of unnecessary or unwanted parentheses (e.g., 1.392, 2.133, 4.168, 7.15, 11.297). On the other hand Mynors is

Apollonius in the use of parenthesis to enliven speeches and fill in background detail, he also employs it in several new or distinctive ways.

One specifically Virgilian touch is the prominent use of parenthesis to convey the narrator's empathetic response to the action being described; in this capacity parenthesis is a component of what Brooks Otis termed Virgil's "subjective style."[33] One form of such involved reaction is the indignant exclamation: Virgil both takes up Catullus' parenthetic *nefas!* (above, p. 148) and adds to it the similar use of *infandum!* (*G.* 1.479, *Aen.* 1.251, cf. also Stat. *Th.* 1.595, 4.204) and *miserum!* (*Aen.* 6.21).[34] Virgil also extends the range of the Homeric θαῦμα ἰδέσθαι (applied to works of craftsmanship remarkable for their material, size, or skill[35]) to events that evoke responses of wonder or terror or pity from narrator or character, *mirabile / miserabile / horrendum uisu / dictu*, cf. *Aen.* 1.111, 439, 2.174 (Sinon), 4.182, 454, 7.64, 8.252 (Evander), 565 (Evander), 9.120 (*mirabile monstrum*), 465, 10.637, 12.252.[36] While the connection between exclamations of this kind and the Homeric phrase has been noted—cf., e.g., Harrison on *Aen.* 10.637—I have not seen it observed that Virgil expresses parenthetically what in Homer is paratactically appended. The difference reflects both the more highly interwoven texture of Virgil's style and also the more overt emotional involvement of the Virgilian narrative voice.[37]

---

noticeably moderate by comparison with Otto Ribbeck's Teubner text of the *Aeneid* (Leipzig 1895), which contains 185 parentheses (Ribbeck marks parentheses in 59 passages where Mynors does not and does not mark them in 14 places where Mynors does).

[33] It is curious that Otis himself does not seem to mention parenthesis in his lengthy chapter of that title (*Virgil: A Study in Civilized Poetry* [Oxford 1963] 41-96), especially since Otis himself was one of the most parenthetic—and subjective—of writers, and might have been expected to notice its effects.

[34] Virgil is in turn the probable model for Ovid's *indignum!* (used mock-solemnly in *Am.* 1.6.1, straight in *Met.* 5.37).

[35] With an interesting difference of usage between *Iliad* and *Odyssey*: in the former three times of armor or weapons (5.725, 10.439, 18.83) and once of Hephaestus' automated tripods (18.377), in the latter once of city walls (7.45) and three times of wool or woven fabric (6.306, 8.366, 13.108).

[36] On such expressions see von Albrecht (above, n. 1) 120, who notes Ovid's even freer and more personal use of them.

[37] A more indirect form of narratorial comment that is functionally parenthetic is the expression of uncertainty about the truth of the incident related, *si uera est fama*

An especially prominent example of narrator's empathy taking the form of a parenthetic comment occurs in the simile at *Aen.* 12.451-57, implicitly describing the fear of the Rutuli as Aeneas returns to the battle after being wounded:

> qualis ubi ad terras abrupto sidere nimbus
> it mare per medium (miseris, heu, praescia longe
> horrescunt corda agricolis: dabit ille ruinas
> arboribus stragemque satis, ruet omnia late),
> ante uolant sonitumque ferunt ad litora uenti:
> talis in aduersos ductor Rhoeteius hostis
> agmen agit.

The emotive language that introduces the parenthesis (*miseris, heu*) nearly merges the narrator's point of view with that of the threatened farmers, while the future indicatives that follow (*dabit, ruet*) stretch the limits of syntax to suggest their terrified thoughts.[38] Like Catullus, Virgil here deploys all the subtlety that Callimachus and Apollonius had brought to the handling of parentheses to produce a far weightier effect.

Given the high proportion of parentheses that occur within speeches in the *Aeneid*, it is natural to ask whether Virgil employs the device as a means of characterization. While I have found no evidence of parenthesis as a distinctive element in the speeches of any single character in the epic, Virgil does seem to use parenthesis, especially in doubled or clustered form, to define the *ethos* of a speaker or of the participants in a dialogue.

Twice in Book 6 doubled parentheses appear in especially solemn pronouncements., with strikingly similar effects. First, the Sibyl's prophecy of Aeneas' future (83-86): *'o tandem magnis pelagi defuncte periclis | (sed terrae grauiora manent), in regna Latini | Dardanidae uenient (mitte hanc de pectore curam), | sed non et uenisse uolent.'* The chiastic symmetry of the parentheses—the first sounding a gloomy antithesis to the optimistic opening, the second offering assurance that is

(*G.* 4.42, *Aen.* 3.551) or *si credere dignum est* (*G.* 3.391, *Aen.* 6.173); such phrases are, however, not generally treated as formal parentheses by editors.

[38] Nothing similar is found in the Homeric model for the simile, *Il.* 4.275 ff. Elsewhere *heu* in a parenthesis appears only in quoted speech, cf. 4.376, 6.150, 11.273.

immediately undercut by the cryptic final phrase—perfectly mirrors the speaker's equivocal outlook.[39] Near the end of the book Anchises' famous prediction of the specifically Roman achievement is also punctuated by two balanced parentheses (847 ff.): *'excudent alii spirantia mollius aera | (credo equidem), uiuos ducent de marmore uultus, |... | tu regere imperio populos, Romane, memento | (hae tibi erunt artes), pacique imponere morem'* etc. Each of the asides is attached to one of the main divisions of Anchises' remarks (*alii—tu*), and each gives added weight to the claim or precept in question: the Greeks will indeed excel in the pursuits of mind and in the arts, but Romans will have their own *artes*, of even greater import and worth.[40] In both passages powerfully consoling or unsettling messages are issued though brief parenthetic phrases, an idiom that effectively suggests the respective speaker's unique authority.

Even more intriguing (in part because apparently unparalleled before or after Virgil) are several examples of parentheses allocated in parallel fashion to the speakers in a dialogue, either to underscore the similarity of ethos between the speakers or to highlight their different emotional attitudes. Of the first kind is the courteous and leisurely exchange between Aeneas and Helenus in Book 3.356 ff., where Aeneas' request for guidance and the prophet's reply are marked by lengthy explanatory parentheses (362-67 *'namque omnis cursum mihi prospera dixit | religio'* etc.; 374-76 *'nam te maioribus ire per altum | auspiciis manifesta fides'* etc.). Similarly in the scene near the end of the poem in which Jupiter and Juno agree to an uneasy truce, Jupiter's claim that Juno must have been responsible for Juturna's intervention (12.798-99 *'aut ensem (quid enim sine te Iuturna ualeret?) | ereptum reddi Turno'*) is acknowledged to be true in a responding phrase that likewise revolves

[39] Austin on 85 *'mitte ... curam'* comments "this new, and comforting, parenthesis balances the previous grim one, neatly arranged to end the line in contrast with the placing of the other at a line-opening." All of Austin's *Aeneid* commentaries are admirably alert to the effects of parenthesis, but his sensitivity was nowhere greater than in the commentary on Book 6; see also the following note and n. 44 below.

[40] Austin again observes the connection (n. on 852), but his comments here seem to me to miss the mark: he says of *'credo equidem'* "the parenthesis makes the structure momentarily informal, giving an impression of lively talk and gesture" and of *'hae ... artes'* "the parenthesis ... gives an illusion of modest, almost apologetic explanation for the 'only thing' left to Rome."

around a parenthesis (813 f. *'Iuturnam misero (fateor) succurrere fratri | suasi'*).

Contrasting parentheses figure in two of the *Aeneid*'s most dramatic dialogue scenes. In the confrontation between Dido and Aeneas in Book 4 each of the characters' opening speeches contains two parentheses, which resemble each other in length and content but which are sharply distinct from those in the opposing speech. Dido's asides both fill an entire line and offer calculatingly poignant reasons for her questions and appeals: 315 *'quando aliud mihi iam miserae nihil ipsa reliqui'*, 324 *'hoc solum nomen quoniam de coniuge restat'*. Aeneas' are short and urgent, the first bolstering his denial that he wished to flee in secret (337-38 *'neque ego hanc abscondere furto | speraui (ne finge) fugam'*) and the second swearing to the reality of Mercury's recent appearance (356-58 *'nunc etiam interpres diuom Ioue missus ab ipso | (testor utrumque caput) celeris mandata per auras | detulit'*). Dido's furious response to Aeneas' speech is also summed up in a parenthetic aside, 376 *'heu furiis incensa feror!'*[41]

Three-way conversations are rare in epic, and although the brief scene involving Turnus, Latinus, and Amata that opens Book 12 has a Homeric analogue in the attempts of Priam and Hecuba to keep Hector from fighting Achilles (*Iliad* 22.25-130), Virgil creates a much stronger sense of the characters occupying the same space and reacting to one another's words. Parenthesis is one means by which their speeches are linked while at the same time given distinct qualities. Turnus' arrogant confidence emerges in his eagerness to meet Aeneas in single combat: 14-15 *'aut hac Dardanium dextra sub Tartara mittam | desertorem Asiae (sedeant spectentque Latini)...'* The verb *spectent* may even recall Turnus' shocking wish before his duel with Pallas that Evander might be present to witness his son's death (10.443 *'cuperem ipse parens spectator adesset'*). Latinus responds *sedato ... corde* (18), but grows agitated at the thought that he might incur blame if Turnus were to die—an outcome that elicits from him a pious apotropaic aside: 40-42 *'quid consanguinei Rutuli, quid cetera dicet | Italia, ad mortem si te (Fors dicta refutet!) | prodiderim...?'* Finally, Amata's overheated and self-

[41] It may be relevant that Virgil opens this scene with a narrator's parenthesis, 296-97 *At regina dolos (quis fallere possit amantem?) | praesensit* etc.

regarding appeal is prefaced by a correspondingly melodramatic parenthesis: 56-60 *'Turne, per has ego te lacrimas, per si quis Amatae | tangit honos animum (spes tu nunc una, senectae | tu requies miserae, decus imperiumque Latini | te penes, in te omnis domus inclinata recumbit) | unum oro'* etc.[42]

Most of the foregoing discussion has tried to place Virgil's practice against the background of earelier poetry; to conclude I mention an apparent Virgilian innovation that was taken up by his immediate successor in Latin epic, Ovid.[43] There is a kind of parenthetic aside that is both frequent in Ovid and also typical of his ironic use of the device, in which the narrator breaks into a character's words to fill in or underline a detail. So, for example, when Jupiter invites the diffident Io to join him in a shady grove in *Met.* 1.589-92 *'o uirgo Ioue digna tuoque beatum | nescioquem factura toro, pete' dixerat ' umbras | altorum nemorum' (et nemorum monstrauerat umbras) | 'dum calet'* etc., the parenthesis adds vividness to Jupiter's words by making us see their accompanying gesture, and in so doing allows us to savor the oily smoothness of his seductive routine. Virgil employs a precisely similar parenthesis, describing the action that accompanies the speaker's interrupted words, when the Sibyl shows Charon the golden bough and compels him to take Aeneas and herself into his boat: *Aen.* 6.405-7 *'si te nulla mouet tantae pietatis imago, | at ramum hunc' (aperit ramum qui ueste latebat) | 'agnoscas.'* The *Aeneid* contains few lighter moments, but this is surely one of them; the retarding parenthesis, though formally in the narrator's voice, perfectly captures the Sibyl's amused mastery of the situation.[44]

---

[42] The words *spes ... recumbit* do not appear as a parenthesis in Mynors and Geymonat, but were so treated by Ribbeck and in R. D. Williams's Macmillan edition (London 1973). On the hyperbolic emotionalism of Amata's lines see Ralph Johnson, *Darkness Visible* (Berkeley 1976) 55-56, who makes the unforgettable comment "Amata ... cannot hide the fact that she is, whatever else she is, a mezzo-soprano."

[43] Caution is needed since the tecnhique in question might have been used by Callimachus in part of his work no longer extant. The similarities between the Virgilian and Ovidian passages, however, at least support a claim of direct influence.

[44] Austin *ad loc.* remarks "an ingenious parenthesis, investing the crucial moment with high drama." This is perhaps an overly serious reading, but "ingenious" (a quintessentially Ovidian epithet) is just right. Austin's excellent note goes on to document Ovid's fondness for this type of parenthesis; to his instance from the

Ovid responded to the humorous aspect of the Virgilian scene in a corresponding episode of the *Metamorphoses*, when Perseus is refused entrance to the grove of the Hesperides by Atlas (4.653-656): *uiribus inferior (quis enim par esset Atlantis | uiribus?) 'at quoniam parui tibi gratia nostra est, | accipe munus' ait laeuaque a parte Medusae | ipse retro uersus squalentia protulit ora.*

The transition from Virgil to Ovid is even more clearly illustrated by Virgil's only other example of this form of parenthesis, when Latinus swears that he will never break the terms of his treaty with Aeneas just as his staff will never put forth leaves (12.206-207): *'ut sceptrum hoc' (dextra sceptrum nam forte gerebat) | 'numquam fronde leui fundet uirgulta'* etc.[45] In the *Ars Amatoria* Ovid ironically transposes this dramatic gesture to an erotic context, in which Ulysses draws with a stick in the sand to help Calypso envision the battlefield at Troy (2.131-132): *ille leui uirga (uirgam nam forte tenebat) | quod rogat in spisso litore pinguit opus.* But Ovid also turns this sort of irony upon himself: an aside like that at *Met.* 7.659-660 *'cum primum, qui te feliciter attulit, Eurus' | (Eurus enim attulerat) 'fuerit mutatus in Austros'* seems to mock the narrator's insistence on supplying information already quite clear from the speaker's words. As with other aspects of Ovid's narrative technique (transitions, for example), an occasional ludicrously inept specimen serves to set off the unobtrusive skill with which the device is normally handled.

The attention here devoted to a detail of Greek and Latin poetic style could be aptly described as *in tenui labor.* But the scholar whom we honor with these essays knows that studies of this kind require no apology, and that when applied to highly sophisticated poets they can afford both enjoyment and instruction.

Metamorphoses may be added 5.280 and 282, 7.660 (discussed above), 8.575, 11.693, 12.88, 13.132-33, 459.

[45] Ribbeck's text contains two more cases, 6.573-74 *tum demum ... portae*, and 9.250-51 *sic memorans ... rigabat*, but neither is compelling. In several places where Ribbeck alone marks a parenthesis the effect would be to make Virgil more "Ovidian"; cf. most conspicuously 8.223-24 *fugit ilicet ocior Euro | speluncamque petit (pedibus timor addidit alas)*, compare *Met.* 4.528 *occupat hunc (uires insania fecerat) Ino.*

MNEMOTECHNICS AND THE RECEPTION OF THE *AENEID* IN LATE ANTIQUITY AND THE MIDDLE AGES

JAN M. ZIOLKOWSKI

In the *Institutio oratoria* Quintilian advises people who lack both a strong memory and an aptitude for extemporaneous speaking to apply themselves to literature, not speech-making:[1]

> But even this is no remedy for a weak memory, except for those who have acquired the art of speaking extempore. But if both memory and this gift be lacking, I should advise the would-be orator to abandon the toil of pleading altogether and, if he has any literary capacity, to betake himself by preference to writing. But such a misfortune will be of but rare occurrence.

Although Quintilian may not have been justified in linking weak memories and literary pursuits (certainly not in the case of the man whom this volume honors!), even so I will provide numerous *aides-mémoire* in the form of quotations as I explore the effects of formal memory training upon the reception of the *Aeneid* in late antiquity and the Middle Ages.

Beginning with the theory of memory, I will first trace how Simonides' famous anecdote about the origin of formal mnemotechnics survived into late antiquity. Then I will consider briefly the standing of formal mnemotechnics in the Middle Ages. From theory I will move to practice, by examining the role of memory training in ancient and medieval school training. In this examination I will focus on the effects that formal memory training and the prestige of memory had on the remembrance and interpretation of the *Aeneid*. Throughout I will set my sights on the problematic period from late antiquity through the twelfth century A.D. Frances Yates, in her still-essential book on *The Art of Memory*, states at one juncture that: "If Simonides was the inventor of

[1] Quint. *Inst.* 11.2.49 (trans. H. E. Butler, Loeb Classical Library): *sed ne hoc quidem infirmae memoriae remedium est nisi in iis qui sibi facultatem aliquam dicendi ex tempore parauerunt. quod si cui utrumque defuerit, huic omittere omnino totum actionum laborem, ac si quid in litteris ualet ad scribendum potius suadebo conuertere: sed haec rara infelicitas erit.*

the art of memory, and 'Tullius' its teacher, Thomas Aquinas became something like its patron saint."[2] Because the stimulating recent publications of Lina Bolzoni and Mary Carruthers have dealt mainly with the later Middle Ages, I will restrict myself to the centuries before Thomas underwent canonization—a period that has received attention from M.T. Clanchy, from Janet Coleman, and even more from Patrick Geary, but with a view toward history rather than the interpretation of literature.[3]

The Simonidean explanation for the foundation of a formal mnemotechnics—which served the quasi-forensic purpose of reconstructing an accident scene—reached the Latin West through several intermediaries. First came classical sources, such as Cicero's *De oratore* and Quintilian's *Institutio oratoria*, neither of which circulated extensively in the early Middle Ages.[4] Second was the three-book treatise on rhetoric by C. Chirius Fortunatianus, a rhetorician who probably lived in the fourth century. In contrast to the classical sources on which it drew, this treatise seems to have exerted a strong influence in the early Middle Ages, to judge from its survival in five seventh-, eighth-, and ninth-century manuscripts.[5] Third, and by far the most important, was that bizarre compendium of the liberal arts produced in the first quarter of the fifth century, Martianus Capella's *De nuptiis Philologiae et Mercurii*.

---

[2] See Frances Yates, *The Art of Memory* (Chicago 1966) 82.

[3] L. Bolzoni, *La cultura della memoria* (Bologna 1992); M. Carruthers, *The Book of Memory: A Study of Memory in Medieval Culture* (Cambridge 1990); M. T. Clanchy, *From Memory to Written Record: England, 1066-1307* (Cambridge, Mass. 1979); Janet Coleman, *Ancient and Medieval Memories: Studies in the Reconstruction of the Past* (Cambridge 1992); and Patrick J. Geary, *Phantoms of Remembrance: Memory and Oblivion at the End of the First Millennium* (Princeton 1994). As Geary points out (p. 27), the two most recent of these excellent studies omit consideration of the early Middle Ages—and I would add that that they pay less attention to belles-lettres than to certain other genres. The most important older work on mnemotechnics in the Middle Ages is Helga Hajdu, *Das mnemotechnische Schrifttum des Mittelalters* (Vienna 1936).

[4] Cicero, *De oratore* 2.86.351-354 and Quintilian, *Institutio oratoria* 11.2.11-17.

[5] Manfred Fuhrmann, "C. Chirius Fortunatianus," in *Der Kleine Pauly. Lexikon der Antike*, ed. Konrat Ziegler and Walther Sontheimer 5 vols. (Munich 1979), 1: 1148.57-1149.5, and D. Matthes in *Lustrum* 3 (1958) 77-78.

Because of their later significance, the passages in Fortunatianus and Martianus Capella merit close scrutiny. Although they are substantially the same, they contain several striking differences:

> Fortunatianus
>
> Who first demonstrated the art of memory? Simonides. What do we gain from Simonides' feat concerning the banquet? That the memory is aided when places are assigned in the mind. Then we will assign some kind of sign in its likeness for what has been written or thought. Who used to employ this skill? Charmadas [fl. c. 107 B.C.] and Metrodorus of Scepsis [fl. c. 70 B.C.]. Which is the superior and simpler principle of memory? That a speech should be learned in parts, if it is on the long side. Why so? Because certain limits must be given, so that first continuous and repeated meditation may join together the verbal context, which is most difficult, and then an order when recalled may join together the parts themselves; and let us record some marks beside those parts which have proved more difficult, and let us repeat them often, and let us learn them from wax tablets on which we have written them. How should we learn them? In a low voice or rather a murmur, in the same manner so that we may often test whether we have retained them. What assists the memory most? Division and composition: for order conserves memory powerfully, as we gathered from Simonides' banquet. What is the greatest art of memory? Practice and toil. How do we practice it? First with poems, then with speeches, and finally with more difficult works, such as legal writings. Why do we retain more easily and securely those things upon which we meditate at night? Because our attention is never distracted and after this thought or reading we think over nothing further in our mind, but at once rest. Should we always memorize word for word? If time allows; otherwise, we will retain only the substance and then extemporize the words to suit them. What else shall we observe in this? We must not fall into the habit of making allow-

ances for ourselves; for it is a great fault to be corrected and, when you are speaking, to consult a book.[6]

Martianus Capella

Now order brings in the precepts for memory which is certainly a natural (gift) but there is no doubt that it can be assisted by art. This art is based on only a few rules but it requires a great deal of exercise. Its advantage is that it enables words and things to be grasped in comprehension quickly and firmly. Not only those matters which we have invented ourselves have to be retained (in memory) but also those which our adversary brings forward in the dispute. Simonides, a poet and also a philosopher, is held to have invented the precepts of this art, for when a banqueting-hall suddenly collapsed and the relatives of the victims could not recognise (the bodies), he supplied the order in which they were sitting and their names which he had recorded in memory. He learned from this (experience) that it is order which sustains the precepts of memory. These (precepts) are to be pondered upon in well-lighted places in which the images of things are to be placed. For example (to remember) a wedding you may hold in mind a

---

[6] Fortunatianus, *Ars rhetorica* 3.13-14 (following *Institutio* 11.2), ed. Carolus Halm, *Rhetores Latini minores* (Leipzig 1863) 129.4-28; the translation is my own: *Artem memoriae quis primus ostendit? Simonides. Ex Simonidis facto de convivio quid adsumimus? iuvari memoriam signatis in animo sedibus. Tunc ad scripta vel cogitata aliquid signi ad eius similitudinem conlocabimus. Hac exercitatione qui utebantur? Charmadas et Scepsius Metrodorus. Quae melior et simplicior memoriae ratio est? ut si longior fuerit oratio, per partes ediscatur. Cur ita? quoniam dandi sunt quidam termini, ut contextum verborum, qui est difficillimus, continua et crebra meditatio, partes deinde ipsas repetitus ordo coniungat, et his, quae difficilius haeserint, aliquas adponamus notas et eas saepe repetamus, et isdem ceris ediscamus, quibus scripsimus. Quem ad modum ediscendum est? voce modica et magis murmure, item ut saepe nos temptemus an teneamus. 14. Quid vel maxime memoriam adiuvat? divisio et conpositio: nam memoriam vehementer ordo servat, sicut accepimus de Simonidis convivio. Quae est maxima ars memoriae? exercitatio et labor. Quem ad modum hanc exercebimus? primo poematibus, dehinc orationibus, novissime operibus durioribus, qualia scripta sunt iuris. Cur ea quae nocte meditamur, facilius ac firmius tenemus? quoniam nusquam avocatur intentio et post istam cogitationem vel lectionem nihil amplius animo cogitamus, sed statim capimus quietem. Semper ad verbum ediscendum est? si tempus permiserit: sin minus, res ipsas tenebimus solas, dehinc his verba de tempore accommodabimus. Quid aliud in hac observabimus? ne consuescamus nobis ignoscere: nam et admoneri et, cum dicas, ad librum respicere vitiosum est.*

girl veiled with a wedding-veil; or a sword, or some other weapon, for a murderer; which images as it were deposited (in a place) the place will give back to memory. For as what is written is fixed by the letters on the wax, so what is consigned to memory is impressed on the places, as on wax or on a page; and the remembrance of things is held by the images, as though they were letters. But, as said above, this matter requires much practise and labour, whence it is customarily advised that we should write down the things which we wish easily to retain, so that if the material is lengthy, being divided into parts it may more easily stick (in memory). It is useful to place *notae* against single points which we wish to retain. (When memorising, the matter) should not be read out in a loud voice, but meditated upon with a murmur. And it is obviously better to exercise the memory by night, rather than by day, when silence spreading far and wide aids us, so that the attention is not drawn outward by the senses. There is memory for things and memory for words, but words are not always to be memorised. Unless there is (plenty of) time for meditation, it will be sufficient to hold the things themselves in memory, particularly if the memory is not naturally good.[7]

---

[7] Martianus Capella 5.538-539, ed. Adolf Dick and rev. Jean Préaux (Stuttgart 1978) 268.13-270.2 (trans. Yates, pp. 51-52): *nunc ordo praecepta memoriae subministrat, quam quidem constat esse naturalem, sed et arte adiuuari posse non dubium est. haec autem ars breuibus praeceptis, sed magna exercitatione formatur, cuius partis hoc munus est, ut non tantum firma, uerum etiam celeri comprehensione res uerbaque percipiantur. tenenda uero sunt non ea tantum, quae a nobis inuenta sunt, sed etiam quae ab aduersario in agendo tractata sunt. Simonides huius rei praecepta inuenisse perhibetur poeta idemque philosophus; cum enim conuiuii locus subito corruisset, nec possent propinqui obtritos internoscere, discumbentium ordinem nominaque memoria recordante suggessit. quo admonitus intellexit ordinem esse, qui memoriae praecepta conferret. is uero in locis illustribus meditandus est, in quibus species rerum sententiarumque imagines collocandae sunt, ueluti nuptiarum uelatam flammeo nubentem aut homicidae gladium uel arma detineas, quas species locus tamquam depositas memoriae reddat. nam sicut id, quod conscribitur, cera continetur et litteris, sic quod memoriae mandatur, in locis tamquam in cera paginaque signatur; imaginibus uero quasi litteris rerum recordatio continetur. sed, ut diximus, magnam exercitationem res laboremque conquirit, in qua illud obseruari compertum est solere, ut scribamus ipsi quae facile uolumus retinere; deinde ut, si longiora fuerint, quae sunt ediscenda, diuisa per partes facilius inhaerescant; tum apponere notas rebus singulis oportebit in his, quae uolumus maxime retinere; nec voce magna legenda sunt, sed murmure potius meditanda; et nocte magis quam interdiu maturius excitari*

Both writers credit Simonides with having originated formal mnemonics and both recommend mastering long texts by dividing them into small sections, copying them out and annotating them, reading them aloud quietly, and working at night when there are no distractions—all *modi operandi* that remain very familiar to active scholars today! Martianus Capella accords considerably more space to describing memory through images, which Fortunatianus leaves only implicit. Both distinguish between memory for things and memory for words, but neither makes much of an issue of poetry: Fortunatianus lists poetry first when he discusses the materials upon which we should train our memories, and Martianus indicates that Simonides was a poet, but beyond that they say nothing specific about the memorization or composition of verse.

Even more widely current than the anecdote about Simonides was vague information about the different techniques of ancient mnemonics. It is interesting to note the clustering of such information in two groups. In the first we find ancient sources such as the *Rhetorica ad Herennium*, Cicero's *De oratore*, and Quintilian's *Institutio oratoria* (as well as the paragraphs already cited from Fortunatianus and Martianus Capella).[8] In the second we encounter sources from the mid-twelfth century and later, such as Thierry of Chartres' commentary on the *Ad Herennium*, John of Salisbury's *Metalogicon*, Geoffrey of Vinsauf's *Poetria nova*, and John of Garland's *Parisiana poetria*.[9] With the exception of Thierry, all of these twelfth- and early thirteenth-century authors who refer to mnemotechnics after this seven-century gap treat it skeptically, as if they were familiar with it only from reading ancient sources and as if they had doubts that it could be applied successfully in practice. The relevant passage from John of Salisbury typifies this guarded outlook:

memoriam manifestum est, cum et late silentium iuuat, nec foras a sensibus auocatur intentio. est quidem memoria rerum atque uerborum, sed non semper ediscenda sunt uerba; nisi spatium meditandi tempus indulserit, sat erit cui res ipsas animo tenuisse, praesertim si nihil eius munere naturali prouenerit.

[8] See *Rhetorica ad Herennium* 3.16.28-24.40; Cicero, *De oratore* 2.86-88.351-358; and Quintilian, *Institutio oratoria* 11.2.1-51.

[9] Geoffrey of Vinsauf, *Poetria nova* 1969-2030, in Edmond Faral, *Les arts poétiques du XIIe et du XIIIe siècle. Recherches et documents sur la technique littéraire du moyen âge*, Bibliothèque de l'École des hautes études (Paris 1924) 193-262 (here: 257-259), and John of Garland, *The Parisiana Poetria* 2.87-115, ed. and trans. Traugott Lawler (New Haven and London 1974) 36-39.

> Seneca glibly promised to impart the art of memorization, of which I certainly wish I were a master; but as far as I know, he did not actually teach it. Tullius seems to have applied himself to this in his *Rhetorical Questions*, but the latter are not of much help to men like me. There are extant some things, it is true, which we can scarcely apprehend, but about these we are very little concerned. On the other hand, rules concerning similar forms and inflections, etymologies, definitions of terms that need explanation, and differences, those pointing out the faults of barbarisms, solecisms, and other grammatical errors to be avoided, those clarifying the question as to what forms of metaplasms, schemata, and tropes are permissible and ornamental, and those explaining prose, enunciating the laws of poetry, and stating cases, as well as the method to be followed in historical and fictional narratives,—all must be extremely advantageous. If anyone wants the definitions and forms of the above, he has but to peruse the books of the aforesaid grammarians.[10]

Even Thierry maintains that the memory system proved to be serviceable only for remembering verses from poets.[11]

To me the evidence implies that from the sixth through the early twelfth century—loosely speaking, during the entire pre-Gothic period—most memorizing followed the monastic mode of *ruminatio* and that it aimed at *memoria uerborum* rather than *memoria rerum*.[12] Many of the metaphors that characterized memory remained the same in medieval literature as they had in ancient times—memory was a store-

---

[10] John of Salisbury, *Metalogicon* 1.20, Corpus Christianorum Continuatio Mediaevalis 98 (Turnhout 1991), ed. J.B. Hall with K.S.B. Keats-Rohan, p. 47.28-40; trans. Daniel D. McGarry, *The Metalogicon of John of Salisbury: A Twelfth-Century Defense of the Verbal and Logical Arts of the Trivium* (Berkeley 1955) 59-60.

[11] Thierry of Chartres, Commentary on *Ad Herennium*, ed. Karin M. Fredborg, *The Latin Rhetorical Commentaries* (Toronto 1988) 306.20-21: *Nota ad carmina poetarum in memoria retinenda verborum memoriam plus quam ad causas valere.* See Carruthers (above, n. 3) 151 and 323, n. 49.

[12] On *ruminatio* see Jean Leclercq, *Love of Learning and the Desire for God*, tr. Catharine Misrahi (New York 1961) 16-22 and 89-90.

house or treasure-house,[13] wax-tablet, belly, bookchest, chest, mirror,[14] and cell[15]—but gone temporarily are explicit expressions of the blueprints for memory that the ancient mnemotechnicians had advocated. Representative of the less sophisticated—or at least less visual—mechanisms of recall that held sway during this half millennium or so is a declaration by Alcuin in a dialogue "Concerning Rhetoric and the Virtues." Asked by Charlemagne to tell him how to obtain and increase memory, Alcuin replies with schoolmasterly common sense: "We have no other precepts about it, except exercise in memorising, practice in writing, application to study, and the avoidance of drunkenness which does the greatest possible injury to all good studies..."[16]

The citation from the *Metalogicon* suggests that one reason for this shift lies in the fundamentally *grammatical* nature of early medieval education, as opposed to the prestige that had attached to *rhetoric* in antiquity. The sort of memory that was esteemed in the Middle Ages was precisely the verbatim recall which Quintilian prescribed for the preschool training of orators: heavy memorization of poetry, especially poetry with moral meaning. To quote the *Institutio oratoria*, "at the tender age of which we are now speaking, when originality is impossible, memory is almost the only faculty which can be developed by the teacher."[17]

Rather than enunciating a new idea of his own, Quintilian was codifying a wisdom that had been in place for centuries before him and that remained in force for centuries to come. From Diogenes of Sinope ("the Cynic") in the fourth century B.C. on, boys were encouraged to memo-

---

[13] Quint. *Inst.* 11.2.1-2 and *Rhet. ad Herenn.* 3.16.28

[14] On these metaphors see Christel Meier, "Vergessen, Erinnern, Gedächtnis im Gott-Mensch-Bezug. Zu einem Grenzbereich der Allegorese bei Hildegard von Bingen und anderen Autoren des Mittelalters," in *Verbum et Signum. Beiträge zur mediävistischen Bedeutungsforschung. Studien zu Semantik und Sinntradition im Mittelalter*, ed. Hans Fromm, Wolfgang Harms, and Uwe Ruberg (Munich 1975) 143-194 (here: p. 187).

[15] On the *cellula* of memory, see Geoffrey of Vinsauf, *Poetria nova* 1972, ed. Faral, p. 258.

[16] Ed. and trans. W.S. Howell, *The Rhetoric of Charlemagne and Alcuin* (Princeton 1941) 136-139 (cited by Yates [above, n. 2] 53).

[17] 1.1.36.

rize, first and foremost poetry.[18] (Remember how Fortunatianus recommended memorizing, *first* poems, *then* orations, and *finally* legal documents.) The boys achieved their rote learning initially through copying out selections and later through thoroughgoing study—the reading aloud, memorizing, and explicating—of full texts. Sometimes, through dictation, they were forced to incorporate writing into the process of memorization.[19] On other occasions they had to repeat lines of text after the master until they had learned them by heart.[20]

Although many different poets were used in elementary education, Virgil enjoyed a special status. According to Suetonius' *De grammaticis*, Quintus Caecilius Epirota took Virgil and other modern poets into his school very early—perhaps even before the *Aeneid* was published.[21] By the time Juvenal writes his seventh satire, Virgil and Horace are already so well established that their soot-blackened busts are regular fixtures in the schoolroom.[22]

A facility for remembering Virgil was valued outside the classroom, among adults. Our most striking example of this aptitude and of this esteem appears in Augustine's treatise "On the nature and origin of the soul":

> A certain man who from his youth has been a friend of mine, named Simplicius, is a person of accurate and astonishing memory. I once asked him to tell me what were the last lines but one of all the books of Virgil; he immediately answered my question without the least hesitation, and with perfect accuracy. I then asked him to repeat the preceding lines; he did so. And I really believe that he could have repeated Virgil line

---

[18] Stanley F. Bonner, *Education in Ancient Rome, From the elder Cato to the younger Pliny* (Berkeley 1977) 39-40.

[19] Bonner (above, n. 18) 177.

[20] This practice, which the Greeks designated *apostomatizein*, is alluded to by Horace, *Epistles* 1.18.12-14: see Bonner (above, n. 18) 177.

[21] Suetonius, *De grammaticis* 16.3 (trans. J. C. Rolfe, Loeb Classical Library): "He is said to have been the first to hold extempore discussions in Latin, and the first to begin the practice of reading Vergil and other recent poets..."

[22] Juvenal 7.226-27: "if only it does not pass away without reward to have sniffed as many lamps as there were boys who stood about (in your class), when the whole bust of Horace was discolored and soot was clinging to black Vergil."

> after line backward. For wherever I wished, I made trial whether he could do it, and he did it. Similarly in prose, from any of Cicero's orations, which he had learnt by heart, he would perform a similar feat at our request, by reciting backwards as far as we wished. Upon our expressing astonishment, he called God to witness that he had no idea of this ability of his previous to that trial. So far, therefore, as memory is concerned, his mind only then learnt its own power; and such discovery would at no time be possible except by trial and experiment.[23].

This recollection reveals a provocative affinity to the Caedmon episode in Bede: with the help of God, a man displays publicly a knack for poetry that he professes never to have known before.[24] In this case the talent is a capacity to recite the texts of Virgil and Cicero backwards and forwards and to scan them without becoming confused.

Although such total recall of Virgil was surely exceptional, the esteem in which Virgil was held ran fairly constant. The rationale for devoting close attention to a pagan poet was that the *Aeneid* contained moral lessons worthwhile even for Christians. Interestingly for our purposes, in the boldest attempt to derive allegorical meaning from the *Aeneid* the main message of the first six books was connected with memory: the premier text for memorization was explained as describing the maturation of wisdom or philosophy over the course of human life, and in this development memory took a privileged place. Indeed, the

[23] Augustine, *De natura et origine animae* 4.7.9, ed. Carolus Urba and Iosephus Zycha, Corpus Scriptorum Ecclesiasticorum Latinorum 60 (Vienna 1913) 389, lines 7-18; trans. Philip Schaff, *A Select Library of the Nicene and Post-Nicene Fathers of the Christian Church* vol. 5 (New York 1887) 358: *amicus quidam meus iam inde ab adulescentia, Simplicius nomine, homo excellentis mirabilisque memoriae, cum interrogatus esset a nobis, quos uersus Vergilius in omnibus libris supra ultimos dixerit, continuo celeriter memoriterque respondit. quaesiuimus etiam superiores ut diceret: dixit. et credidimus eum posse retrorsus recitare Vergilium; de quocumque loco uoluimus, petiuimus ut faceret: fecit. prosa etiam de quacumque oratione Ciceronis, quam memoriae commendauerat, id eum facere uoluimus: quantum uoluimus sursum uersus secutus est. cum ammiraremur, testatus est deum nescisse se hoc posse ante illud experimentum. ita, quantum ad memoriam pertinet, tunc se eius animus didicit et quandocumque disceret, nisi temptando et experiendo non posset.* The passage has also been translated by Carruthers (above, n. 3) 19.

[24] See Bede, *Ecclesiastical History of the English People* 4.24, ed. and trans. Bertram Colgrave and R.A.B. Mynors, Oxford Medieval Texts (Oxford 1969).

commentator (who may or may not have been Bernardus Silvestris) eventually consigns philosophy to the chamber of memory in the mind. This chamber occupies one entire unit in a mind that the commentator conceives as being tricameral:

> For there are three things which furnish perfect wisdom: wit, the power of discovering; reason, the power of discerning what is discovered; memory, the power of retention. In the brain there are three chambers which others call ventricles: in the anterior chamber is the exercise of wit; in the middle, the exercise of reason; in the posterior, the means of memory.[25]

So obsessed with the issue of memory is the commentator that once he has brought Aeneas and the Sibyl to the cells of memory, he seems convinced that he has completed his philosophical task and therefore he concludes his commentary exactly at the moment when Aeneas has planted the bough—which the commentator takes to be philosophy—across the threshold of memory.

> *Portas*: the three chambers. We come to heavenly contemplation through these (as we said before) by exercising wit, reason, and memory… *Hec ubi*: at the gates, since Aeneas and the Sibyl are presently in the cells of memory… But since committing to memory follows discovery by wit, Aeneas places the branch—philosophy—across the threshold, the rear chamber. Here end the glosses on the *Aeneid* according to the integument.[26]

---

[25] *The Commentary on the First Six Books of the "Aeneid" of Vergil Commonly Attributed to Bernardus Silvestris* 6.68, ed. Julian Ward Jones and Elizabeth Frances Jones (Lincoln, Nebraska 1977) 47.2-6; trans. Earl G. Schreiber and Thomas E. Maresca, *Commentary on the First Six Books of Virgil's "Aeneid"* (Lincoln, Nebraska 1979) 47 (see also pp. 57, 63): *Tria namque sunt que sapientiam perfectam reddunt, ingenium, scilicet vis inveniendi, ratio vis discernendi inventa, memoria vis conservandi. In cerebro autem sunt tres cellule quas alii ventriculos vocant: prima anterior in qua est ingenii exercicium; secunda est mediana in qua est vis rationis; tercia est postica in qua est vis memorie.*

[26] *Commentary* 6.631-632 and 636, ed. Jones and Jones, p. 114.13-15 and 15-16, and 115.4-5; trans. Schreiber and Maresca, p. 81: *PORTAS, id est cellulas. Per has enim ut supradiximus exercendo ingenium, rationem, memoriam celestia contemplatione ingredimur... HEC UBI, id est in quibus portis, quia in cellula memorie... Sed quia sequitur ingenii inventionem memorie mandatio, addit RAMUM: philosophiam. ADVERSO LIMINE: cellula postica. Expliciunt glosule*

However strained such an interpretation of the *Aeneid* may seem to us, in putting it forth the commentator was in good company, since both Fulgentius (who flourished around 500) and Cristoforo Landino (who lived from 1424-1498) follow a similar procedure.[27]

Since memory so engaged the authors who have been considered in this essay, it is entirely consistent to discover that forgetting was also a topic to which they were by no means oblivious. At the end of the Fortunatianus passage cited above, the rhetorician professes the belief that it is a flaw not to remember properly. Christians saw the *uitium* as being no mere peccadillo but actually a vice, because of the dialectic relation-

---

Eneidos secundum integumentum. Compare *Commentary* 6.69-70, trans. Schreiber and Maresca, p. 48: "*Templa*: what memory has preserved. *De marmore*: here Virgil notes the origin of memory. Indeed, wit and memory have diverse origins. For wit arises from a fiery nature, memory from a cold nature. All velocity arises from fire, whence we see that sanguine people are swift but heavy. Wit is speed in seizing the knowable. And so it has its origins in heat. In contrast, memory is the slowness and retention of what is comprehended. For all slowness arises from cold. Whence we see melancholic people are slow but thin. Martianus allegorically notes these diverse origins when he says that Urania resides in *Aplanes* (the firmament), that is, by seizing strength in fire; Polyhymnia resides in Saturn, that is, great memory resides in coldness. Virgil notes the same things here when he says *templa de marmore*, 'marble temples,' that is, the preservation of memory by tenacious cold."

27 Bernard's knowledge of Fulgentius is evident in the *Commentary* 6.138, trans. Schreiber and Maresca, p. 58, where the altar of Proserpina "is the posterior cell of memory in which, as noted earlier, are coldness and slowness, which come from the moon and which serve memory. And therefore it is said that the branch is sacred to this goddess since in that cell philosophy is entrusted to memory." Compare Fulgentius, *Virgiliana continentia* 160, ed. Rudolf Helm and rev. Jean Préaux (Stuttgart 1970) 101.15-102.4; *Fulgentius the Mythographer*, trans. Leslie George Whitbread (Columbus, Ohio 1971) 131-132: *Deinde ramum aureum postibus deuotis infigit et ita Elisium ingreditur, quo clareat, dum perfectionem omisso iam labore discendi ... memoriae quae in cerebro est sicut in postibus perpetue infigenda. Elisium ingreditur campum—elisis enim Grece resolutio dicitur—id est feriatam uitam post magistrianum timorem. Sicut enim inferni Proserpina regina est, ita scientiae regina memoria est, quae in elisis proserpens dominatur perenniter mentibus. Huic ergo doctrinae aureum ramulum dedicatur; de qua memoria Cicero dicere solitus solitus erat tesaurum scientiae* ("Next he fastens the golden bough on the dedicated gateposts [dedicated to learning] and so enters Elysium, where, the labor of learning now over, he celebrates the perfecting of memory, which is to be fastened on the brain as enduringly as the golden bough on the gateposts. He enters the Elysian fields—*elisis* in Greek means release—that is, the liberated way of life after finishing with fear of teachers. As Proserpine is queen of the lower world, so the queen of knowledge is memory, which as it advances reigns forever supreme in liberated minds. In this way is the golden bough dedicated to learning. Cicero used to say that memory was the treasure house of wisdom").

ship between memory and oblivion.[28] The Vulgate Bible offers approximately 600 instances of remembering or forgetting, among which are verses that present eternal memory as the reward for good life and oblivion as the fate of those who have offended God.[29]

The counterpoint between memory and oblivion attracts the remark of Bernardus (or pseudo-Bernardus) Silvestris repeatedly in the commentary on Virgil. In two passages, he envisages memory as the burial of virtue and knowledge in the mind (presumably in anticipation of resurrection, on analogy with a Christian buried in the earth!), as opposed to the permanent oblivion to which vice should be consigned:[30]

> *Vita*: philosophy understands two lives, two deaths, and two graves in poetic fictions... The first burial is the engraving of thoughts in firm memory, in which virtue and knowledge are to be buried. The other grave is the one in which Misenus and Palinurus are to be buried: the enwrapping in oblivion.[31]

---

[28] Thus there is an opposition between on the one hand *oblivio* and *obliviscor* and on the other words such as *memini, memor, memoria, memoriale, reminiscor, rememoror, recordatio,* and *recordor.* For the conceptual framework, see Alan of Lille, *Liber in distinctionibus dictionum theologicalium*, ed. J.-P. Migne, *Patrologiae cursus completus, Series latina*, 221 vols. (Paris, 1844-1864), vol. 210 (1855), columns 855BC: Memoria, *proprie. Dicitur vita aeterna; unde in Psalmo:* In memoria aeterna erunt justi. *Dicitur fama vel gloria, unde David:* Periit memoria eorum cum sonitu. *Dicitur poena memorialis, unde David:* Redeat in memoriam iniquitas patrum ejus, *id est puniatur poena memoriali.* And: Memor, *proprie, adimplens promissum; unde David:* Memor esto verbi tui servo tuo. *Dicitur judicans vel damnans, unde Psalmista:* Nec memor ero nominum eorum, *id est non judicabo eos in die judicii, dicens:* Ite, maledicti, in ignem aeternum. *Dicitur aliquis gratiam conferens, unde David:* Memor ero Raab et Babylonis, *id est peccatoribus qui significantur per Raab et Babylonem conferens gratiam.*

[29] Compare for example Wisdom of Solomon 4.1 *inmortalitas est enim in memoria illius* with Psalm 87.13 *iustitia tua in terra oblivionis.*

[30] *Commentary* 6.331-334 and 6.374-383, trans. Schreiber and Maresca, pp. 78 and 81: "*Mortis honore*: burial. We have called the grave the memory of virtue and knowledge, the forgetting of vice. And note that these two men lack the honor of death (*honore mortis carere*), that is, proper burial, since we commit virtue and knowledge to memory with greater difficulty than we do the filth of vice--in wicked things we are mutable" and "*Inhumatus*: 'unburied,' unless you wrap virtue in oblivion... *Tumulum*: oblivion."

[31] *Commentary* 6.306, ed. Jones and Jones, p. 79.1-2 and 8-10; trans. Schreiber and Maresca, pp. 75-76: *VITA: Duas vitas et duas mortes et duas sepulturas in integumentis philosophia intelligit... Una sepultura est cognitorum intimatio in firma*

> *Sepulti*: they are enwrapped—their vices in oblivion, their virtues in memory. *Ossa quierunt sedibus*: virtues were placed in the treasury of memory.[32]

In another two he condemns oblivion as the forgetfulness of a soul that allows its memory of God to lapse:

> Aeneas buries his father in anger when God is given over to oblivion. For the angry are almost apostates. The grave is a kind of oblivion.[33]

> Four rivers are designated in the infernal regions: Phlegethon, the fires of the angry; Lethe, the oblivion of the mind forgetful of the majesty of its own divinity; Styx, hatred; and Acheron, sorrow.[34]

The preoccupation with memory affected the reception of Virgil not just by commentators but also by practicing Latin poets, since together with a tendency to view reading as a memory act came an inclination to conceive of composition in similar terms. At its most extreme, the sort of memory that Simplicius demonstrated had a compositional correspondent in the *cento*, since it is likelier that the *centos* were woven together by memory than cobbled together half line by half line by constant reference to a written text.[35] Etymologically a word for a patch-

---

memoria in qua sepeliende sunt virtutes et scientie. Altera sepultura est in qua Misenus et Palinurus sepeliendi, scilicet involutio in oblivionem.

[32] *Commentary* 6.326, ed. Jones and Jones, pp. 80.23-81.1; trans. Schreiber and Maresca, p. 77: *SEPULTI sunt: id est, involuta sunt vicia sua in oblivione, virtutes in memoria. OSSA: virtutes QUIERUNT: locate fuerunt SEDIBUS: thesauris memorie.*

[33] *Commentary* 3.707, ed. Jones and Jones, p. 23.15-16; trans. Schreiber and Maresca, p. 24: *In iracundia pater sepelitur cum oblivioni deus datur. Iracundi enim pene apostantur. Sepultura quedam oblivio est.*

[34] *Commentary* 6.1, ed. Jones and Jones, p. 29.12-14; trans. Schreiber and Maresca, p. 32: *Quattuor etiam fluvios in eis assignabant: Flegetontem ardores irarum, Leten oblivionem mentis maiestatem sue divinitatis obliviscentis, Stigem autem odium, Acherontem tristiciam.*

[35] On the cento, see Joseph Octave Delepierre, *Tableau de la littérature du centon chez les anciens et chez les modernes*, 2 vols. (London 1874-1875); Filippo Ermini, *Il centone di Proba e la poesia centonaria latina* (Rome 1909); and K. H. Schelkle, "Cento," in *Reallexikon für Antike und Christentum*, ed. by Theodor Klauser (Stuttgart 1950), vol. 2 (1954) 972-973.

work robe, the *cento* makes the purple patch into a raison d'être: the composition of poetry becomes itself a mnemonic tour de force.

The cento offered one workable solution for Christians who could not dispense altogether with the classical educational system erected around canonical authors—most particularly, Virgil—but who could not accept pagan writings without somehow retooling them for Christian purposes. They achieved an accommodation with Virgil's oeuvre sometimes by arguing that the poet had been a covert or unwitting Christian. At other times they interpreted his works—regardless of what they maintained were his own religious convictions—as containing moral or even theological truth. Finally, they took the even more active measure of centoizing: of taking lines and half lines from Virgil which they reassembled so as to express Christian concerns. A closely related but less drastic means of creating Christian literature with the help of Virgil was to write a Bible epic, in which the poet paraphrased parts of the Bible in language suited to the Latin epic tradition.

The cento and the Bible epic would not have experienced the vogues that they did if poets and their readers had not possessed a deep and intimate memory of Virgil's style and content. In a classic book entitled *Vergil in the Middle Ages*, Domenico Comparetti wrote that the idea for the Virgilian centos "could only have arisen among people who had learnt Virgil mechanically and did not know of any better use to which to put all these verses with which they had loaded their brains."[36] Some of the poets merited such harsh disparagement, but in other cases we must be careful not to vituperate poets for having adopted aesthetic values which differ from our own but which served them well; for memory, especially as transmitted through schools where for more than a millennium Virgil was recognized as the most memorable of poets, contributed powerfully to the expression of late antique and early medieval aesthetics—and it is by no means clear that we are better off without it.

Even once the imagic systems of Cicero and pseudo-Cicero returned to prominence in the late twelfth and early thirteenth centuries, memory of Virgil remained an essential for both the interpretation and

---

[36] *Vergil in the Middle Ages*, trans. E.F.M. Benecke, 2d ed. (London and New York 1908; repr. Princeton 1997) 53.

the production of literature. A graphic example of his continued salience would be the *rota Vergilii*, a famous "memory grid" recorded by John of Garland so that his readers could use the distinct differences between the *Eclogues*, *Georgics*, and *Aeneid* as an organizing principle for their recollections of the three styles and the features appropriate to them.[37] The concentric circles of John of Garland's schema bring us, literally, full circle to the formal mnemotechnics with which this walk down memory lane began. As we have seen, the urgency of remembering Virgil and the challenge of understanding his various pronouncements on memory remained intense from late antiquity through the twelfth century, both when verbatim memorization prevailed and when formal mnemotechnics based on classical memory systems held sway.

[37] See John of Garland, *Parisiana Poetria* 2.116-123, ed. and trans. Lawler, pp. 38-41.

VITAE VERGILIANAE IN UNPUBLISHED VIRGILIAN COMMENTARIES (SAEC. XV AND XVI)*

VIRGINIA BROWN

The "congeries of fact and fiction" that makes up the fourth-century Vita Suetonii / Donati (VSD),[1] the oldest extant biography of Virgil, is to be found as well in other lives of the poet. Separation of reality from legend becomes possible only after the sources have been assembled. But this is not an easy thing to do for Virgil simply because the material is so abundant and diverse.

A most convenient collection of the testimonia of Virgil's contemporaries and embellished, often fabulous, accounts of later authors was recently published in the *Enciclopedia Virgiliana*.[2] The range of material included is truly admirable, especially for the Middle Ages. A much smaller section for the fifteenth century (six Latin lives) and nothing for the sixteenth century reflect the obvious difficulty of extracting relevant texts from mostly unpublished sources.

One such source must be the manuscripts which preserve the poems of Virgil. The tremendous popularity implied by his continuing status as a school author is attested by the survival of more than a thousand witnesses; of these at least two thirds are humanist, i.e., copied in the fifteenth and sixteenth centuries.[3] Probably almost half of the humanist

---

* Part of the research presented here was derived from my work on the article on "Virgil" for the *Catalogus translationum et commentariorum* and was facilitated by a grant, for the latter project, from the Research Division of the National Endowment for the Humanities, an independent Federal agency. I am happy to acknowledge this assistance. I should also like to thank Prof. A.C. de la Mare for her help in identifying and dating some Italian manuscripts (serial nos. 9, 10, 18, 23 below).

[1] W. Clausen, *Virgil's Aeneid and the Tradition of Hellenistic Poetry* (Berkeley 1987) 126 n. 8. Names and texts of other published lives are cited, unless otherwise specified, from K. Bayer, *Vergil-Viten*, in J. and M. Götte, eds., *Vergil. Landleben: Bucolica, Georgica, Catalepton, Vergil-Viten*, 2nd ed. (Munich 1978) 211-395, with list of sigla on pp. 657-58.

[2] G. Brugnoli–F. Stok, "Fontes ad vitam Vergilii pertinentes," *Enciclopedia Virgiliana* 5** (Rome 1991) 427-540.

[3] For lists of the manuscripts containing Virgil's work see: M. Geymonat, "Codici," *Enciclopedia Virgiliana* 1 (Rome 1984) 831-38; G.C. Alessio, "Medioevo.

codices show signs of use by readers, and many contain an accessus to a commentary which, in a *catena* or gloss format, may accompany the text.

Here, obviously, there are many opportunities for biographical discoveries, and the list of thirty-five new lives given below is intended as a modest supplement to the *Enciclopedia Virgiliana.* Stringent criteria of selection have kept the number of texts with relevant information from being much higher; for purposes of this article I have included only those *vitae* which seem to be complete and are one of the components of an accessus to an unpublished commentary or complete *vitae* not formally part of an accessus but evidently connected with the unpublished commentary which follows in the manuscript. Necessarily excluded, therefore, are those *vitae* lacking the accessus context or commentary link, incomplete *vitae* in a truncated accessus, and the very large number of glosses with interesting biographical interpretations.[4]

Broadly speaking, the scribes of the *vitae* come from four regions (for convenient reference the serial number in the list below will be cited). Nearly half (nos. 1, 4, 6-8, 11-14, 19, 20, 25, 27, 29, 33-35) are "Germanic" (taken very broadly and including Austria, Switzerland, etc.). Ten *vitae* (nos. 9, 10, 17, 18, 23, 24, 26, 28, 31, 32) were written in Italy, and there are seven witnesses from France (nos. 2, 3, 5, 15,

---

Tradizione manoscritta," ibid. 3 (Rome 1987) 432-43; A. Petrucci, "Papiri," ibid. 3 (Rome 1987) 964-65.

[4] I have noted surprisingly few instances of seemingly new, complete *vitae* appearing on their own, namely:

Copenhagen, Det Kongelige Bibliotek Thott 311 8°, fols. 217r-218v, saec. $XV^{3/4}$ (Florence), perhaps copied by Alessandro Braccesi. *Inc.* "Publius Maro Virgilius ortus patre Marone matre Maia, Cn. Pompeio Magno et M. Licinio consulibus, urbe Mantua, que civitas est Venetie. Virgilius autem appellatus vel quia mater virgam lauream parere somniasset vel quia virga populea (fol. 217v) ubi ortus esset terre compacta... *expl.* (fol. 218v) Leges eum summum poetam Latinorum appellant, et antiqui Romani nullum bene institutum (?) credebant nisi prius in Virgilio esset eruditus et eius poesim totius humane vitae speculum esse arbitrabantur. Quare lege foeliciter."

Munich, Bayerische Staatsbibliothek Clm 8042, fol. 4v, saec. XVI (Germany). *Inc.* "Natus est poeta noster Virgilius in vico Ande qui non longe abest ab urbe Mantua. Unde et andinus vates appellatur a Sylio Italico... *expl.* Periit Tarenti ut Servius refert, ut alii volunt Brundusii, contracta agritudine ex itinere. Paulo mox delatum est illius cadaver Neapolim ibique sepultum addito hoc carmine: 'Mantua me genuit ... cecini pascua rura duces'."

21, 22, 30). There seems to be only one representative from England (no. 16). Diversity characterizes the milieux in which the new humanist *vitae* originated: five are explicitly connected with University lectures at Cracow (no. 4), Leipzig (nos. 12, 29) and Paris (nos. 5, 21, 22). At the other end of the spectrum are those commentaries (nos. 16, 18) whose contents indicate that they were intended for students at a much lower level.

The derivative aspect, generally speaking, of *vitae vergilianae* composed after the VSD is well known.[5] This is certainly true of the humanist *vitae* which were also subject to the influence of Donatus Auctus; the impact of both is immediately evident in our nos. 1, 8, 9, 14, 15, 19, 23, 25, 28, 31-34. Yet it is important not to neglect the later lives: they tell us who was reading Virgil, when, where and how, and so are important for the history of culture and education; this in turn leads to the detection of patterns of transmission and intellectual shifts.[6]

What, then, do the new humanist *vitae* listed below have to offer? An admittedly sketchy survey suggests the following:

(a) the general setting of the accessus in which each *vita* appears confirms a preference for the Servian type of prologue. Only in no. 6 do we find the Aristotelian type of accessus which developed during the thirteenth and fourteenth centuries and specified four causes (efficient, material, formal, final).[7] Indeed, the *vita* of no. 6 is so medieval in character, with its emphasis on the miraculous deeds performed by Virgil, that it may antedate the fifteenth century although it is presently known to survive only in humanist witnesses.

---

[5] See, e.g., N.M. Horsfall, "Virgil: His Life and Times," in Horsfall, ed., *A Companion to the Study of Virgil* (Leiden 1995) 4.

[6] For different ways in which four humanists (Petrarch, Sicco Polenton, the anonymous compiler of Donatus Auctus, and Pomponio Leto) grappled with the problem of fact and fiction in Virgilian biography, see F. Stok, "Virgil between the Middle Ages and the Renaissance," *International Journal of the Classical Tradition* 1 (1994) 15-22, with further references in the notes to discussions of medieval lives. I regret that I was not able to consult G. Brugnoli–F. Stok, eds., *Vitae Vergilianae* (Rome 1994), cited on p. 15 n. 1 of Stok's article.

[7] A.J. Minnis, A.B. Scott, D. Wallace, edd., *Medieval Literary Theory and Criticism c. 1100 – c. 1375. The Commentary – Tradition*, rev. ed. (Oxford 1991) 2-4 and passim.

(b) no. 6, however, could also serve as a salutary reminder that humanists from all regions knew and continued to consult medieval sources. This is clear in the case of no. 2 where Vincent of Beauvais is specified as one of the sources (a major unnamed source in the same Vita is the accessus attributed to Anselm of Laon [d. 1117]).

(c) it is satisfying to be able to identify some of the new *vitae* as the work of known humanists, namely, Cristoforo Landino (nos. 9 [autograph, perhaps hitherto unidentified], 28), Johannes Albinus (no. 12), Johannes Kegelius (no. 14), Gaspar Tribraco de' Trimbocchi (no. 24), and Giovanni Piedemonti (no. 26).[8] In this regard, it may be helpful to mention that something more can be said about three anonymous *vitae*, long known and published from humanist manuscripts: the Expositio Monacensis III is the accessus to "Anselm of Laon's" commentary on the *Aeneid*;[9] the Vita Monacensis II is found in a longer form in no. 6;[10] the Vita Monacensis III may be an excerpt from Jacques Legrand (Jacobus Magni, died c. 1425), *Sophologium* II.5 (p. 270a, edition of Lyons, 1585).[11]

(d) despite the VSD / Donatus Auctus background and looming presence, many of the *vitae* report new information. While they may not be authentic, such details can be illustrative of the environment in which individual humanists worked. For example, Cristoforo Landino (no. 9)

[8] An "Oratio Virgilii", which includes a vita (no. 31), was attributed to Gasparino Barzizza by R.G.G. Mercer, *The Teaching of Gasparino Barzizza with Special Reference to His Place in Paduan Humanism* (London 1979) 82-3 and n. 46. Closer inspection, however, shows that the "Oratio Virgilii" was delivered by an anonymous *adulescens* at the instigation of his teacher ("Magister Philippus").

[9] C. Baswell, *Virgil in Medieval England. Figuring the Aeneid from the Twelfth Century to Chaucer* (Cambridge 1994) 313-14 has edited the text from twelfth–century witnesses.

[10] In analyzing the character and sources of the Vita Monacensis II, W. Suerbaum, "Von der Vita Vergiliana über die Accessus Vergiliani zum Zauberer Virgilius. Probleme–Perspektiven–Analysen," *ANRW* 31.2 (Berlin—New York 1981) 1241-53 recognized that the text printed by Bayer, pp. 370-77 was actually composite, the last section (Bayer, pp. 374.72-376.92) coming from Conrad of Mure, *Fabularius*. The Vita in our no. 6 continues instead with the episode of the *Salvatio Romae* and the account of the heckler Philistus (from Donatus Auctus [Bayer, p. 366.201-219]). Suerbaum describes the Vita Monacensis II as a genuine Vita and not an excerpt from an accessus (p. 1241, "eine echte Vita, kein Accessus"). He may still be right since the sentence preceding the Vita could imply an insertion.

[11] Dr. Sigrid Krämer kindly suggested this work as a possible source.

notes that the grateful Neapolitans gave Virgil a magnificent funeral. Not attested in the VSD or, seemingly, in Late Antique and medieval lives, this observation stems possibly from his own attendance at funerals of illustrious Florentines. Complete editions of the new lives, now in preparation by the present writer, will make accessible other interesting differences and insertions.

* * *

In transcribing from microfilm the *incipits* and *explicits* of the texts given below, I have preserved manuscript orthography and slips in grammar and syntax, with angle brackets enclosing my occasional additions and question marks after those words which could not be deciphered with surety. Capitalization and punctuation, however, observe modern conventions. The *incipit* of the accessus (where there is one) is also given; the *explicit* is always that of the Vita (this sometimes coincides with the conclusion of the accessus). There is a brief indication of the Virgilian content of each manuscript.

Bibliography has been kept to a minimum, and these abbreviated titles have been used:

Cat. Monac. = C. Halm et al., *Catalogus codicum latinorum Bibliothecae Regiae Monacensis*, 5 vols. in 7 parts (Munich 1868-81).

Iter = P. O. Kristeller, *Iter italicum. A Finding List of Uncatalogued or Incompletely Catalogued Humanistic Manuscripts of the Renaissance in Italian and Other Libraries*, 6 vols. (London—Leiden 1963-92).

Pattie = T. S. Pattie, "Latin Manuscripts of Virgil in the British Library," in R. D. Williams-T. S. Pattie, *Virgil. His Poetry through the Ages* (London 1982) 125-38.

(1) AUGSBURG, Staats- und Stadtbibliothek 4° Cod. 21. Saec. XV. Germany.

fol. 10v: accessus to the *Georgics*.

fol. 11r-v: accessus to the *Aeneid*.

fols. 21r-23r: accessus, with Vita, to the *Eclogues*. *Inc*. (fol. 21r) Solent in inicio auctorum sex (*corr. ex.* sep) requiri: titulus operis, vita poete, qualitas carminis, intencio scribentis, numerus librorum... Vita poete inquiritur a quo scilicet et natus et unde et ubi studuerit et quando studuerit et quales habuisset mores. Virgilius natus fuit patre Virgilio et Maia matre ciuis mantuanus. Patrem eius quendam (find *canc.*) fingulum dicebant et uiatoris saepius mercennarium et curam apum tandem a cive quodam ma<n>tuano in silva accepisse... *expl.* (fol. 22r) Quarta egloga et disticon: "Sicilides Muse paulo maiora canamus."

Bibl.: *Iter* 3.455a-b.

(2) BERNE, Burgerbibliothek 451. Copied in 1457 by Johannes Chauflon (cf. fol. 180r). France.

fols. 1r-180r: *Georgics*, with a commentary preceded by an accessus including a Vita. *Inc*. (fol. 1r) Quoniam divino confisi iuvamine iuxta virium facultatem, exilem licet ac tenuem, ad quantulamcumque iuvenum instruccionem lecturam Georgicorum Publii Virgilii Maronis discurrere statuimus, in primis ideo nonnulla haud incongrue percunctari libet. Et ut imitabilem priscorum morem aliqualiter observemus, iuxta Servium septem que solent inquiri responsionibus debitis occurrunt terminanda. Primum est vita poete; secundum titulus operis; tertium carminis qualitas; quartum intencio scribentis; quintum numerus librorum; sextum ordo librorum; septimum explanacio que est ipsa narracio. Circa vitam poete queritur a quo sit natus, ubi sit natus, ubi studuerit, quales mores habuerit, et quid composuerit. Virgilius itaque fuit natus patre Figulo matre Maya. Unde Figulus non est nomen professionis sed proprium... *expl.* (fol. 2r) Anno Augusti 25.° Virgilius Brundusii moritur Censio Saturnino et Lucrecio Cinia (?) consulibus. Ossa eius Neapolim translata in secundo ab urbe miliario (*corr.* ?) sepeliuntur, titulo redimita sequenti: "Mantua me genuit ... cecini pascua rura duces." Pascua id est Bucolica; rura id est Georgica; duces id est Eneida. Hec Vincencius. Et hec de vita poete sufficiant.

Bibl.: B.M. von Scarpatetti et al., ed., *Kataloge der datierten Handschriften in der Schweiz in lateinischer Schrift vom Anfang des*

Mittelalters bis 1550, vol. 2 (Dietikon–Zürich 1983) 23, no. 51 and pl. 377 (fol. 1r).

(3) BRUSSELS, Bibliothèque Royale 9783. Copied in 1447 at Paris (cf. fol. 25v).

fol. 1r: incomplete accessus, with Vita, to the *Eclogues*. *Inc.* Virgilius natus Mantue civitatis Venecie de humili plebe, patre nomine Figulo, matre nomine Maia. Studuit Cremone vicine Mantue, Neapoli, Mediolani, Athenis. Fuit etiam omni virtute probatus, uno dempto quia totus urebatur et fervebat libidine... *expl.* Et quia post (post quia *ante corr.*) mortem Antonii et Cleopatre, anno videlicet XV imperii Octaviani (Oct- imp- *ante corr.*), Virgilius incepit scribere et XXI annis continuavit, cumque a morte Cleopatre et Antonii usque ad Christi nativitatem fluxerint anni XXVII, sequitur manifeste quod Virgilius mortuus est sex annis ante Christi nativitatem.

fols. 2r-25v: *Eclogues*, with dense glosses.

Bibl.: F. Masai-M. Wittek, eds., *Manuscrits datés conservés en Belgique. Manuscrits conservés à la Bibliothèque Royale Albert Ier Bruxelles*, vol. 3 (Brussels–Ghent 1978) 39, no. 277 and pl. 496 (fols. 25v [detail], 8v).

(4) BUDAPEST, Országos Széchényi Könyvtár Clmae 285. Copied in 1465 at the University of Cracow (cf. fol. 47ra).

fol. 1r: incomplete accessus, with Vita. *Inc.* Pro libelli huius introduccione accipitur dictum Oracy in poetria veteri sic dicentis: "Aut prodesse volunt aut delectari poete"... Aliis itaque dimissis dicendum est de causis et maxime de efficiente. Et fuit Virgilius filius Maronis et Maye civis manthuanus qui primo studuit in Armenia ciuitate Silicie regionis, tandem vero se ad Mediolanum civitatem Apulie transtulit... *expl.* Tandem traxit ad Athenas per certa negocia Augusti expedienda et in quadam nauicula est dimersus; contra quam Ovidius exclamans dicit: "redde nobis Virgilium tibi traditum."

fol. 1r: Walter Burley, *Liber de vita et moribus philosophorum* 104 (life of Virgil).

fol. 1v: Servius, commentary on the *Eclogues* (part of the accessus only).

fols. 2r-47ra: *Eclogues* and *Georgics*, with glosses.

Bibl.: *A Magyar Nemzeti Múzeum Országos Széchényi Könyvtárának*, Címjegyzéke 12. *Codices manu scripti latini*, vol. 1: E. Bartoniek, *Codices latini medii aevi* (Budapest 1940) 263-66.

(5) CAMBRIDGE (Massachusetts), Harvard University, Houghton Library FLS.A100.B565v.

14 small volumes, printed at Paris c. 1542-66 for the use of students, now bound together and containing selections from Virgil, Ovid, Horace, Persius, Statius as well as a few humanist texts. The first six items, all with dense MS. glosses representing the teaching of Parisian masters, contain respectively *Eclogues* 1.1-6.81 ("Ex typographia Thomae Richardi, sub Bibliis aureis, e regione collegii Remensis", 1561), *Georgics* 4 ("sub Phoenice Petrus Gromorsus excudebat", 1542), *Aeneid* 1 ("Ex typographia Thomae Richardi, sub Bibliis aureis, e regione collegii Remensis", 1561), *Aeneid* 6 (no title-page), *Aeneid* 8 (no title-page), and Appendix vergiliana (no title-page).

item 3: *Aeneid* 1. An unidentified master lectured on this text in 1564 (note at top of fol. 2r); the hand of the same glossator entered on the verso of the title-page a brief Vita. *Inc.* (title-page, verso) Publius Vergilius Maro patrem habuit Virgilium et matrem Maiam quae nullius prope fuit fortunae. Verum ob praeclaram ingenii dotem duxit uxorem et diligenter agro colendo operam sumens rem familiarem auxit. Fuit civis mantuanus atque variis in locis literis operam dedit... *expl.* Postea ab Augusto Aeneidem propositam scripsit annis undecim sed non emendavit nec edidit. Unde moriens praecepit incendi. Augustus eam perire noluit; verum Tucae Varro emendandam tradidit iussitque ut nihilum adderent sed solum superflua demerent.

Bibl.: R.E. Stoddard, *Marks in Books, Illustrated and Explained* (Cambridge, Mass. 1985) 30, no. 47 and pl. 47 (item 2, fols. 54v-55r).

(6) CARLSRUHE, Badische Landesbibliothek Ettenheim-Münster 28. Saec. XV. Germany.

fols. 9r-10r: accessus, with Vita, to the *Aeneid*. *Inc*. (fol. 9r) In principio cuiuslibet libri potissimum quatuor solent indagari causae videlicet materialis formalis efficiens et formalis (finalis *Clm 8952*)... Aliter triplex est causa efficiens: incitans movens et perficiens. Causa incitans et acuens Virgilium ad hoc opus compilandum fuit Octavianus. Causa movens fuit hystoria Enee. Causa autem perficiens fuit Virgilius. De cuius vita seu ortu notentur infra scripta etc. Virgilius omnium poetarum clarissimus Mantue est natus ex matre Maia et patre Figulo rustico in quodam loco qui tunc dicebatur Andos nunc vero Pietola... *expl*. (fol. 10r) Cuius finem meminit quinto Eneidorum cum dixit: "Nate dea, quo fortuna trahit retrahitque sequamur. | Quidquid id est omnis superanda fortuna ferendo est" (*Aen*. 5.709-710) etc.

fols. 11v-305v: *Aeneid*, *Eclogues* and *Georgics*, with glosses (often dense).

Other MSS.: Munich, Bayerische Staatsbibliothek Clm 4393, fols. 200v-201r (Vita, incomplete [*ut vid*.]), saec. XV, Germany (= Bayer, Vita monacensis [VM] II, pp. 370.1-374.71).

__________ Clm 8952, fols. 131r-135r (accessus, with Vita [fols. 131v-133r]), saec. XV, Germany.

Turin, Biblioteca Nazionale Universitaria F IV 7, fols. 2r-3r (acephalous accessus added by a saec. XV German hand).

Bibl.: *Die Handschriften der Badischen Landesbibliothek in Karlsruhe*, vol. 9: K. Preisendanz, *Die Handschriften des Klosters Ettenheim-Münster* (Wiesbaden 1973 [reprint, with additional bibliography, of 1932 edition]) 15, 99; *Cat. Monac*. I.2, pp. 186-87 (Clm 4393) and II.1, p. 67 (Clm 8952); G. Mazzatinti, *Inventari dei manoscritti delle biblioteche d'Italia* 28 (Florence 1922) 92, no. 887 (Turin F IV 7).

(7) CRACOW, Biblioteka Jagiellońska 2141. Saec. XV. Germany.

fol. 38r: accessus, with incomplete Vita. *Inc*. Pro huius libri introductione seu recomendatione notetur dictum Horatii in Poetria veteri sic dicentis: "Aut prodesse volunt aut delectari poetae"... Dimissis tribus quod hec circa initia aliorum librorum sepius dicuntur solum de causa efficiente dicendum est. Unde causa effitiens istius libri fuit Virgilius Maronis filius et Megne (*ut vid*.), ciuis mantuanus, de quo habetur talis hystoria valens ad omnes libros suos. Virgilius erat nacione mantuanus de illa civitate Mantua. Pater eius erat Maro nomine et mater Meya... *expl.* Contigit autem illo tempore quod Augustus Cezar Romanis iniit prelium cum Cleopatra rege vel do-.

fols. 39r-64r: *Eclogues*, with glosses (often dense).

Related MS.: Freiburg i. Br., Universitätsbibliothek 1, fol. 1v, 1499 (cf. fol. 84v), Germany.

Bibl.: W. Wisłocki, *Catalogus codicum manuscriptorum Bibliothecae Universitatis Jagellonicae Cracoviensis,* vol. 2 (Cracow 1881) 292; W. Hagenmaier, *Die lateinischen Mittelalterlichen Handschriften der Universitätsbibliothek Freiburg im Breisgau*, vol. 1 (Wiesbaden 1974) 1-2.

(8) CRACOW, Muzeum Narodowe w. Krakowie 3230. Saec. XVI^{2}. Cracow (?).

pp. 25-28: "Compendiosa Periocha vitae Virgilii", argumenta, and comments on *Ecl.* 2.24-32. (The Vita [pp. 25-26] is an abridgment, with a few variants, of the Vita found in the same manuscript on pp. 29-33).

pp. 29-44: accessus, with Vita, and commentary on *Aeneid* 1.1-109. *Inc.* (p. 29) In explicandis omnibus poetis (*corr. ex* autoribus) haec praecipue consideranda sunt: vita poetae, titulus operis, qualitas carminis, scribentis intentio, ordo librorum cum explanatione. De vita Virgilii. Publius Virgilius Maro poetarum latinorum princeps, natus patre Marone figulo et Maia matre in pago Andes non procul a Mantua (quae civitas est Venetiae), multis praesagiis eius felicem indolem portendentibus, quorum non hoc minimum est quod mater eius Maia se ramum laureum enixam somniaverit... *expl.* (p. 33) Periit vero Tarenti in Apuliae civitate. Nam dum Methapontum petit, vitae (*ut vid.*) valetudinem ex solis ardore contraxit. Sepultus est vero Neapoli, in cuius tumulo ab ipso tale compositum est distichon: "Mantua me genuit ... cecini pascua rura duces."

pp. 45-169: Paulus Hanicius Cracoviensis, accessus and commentary on the *Eclogues* (dated "1567" on p. 45).

Bibl.: J. Korzeniowski-S. Kutrzeba, *Catalogus codicum manu scriptorum Musei Principum Czartoryski Cracoviensis*, vol. 2 (Cracow 1913) 132.

(9) FLORENCE, Biblioteca Medicea-Laurenziana 52.32. Saec. XV (1460s). Florence. Autograph of Cristoforo Landino.

fols. 1r-181r: Cristoforo Landino, accessus, with Vita, and commentary on *Aen.* 1.1-7.816.

Inc. (fol. 1r) Idibus Octobribus Pompeio et Crasso consulibus in agro mantuano ex (*corr. ex* et) patre Marone, Maia vero matre mercennariam artem exercentibus natus est Virgilius, non tamen obscuris signis divini ingenii cum parens ipsa adhuc pregnans per quietem laurum floribus et pomis ornatissimum sibi parere videretur... *expl.* (fol. 2r) Quapropter doctissimis (*corr. ex* -mus) id temporis viris emendandum dedit hac condictione ut si quid superfluum videretur demerent, nihil vero adderent. Ex qua re aliud est principium ac ipse Virgilius scripserat; eadem ratione plura carmina inperfecta reperiuntur. Et hec ad vitam poetae sint satis.

Bibl.: A. Field, *The Origins of the Platonic Academy of Florence* (Princeton 1988) see index, p. 291.

(10) FLORENCE, Biblioteca Riccardiana 3007. Saec. XV^2. Florence.

fols. 131r-180r: accessus, with Vita, and commentary on *Ecl.* 1-10. *Inc.* (fol. 131r) Ut res quam dicturi sumus facilius intelligatur, in huius lectionis principio duo proponemus scilicet auctorem et carmen. Auctor huius operis fuit Virgilius Maro mantuanus modicis parentibus natus et praecipue patre, quem quidam opificem figulum, plures Magi cuiusdam viatoris initio mercenarium, mox ob industriam generum tradiderunt. Mater fuit Maia filia Magi... *expl.* Postea sepultus fuit Neapoli. De praestantia Virgilii illud nos admonet quod nullius laude (fol. 131v) crescat nec ullius vituperio minuatur.

fols. 180r-224r: accessus and commentary on *Georg.* 1.1-2.400.

Bibl.: *Iter* 1.224b-225a.

(11) GOTHA, Forschungs- und Landesbibliothek Chart. A 239. Copied by "Caspar Philippi de Dresden" in 1461 (cf. fols. 254rb, 370va). Germany.

fol. 1ra-b: accessus (incomplete), with Vita, to the *Eclogues*. *Inc.* (fol. 1ra) Natus est Vergilius familia mediocri patre Figulo matre Maia Idibus Octobris Gneo Pompeio Marco L. (?) Crasso consulibus in villa que Andries dicitur haud procul a Mantua, avunculo Lucrecio poeta qui eum famosissimum futurum predixit cum a Maia sorore nate de se sompnium virge audisset. Studuit eciam Rome aput Epidium oratorem cum Augusto... *expl.* Usus igitur amicicia Cesaris Augusti et complurium probatissimorum virorum vixit annis LII, profectusque Athenis discendi gratia et fugiendi honorum causa Brundisii id est Calabrie moriatur. Vnde dicticum ipsius compositum est: "Mantua me genuit ... cecini pascua rura duces." Cuius ossa Neapolym translata et in loco qui dicitur via puteolana condita sunt, consulibus Cresencio Laurencioque.

fols. 3r-209v: *Eclogues*, *Georgics*, *Aeneid*, *Moretum*.

fols. 210v-219r: Appendix vergiliana.

fols. 220ra-370va: Servius, commentaries on the *Eclogues*, *Georgics*, *Aeneid*.

fols. 370vb-372vb: Fulgentius, *Expositio virgilianae continentiae*.

Other MSS.: Würzburg, Universitätsbibliothek M. ch. f. 60, fol. 175v, saec. XV, Germany.

Related MSS.: (the *incipit* of this Vita occurs as part of the introductory gloss to the *Moretum*) Göttingen, Universitätsbibliothek Philol. 114, fol. 51r, saec. XV, Germany.

Gotha, Forschungs- und Landesbibliothek Chart. A 869, fol. 22r, saec. XV, Germany.

Würzburg, Universitätsbibliothek M. ch. f. 18, fol. 404r, saec. XV, Germany.

(12) GOTHA, Forschungs- und Landesbibliothek Chart. B 500-501. Lectures delivered at the University of Leipzig in 1580-81 (cf., e.g., Chart. B 500, fol. 143r and Chart. B 501, fol. 87v).

(Chart. B 500) fols. 2r-383r: Johannes Albinus, accessus, with Vita, and commentary on *Aen.* 1-4. *Inc.* (fol. 2r) Quae de Virgilii, quem quidam Platonem poetarum latinorum, nonnulli Jovem etiam appellare non dubitaverunt, vita fortunis morte denique memoriae prodita sunt, illa cum a Servio tum a Tyberio Donato et aliis descripta sunt in medio ut cognosci possint ab omnibus. Libet autem ex his quaedam initio explicationis nostrae repetere ut aditus ad summum et sapientissimum hunc poetam cognoscendum expeditior nobis sit et facilior. Natus igitur esse scribitur humilibus quidem parentibus (nam patrem Maronem alii figulum, alii mercenarium Magi cuiusdam viatoris perhibuerunt) in pago Andes prope Mantuam, quae urbs amoenissima sub ditione hodie est Venetorum… *expl.* (fol. 3v) Mortem obiit Tarentii in Apulia vel, ut alii scribunt, Brundusii, rediens ex Graecia cum aegritudinem ex itinere propter nimium aestum contraxisset anno Urbe condita 735, anno mundi 3937, ante Christum 14, Caio Sentio Saturnino et Spurio Lucretio Vispallone (*add.*, *ut vid.*) consulibus. Sepultus est Neapoli ubi tale habuit epitaphium: "Mantua me genuit … cecini pascua rura duces."

(Chart. B 501) fols. 1r-194r: Johannes Albinus, commentary on *Aen.* 5-6.

Bibl.: *Iter* 3.395b.

(13) KONSTANZ, Bibliothek des Suso-Gymnasiums 13. Saec. XVI. Germany.

pp. 423-424: accessus (incomplete, *ut vid.*); brief Vita. *Inc.* (p. 423) Consuetum est Romanis plerumque tribus vocari nominibus. Quorum primum erat prenomen quod differentiae dignitatis honorisve ratione tribuebatur… (p. 424) Vita Maronis. Vergilius Maro parentibus modicis fuit, patre quidem Marone et matre Maia in pago qui Andes dicitur, a Mantua non procul. Huius mater Maia somniavit se virgam parituram quae usque ad coelum pertingeret… *expl.* Moriens autem quinquagesimo primo suae etatis anno voluit ut ossa eius Neapolim transferentur. Quod et factum est, et a se factum est disticon sepulchro (*corr. ex* - rum) inscriptum est. Epitavium Vergilii: "Mantua me genuit … cecini pascua rura duces."

pp. 425-477: *Georgics* 1.1-2.508, with glosses.

Bibl.: *Iter* 3.596a.

(14) KREMSMÜNSTER, Stiftsbibliothek 59. Saec. XVI^2. Austria.

fols. 43r-46r: Johannes Kegelius, accessus, with Vita, to the *Eclogues*. *Inc.* (fol. 43r) In exponendis authoribus haec consideranda sunt: 1. poetae vita; 2. titulus operis; 3. qualitas carminis; 4. scribentis intentio; 5. et numerus librorum. Vita Vergilii. Publius Vergilius Maro qui, ut apud Graecos Homerus, sic apud Latinos omnium quos anteacta secula tulerunt poetarum facile princeps fuit. Hic Vergilius Maro modicis parentibus, patre videlicet Marone et matre Maia, in pago mantuano qui Andes dicitur, Cn. Pompeio Magno et Marco Licinio Crasso primum consulibus, Idibus Octob. natus est... *expl.* (fol. 44v) Tandem Brundusium appulit ubi paucis diebus post diem suum obiit 10. Cal. Octob. cum ageret annum aetatis suae 52, anno ante natum Christum 16. Eius ossa iussu Augusti Neapolim translata sunt. Decumbens hoc epithaphion sibi ipsi fecit: "Mantua me genuit ... cecini pascua rura boves."

pp. 1-172: *Eclogues*, with glosses (often dense).

Bibl.: *Iter* 3.23a.

(15) LONDON, British Library Add. MS. 10300. Saec. XVI. France.

fols. 96r-97r: accessus, with Vita, to the *Eclogues*. *Inc.* (fol. 96r) Virgilii compendiosa vita. Vergilius Maro modicis parentibus fuit, patre Marone a quo Maro est appellatus et matre Maia, in pago mantuano qui Andes dicitur Idibus Octobris natus. Initia etatis id est usque ad septimum decimum annum... *expl.* Demum hortante Augusto Eneida ad imitationem Homeri undecim annis composuit, sed preventus morte castigare non potuit.

fols. 98r-119r: *Eclogues* 1-3, with commentary; argumenta for *Ecl.* 4-5.

Bibl.: Pattie, p. 125, no. 3.

(16) LONDON, British Library Add. MS. 11959. Saec. XV med. England.

fol. 1r: accessus, with Vita. *Inc.* Virgilius Mantue natus a patre Figulo, matre vero Maya. Studuit Athenis. Plures libros composuit sed tres principales scilicet Bucolica Georgica et Eneyda... *expl.* Postea Eneyda rogatu Augusti duodecim annis composuit, sed morte preventus non perfecit nec correxit. Unde illum (?) moriens (?) incendi precepit. Sed Augustus ne tantum opus periret Tuscam et Varrum duos peritissimos poetas ad corrigendum adhibuit tali lege, ut suo operi nihil adderent sed quecumque superflua abraderent.

fols. 1r-36v: Virgil, *Eclogues* and *Georgics* 1.1-2.13, with commentaries on both works.

Bibl.: Pattie, p. 127, no. 13.

(17) LONDON, British Library Harley 2473. Saec. XV. Italy.

fol. 5r: accessus, with Vita, to the *Aeneid. Inc.* In exponendis auctoribus quamquam multa more poetico investigentur que in indagatione presentis illustris poete investigari auridice (?) merentur, attamen hiis omnibus omissis dimissisque investigatione auctoritateque famosorum experitorum doctorum infra scripta pro aliquali introductione commemorentur. Nam in ipsis poetis exponendis hec summatim investigantur et considerantur scilicet vita poetae, titulus operis, qualitas carminis, stilus, numerus librorum, ordo librorum, explanatio. Virgilii autem vita et condicio fuit talis. Nam ipse Vergilius civis fuit oriundusque civitatis Mantue; patre Figullo et matre Maia ortus et humili loco natus, clarus et illustris evasit... *expl.* Augustus vero ne tantum periret opus iussit ipsum opus per Ovidium et Varronem poetas tunc contemporaneos corrigi et emendari, hac tamen lege ut diminuenda dumtaxat diminuerent (*corr. ex* -tur) sed nichil adderent. Et ob hanc causam in ipso opere semiversus reperiuntur ut "Cursus fuit" et in multis locis. Et ante ipsius operis inchoationem sciens se ab armis incepturum premisit et preposuit hos versus "Ille ego qui quondam gracili" etc.

fols. 5v-168v: *Aeneid*, with glosses.

Bibl.: Pattie, p. 131, no. 42.

(18) MODENA, Biblioteca Estense Campori Appendice 263 (gamma H. 5. 11). Saec. XV. Italy.

fols. 1r-57v: accessus, with Vita, and commentary on the *Eclogues*. *Inc.* (fol. 1r, stained and cropped) In omnibus auctoribus exponendis duo sunt principaliter <inqui>renda: vita in primis poete, deinde origo carminis; quo<d> erit agendum in hoc Buccolicorum opere. Nam primo est vide<re> vitam poete scilicet Virgilii; postea origo carminis sui Buccolicorum. <Fu>it igitur Virgilius venetus nactione, non ex urbe Venetiarum q<uia> nondum erant edificate Venetie, sed quia Venetia est provintia que continebat multas urbes et inter alias Mantuam, dictus est venetus nactione... *expl.* Cum autem < > venire Tarentum nondum finito <E>neide, correptus febre mortuus est. Et antequam moriretur fecit duo carmina videlicet unum disticon. Nam grece < > qui continetur et duobus versibus vocatur disticon. De se dixit sic: "Mantua me genuit ... cecini pascua rura duces." (fol. 1v) <Aliqui>d dicamus de carmine quia satis patet vita sua.

fols. 59r-196v: Benvenuto da Imola, commentary on the *Georgics*.

Bibl.: *Iter* 1.388a.

(19) MUNICH, Bayerische Staatsbibliothek Clm 3111. Saec. XV. Germany.

fol. 75r-v: accessus, with Vita, to the *Eclogues*. *Inc.* (fol. 75r) <V>irgilius Maro parentibus modicis fuit et praecipue patre Marone qui vigulus fuit secundum quosdam; alii vero tradunt eum mercenarium fuisse. Natus est Virgilius in pago qui Anges dicitur, qui est a Mantua non procul... *expl.* Fuit etiam Virgilius verecundissimus et morum honestate preditus ut ex moribus cognomen acciperet: nam dictus parthenneas, omni vita probatus. Scripsit etiam plurimos libros et inter illos scripsit tria opera principalia scilicet Bucolica Georgica et libros Eneidum.

fols. 76r-102v: *Eclogues*, with glosses.

Bibl.: *Cat. Monac.* I.2, p. 76.

(20) MUNICH, Bayerische Staatsbibliothek Clm 15147. Copied at Zwickau, 1500-1501 (cf. fols. <20r>, 151r, 192v). Irregular foliation, as noted below.

<fols. 2r-19v>: printed text of the *Eclogues*, with MS. glosses.

fol. 1r: incomplete accessus, with Vita, to the *Georgics*. *Inc*. Quoniam quidem in exponendis authoribus (ut Servio placet) circa primum maioris operis Virgilii librum quedam ordine suo sunt consideranda: poete scilicet vita, operis titilus, qualitas carminis, scribentis intencio, numerus denique librorum. Que quidem ordine suo enitescunt. Publius Virgilius Maro poetarum heroicorum illustrissimus, parentibus modicis (ut aiunt) enutritus, Gneio Pompeo Magno et Marco Licinio Crasso primum consulibus natus in pago qui dicitur Andes... *expl*. Periit autem Tarenti in Apulie civitate dum Methaponthum exiens videre. Contraxit namque adversam ex solis ardore valitudinem sepultusque Neapuli. Itaque Servius. In cuius tumulo ab ipso editum ponitur disticon.

fols. 2r-48r, 52v-151r: *Georgics* and *Aeneid*, both with glosses.

Bibl.: *Cat. Monac*. II.3, pp. 4-5.

(21) NEW HAVEN, Yale University, Beinecke Library Gb4 212.

14 small volumes, printed at Paris *c*. 1577-80 for the use of students, now bound together and containing selections from Cicero, Caesar, Horace, Isidore and Virgil as well as a few humanist texts. Items 12-14 ("Ex typographia Dionysii a Prato, via Amygdalina, ad Veritatis insigne") contain respectively *Eclogues* 1-10 (1579), *Aeneid* 4 [1580?] and *Aeneid* 3 [1580?]; they exhibit dense MS. glosses and interleaved pages with MS. notes in the hand of "F. Graff" (who identifies himself as the scribe in item 13 and the Parisian master Dionysius Lucanus as the author of the *annotationes* in *Aen*. 4).

item 12: *Eclogues* 1-10. On the unnumbered interleaved folio preceding the printed text there is a brief accessus, with Vita. *Inc*. P. Virgilius Maro humili genere Mantuae natus est in vico Andes nomine, a quo andinus ipse et andiades Musae dictae sunt. Siquidem patrem P. Virgilium Maronem figulum et matrem Maiam habuit... *expl*. Quamobrem Augusto eiusque amicis, quorum opera hac in re adiutus multisque aliis opibus auctus fuerat, gratias acturus triennio haec Bucolica scripsit, tum septennio Georgica, undecim vero annis Aeneiden. Quam morte praeventus cum extrema manu limare non potuisset (sicuti fuerant duo priora opuscula), igni dicaverat, sed iussu Augusti conservata ad nos pervenit. Ille vero obiit Brundusii decimo Cal. Octobris sepultusque est (ut rogaverat) Neapoli in via puteolana.

(22) NEW HAVEN, Yale University, Beinecke Library Gnc60 nj565b.

17 small volumes, printed at Paris *c.* 1541-68 for the use of students, now bound together and containing selections from Cicero, Isidore, Quintilian, Ps.-Seneca, Horace, Ovid and Virgil, as well as a few humanist texts. Item 17 contains Virgil, *Eclogues* 1-10 ("Apud Ioannem Roigny, in via Iacobaea, sub insigni Basilisci & quatuor elementorum", 1548), with dense MS. glosses.

Item 17: *Eclogues* 1-10. Accompanying the printed *Argumentum* (p. 2) is a marginal gloss entitled *Argumentum* and containing a *Vita*. *Inc.* (p. 2) Virgilius tenuem fortunam nactus eam ingenii prestantia vicit. Primum enim Mantuae (ubi natus erat), deinde Mediolani, tum Neapoli humaniores disciplinas didicit ac tandem graecis literis instructus in poesi tantum elaborauit ut apud omnes eius fama brevi pervagata fuerit... *expl.* Atque imprimis (*corr. ex* -mus) fertur in bucolico carmine Theocritum expressise quemadmodum ex utriusque scriptis liquido constat, deinde Hesiodum in tradendos preceptos de agri cultura que Georgica appellantur, postremo Homerum in Aeneide, quibus id authoritatis adeptus est apud Latinos ut poetarum coriphaeus et facile princeps habeatur.

(23) OXFORD, Bodleian Library Add. A. 164. Saec. XV. Italy.

fols. 1r-125v: accessus, with Vita, and commentary on *Georg.* 1.1-4.549 (on fol. iir a late entry reports the author of the commentary as "B. C."). *Inc.* (fol. 1r) Dividimus in duas partes, in auctorem et opus. Auctorem intelligimus ut cognoscamus vitam eius, unde ortus fuerit, quibus parentibus, ubi studuerit. Pater eius Maro (*corr. ex* Virgilius) vocabatur qui erat figulus; mater eius Maia nomine vocabatur. Educatus Cremone maxime studuit Mediolani, postea Neapoli... *expl.* Nam sicut afirmat Plinius et alii scriptores, natura (*corr. ex* -ram) ursi est haec, ut pariat unum frustrum carnis; deinde lingua (fol. 1v) tantum pollit illam ut praebeat ei menbra.

Bibl.: F. Madan, *A Summary Catalogue of Western Manuscripts in the Bodleian Library at Oxford*, vol. 5 (Oxford 1905) 52, no. 24720.

(24) PARIS, Bibliothèque Nationale lat. 7948. Copied in 1461 by "Matheus filius Iohanis Simonis de Canzeleriis" (cf. fol. 162v). Italy. Purchased at Ferrara in 1465 and glossed by Simon De Veteri Castro while he was attending the lectures there of Gaspar Tribraco de' Trimbocchi (cf. fol. 163r).

fols. 1r-162v: *Aeneid*, with frequent glosses; a long marginal note on fol. 1r contains an incomplete accessus with a brief *Vita* (presumably from Tribraco's lectures and transcribed in full as follows). *Inc.* (fol. 1r) In principio cuiuslibet poete ista sunt investiganda: poete vita, titulus operis, qualitas carminis, numerus librorum, ordo ipsorum, explanacio, intencio actoris. Virgilius mantuanus fuit qui, amissis omnibus possessionibus suis que fuerant in predam militibus Octaviani Cesaris, venit Romam ubi in favorem Pollionis et Mecene omnia recuperavit. Unde ad requestam Pollionis composuerit Bucolicas que in triennio complete sunt. Ad invitacionem autem Mecene composuit Georgicas que in septennio complete sunt. Subsequenter composuit hoc opus in quo intendit laudare Augustum Cesarem, sed morte preventus non correxit. Obiit autem in Terento civitate Calabrie et sepultus est Neapolis. Qui cum in extremis laboraret fecit epigramma suum suum (*sic*) in hunc modum: "Mantua me genuit, Calabria rapuere tenet nunc / Parthenope; cecini pascua rura duces."

Bibl.: C. Samaran-R. Marichal, *Catalogue des manuscrits en écriture latine portant des indications de date, de lieu ou de copiste*, vol. 2 (Paris 1962) 443 and pl. 129 (detail of fol. 162v).

(25) PARIS, Bibliothèque Nationale lat. 8262. Saec. XV. Germany.

fols. 346v-350r: Appendix Vergiliana; *Moretum*.

fols. 350v-352v: accessus, with Vita, to the *Eclogues*. *Inc.* (fol. 350v) In ingressu interpretacionis carminis buccolici duo principalia sumatim perstringenda sunt: unum de autore eius; secundum de ipso carmine... Virgilii vita hec est. Virgilius Maro patre Virgilio matre Maya natus civis mantuanos (*ut vid.*), que civitas est Veneciarum. Diversis in locis studiis operam vehementissimam adhibens, omnium poetarum optimus (*corr. ex* -mos) decreto senatus et inperiali edicto est iudicatus (iu *corr.*)... *expl.* (fol. 352r) Vendicat autem Virgilius huic carmini nomen, u<t> Servius habet, a custodia boum. Precipua etiam sunt animalia apud rusticos boves, licet et hoc nomen ante apud Theocritum habitum fuerat.

fols. 359r-382r: *Eclogues*, with dense glosses.

Other MSS.: Cracow, Biblioteka Jagiellońska 2141, fol. 38r (incomplete *Vita*), saec. XV, Germany.

(26) PARIS, Bibliothèque Nationale lat. 8413. Saec. XV (1460s?). Italy (Chieti?). Fols. 1r-73r, 105v-175r, 204r-221v copied by Pietro di Celano (cf. fols. 7v, 44r, 71v).

fols. 107r-164v: Giovanni Piedemonti, accessus, with *Vita*, and commentary on *Aen.* 1.1-535. *Inc.* (fol. 107r) Declaratio vite Virgilii per eloquentissimum preceptorem meum Ioannem P. in civitate theatina edita incipit. Cum de arte poesiae, ut refert Plinius de naturali istoria, dubium sit et magna inter doctores comptentio, tamen apud multos legimus eam habuisse ortum ante bellum troianum et, ut refert Aristoteles, poesis orta fuit cum congregatio aliqua popularis esse cepit. Verum aput Latinos Maronem poetam famosissimum esse dicunt, cuius quidem vite declaratio impresenti nobis intentio est. Is enim habuit parentes rusticos, natus in quodam oppido vel villa que dicitur Antes. Pater fictilia primum fecit vasa, demum viatoris mercennarius... *expl.* (fol. 109r) Sunt enim tres stili: humilis mediocris et grandilogus. Grandilogus est altis sententiis et orationibus plenus. Quod in hoc opere enarrare (?) contingit, et incipit a vocabulo de quo intendit ut Lucanus: "Bella per Eumathios". Hic vero: "Arma virumque cano" etc.

Bibl.: C. Samaran-R. Marichal, eds., *Catalogue des manuscrits en écriture latine portant des indications de date, de lieu ou de copiste*, vol. 3 (Paris 1962) 51, 621 and pl. 132 (detail of fol. 187r); *Iter* 3.224b-225a, with further bibliography.

(27) PRAGUE, Knihovna metropolitní kapituly L. 84 (1332). Saec. XV. Germany.

fol. 86v: accessus, with Vita. *Inc.* Nota in libris reperitur vetustis quod in cuiuslibet libri principio ista sunt premittenda, videlicet que sit vita poete, que intencio, que materia, cui parti philosophie supponatur, que utilitas carminis. Vita siquidem poete ex isto potest habere exordium (?). Auctor libri huius fuit Virgilius Maronis et Magie civis mantuanus. Qui primo studuit in Cremona civitate Sicilie... *expl.* Tandem traxit se Athenas pro quibusdam negociis Augusti expediendis et in quadam mavicula est dimersus. Quam Ovidius inclamat dicens "Perfida redde nobis Virgilium tibi creditum". Et hec est vita poete.

fols. 87r-117v: *Eclogues*, with dense glosses.

Bibl.: A. Podlaha, *Soupis rukopisů Knihovny Metropolitní Kapitoly Pražské*, vol. 2 (Prague 1922) 248-49.

(28) ROME, Biblioteca Casanatense 1368. 1463 (cf. fol. 280v). Copied in Florence by Giovanni di Bartolomeo Guidi.

fols. 1r-280v: Cristoforo Landino, accessus, with Vita, and commentary on *Aeneid* 1.1-7.817. *Inc.* (fol. 1r) Cum superius non pro rei dignitate sed pro nostri ingenioli facultate poeticae artis laudes prosecuti sumus, nunc consequens videtur ut primo (*corr.*) in loco poetae vitam breviter perstringamus, deinde voluminis titulum addamus, tertio vero in loco quae sit ipsius carminis species et forma, postremo consilium et intentionem nostri poetae et librorum numerum et ordinem... Natus est igitur Virgilius Idibus Octobribus Pompeio et Crasso primum consulibus, humili sane loco, humilibusque parentibus, quippe qui in agro mantuano in pago qui Andes dicitur... *expl.* (fol. 2r) Tuccae Varoque ea tempestate vatibus egregiis ea lege emendandum tradidit ut si quid videretur supervacaneum demerent, nihil autem adderent. Vnde a principio nonnulla carmina sublata et in secundo etiam et multi versiculi qui imperfecti reperiuntur mende (?) sunt, ut infra suis locis patebit. Haec autem sunt quae habui de vita nostri poetae breviter dicenda.

Bibl.: A. Field, "A Manuscript of Cristoforo Landino's First Lectures on Virgil, 1462-63 (Codex 1368, Biblioteca Casanatense, Rome),"

Renaissance Quarterly 31 (1978) 17-20; idem, *The Origins of the Platonic Academy of Florence* (Princeton 1988), see index, p. 292.

(29) STOCKHOLM, Kungliga Biblioteket Va 31. Copied in 1467 at the University of Leipzig (cf. fols. 159r, 250v).

fols. 1r-193v: *Aeneid* and *Georgics*, both with glosses.

fol. 194r: accessus, with Vita, to the *Georgics* and a comment on *Georg*. 1.1. *Inc.* Virgilius Maro in operibus suis diversos quippe autrios (*sic*) sectatus est, et quidem in Bucolicis Theocritum siracusanum, in Eneyde Homerum, in Georgicis Esiodum ascreum... Ortus est autem Virg<i>lius a patre Virg<i>lio figulo et matre Maya. Pater autem oculis captus est et excecatus... *expl.* Augusto autem ad Romam redeunti ex victoria, occurrit Augusto et in via febre tactus interiit. Deducta sunt autem ossa eius in Calabriam et inde ad secundum lapidem, id est secundum miliare (quia termini miliarium lapide erecto signabantur), in via neopolitana sepultus est.

Bibl.: M. Hedlund, *Katalog der datierten Handschriften in lateinischer Schrift vor 1600 in Schweden*, vol. 2 (Stockholm 1980) 34 and pl. 89 (fol. 159r).

(30) TURIN, Biblioteca Nazionale Universitaria G III 25. 1471. Copied by "Wilhelmus Schulteti dictus de Mechlinia | presbiter Cameracensis dyocesis possessor" who began his task at Brussels (cf. fol. 1v).

fol. 7r-v: accessus, with Vita, to the *Eclogues*. *Inc.* (fol. 7r) Siquidem Virgilius nobilis poeta, qui presentis operis cuius lectioni intendimus autor extitit facundus, adeo morum honestate fuit preditus ut Parthemias diceretur, id est dei vita probatus, excepto tantum quod unico laborabat morbi inquinamento quia inpacientis legitur fuisse libidinis... *expl.* Et postea scripsit Eneyda ab Augusto sibi propositam et iniunctam, quam infra decennium complevit. Sed morte preventus nec correxit nec edidit; unde moriens incendi eam precepit. Sed de hac non est presens speculacio.

fols. 8r-67v, 71r-279v: *Eclogues*, *Georgics*, *Aeneid*, all with glosses.

Bibl.: G. Mazzatinti, *Inventari dei manoscritti delle biblioteche d'Italia* 28 (Florence 1922) 106, no. 1022.

(31) VATICAN CITY, Biblioteca Apostolica Vaticana Chigi I VII 266. Copied by Bernardinus Bornada in 1486 at Brescia (cf. fol. 63v).

fols. 37r-38v: "Oratio Virgilii" (= *declamatio*), with Vita. *Inc.* (fol. 37r) Non putarem, viri optimi et homines doctissimi, quod ea mentes vestras de me caepisset opinio... (fol. 37v) Publius Virgilius Maro mantuanus humilibus parentibus fuit et praesertim patre quem nonnulli opificem figulum, Magi cuida<m> matris (?) mercenarium fuisse crediderunt... *expl.* (fol. 38v) Cuius ossa Neapolim translata sunt tumuloque condita, in quo disticum fecit: "Mantua me genuit ... cecini pascua rura duces."

Other MSS.: Naples, Biblioteca Nazionale IV C 20 bis, fol. 1r-v, saec. XV, Italy (begins with the same Vita and continues with the remainder of the *declamatio* and a commentary [fols. 1v-35v] on Aen. 1.1-2.251-252).

Bibl.: *Iter* 2.486a-487b (Chigi I VII 266); *Virgilio. Mostra di manoscritti e libri a stampa. Catalogo* (I Quaderni della Biblioteca Nazionale di Napoli, Serie V 5; Naples 1981) 32, no. 36 (Naples IV C 20 bis).

(32) VERONA, Biblioteca Capitolare CCLXXXI (351). Saec. XV ex. Italy. Copied by "Bartholomaeus Arduinus Regiensis" (cf. the saec. XVI entry on fol. 66r).

fols. 1r-66r: accessus, with Vita, and commentary on the *Eclogues*. *Inc.* (fol. 1r) Virgilius Maro Cneo Pompeio et M. Licinio Crasso primum consulibus ex ignobili stirpe in lucem editus, mantuanus extitit ex patre figulo qui Maro denominatus fuit, a quo ipse Maro appellatus est... *expl.* (fol. 1v) Augustus vero cum cognovisset homine<m> facetum, convocavit ad se hominem et oculis suis intuens oculos Virgilii dixit se filium esse non pis<toris> sed magnifici principis. Fecit (?) et multa alia ut vix commemorari (?) posset. Ut inquit Macrobius, adeo laus Maronis crevit ut iam desinat esse laus.

Bibl.: *Iter* 2.298b.

(33) VIENNA, Österreichische Nationalbibliothek 3230. Saec. XV/XVI. Germany.

fol. 62v: "Compendiaria vita". *Inc.* Virgilius Maro Cn. Pompeio Magno et Marco Licinio Crasso primum consulibus Iduum Octobrium die natus est in vicco qui Andes dic<i>tur estque a Mantua non procul, parentibus modicis et sortis humilioris, patre Marone et matre Maia... *expl.* Mortuus est Brondusii rediens cum Augusto ex Grecia ac Megaro langore contracto. Sed ossa eius Neapolim (hoc enim optaverat) translata sunt sepultaque via puteolana cum tali distico quod et ipse extrema iam valitudine fecerat: "Mantua me genuit ... cecini pascua rura duces."

fols. 63r-89r, 90v-152v, 154v-166v: *Georgics* and *Aeneid*, both with glosses; Appendix Vergiliana, with glosses.

Bibl.: *Tabulae codicum manu scriptorum praeter graecos et orientales in Bibliotheca Palatina Vindobonensi asservatorum*, vol. 2 (Vienna 1868; rpt. Graz 1965) 239.

(34) WARSAW, Biblioteka Narodowa III 3446. 1493 (cf. fol. 23r). Germany.

fols. 12r-23r, 24v-39v: *Aeneid* 4; excerpts from Servius, commentary on the *Aeneid* (accessus); *Eclogues*, with glosses.

fol. 24r: Vita. *Inc.* Virgilii vita hec est. Ortus fuit patre Virgilii, matre vero Maya. Fuit civis mantuanus; que civitas est Venetencium. Adeo autem verecundissimus fuit ut ex moribus cognomen acciperet. Nam dictus est Pa<r>tenias... *expl.* Augustus vero ne tantum opus perieret, Dutam et Pharram iussit emendare ut superflua demerent, nil tamen adderent. Unde et semiplenos eius invenimus versiculos. Ideo dicit "cecini duces" in eo versu qui recitatus est: "Mantua me genuit" etc.

(35) WROCŁAW, Biblioteka Kapitulna 81n. Saec. XVI. Germany.

<fols. 1r-6v>: accessus, with Vita, to the *Aeneid. Inc.* <fol. 1r> In explicatione authorum quaedam commemorari solent quibus praecipua capita tocius operis continentur... De titulo operis et primum de authore. <fol. 1v> Autorem scimus esse horum poematum ex inscriptione libri Publium Vergilium Maronem tribus nominibus romana consuetudine

vocatum... Natus est Vergilius in foro Andis non longe a Mantua distans (unde antinus vates dicitur) exiguis parentibus, patre Marone et Maia matre... *expl.* <fol. 2v> Cum plerique oratores in uno tantum dicendo genere excellerent, ipsius autem vacundia adeo esset multiplex ut simul omnia complexus esse videatur. Haec de autore breviter commemorata sufficiant.

<fols. 6v-9r>: argumentum (*Aen.* 1); commentary on verse argumentum and *Aen.* 1.1-2.

Bibl.: *Iter* 4.420a.

III

Style and Tradition

HOMER AND HITTITE REVISITED

CALVERT WATKINS

The Oxford Classicist Jasper Griffin in a recent issue of the *New York Review of Books* (June 20, 1996) critically analyzes several recent volumes dealing with various approaches and responses to "the Greek experience", notably *Not Out of Africa* and *Black Athena Revisited*, with Mary R. Lefkowitz as author and coeditor respectively. Griffin rightly concludes his remarks on these two volumes with the observation that

> In recent years there has been a great movement among classical scholars in the direction of finding sources and models for Greek literature and thought, not indeed in Egypt, but in the ancient societies of the Near East. The reason why there is much more of this research now than there was in the nineteenth century is not simply that earlier scholars concealed the evidence because they were all racist and anti-Semites. It has vitally to do with the fact that we can now read so much that they could not. Nineteenth-century scholars were not in a position to assess the significance for early Greek literature of the literatures of Ugarit (in Syria) and of the Hittites (in Anatolia), because they had not yet been deciphered. When they were, classicists immediately began to work on their implications.

This development is surely to be welcomed, even if classicists do not seem to be trampling each other in their haste to get at the Hittite texts. It is also to be welcomed when Hittite specialists cast their eyes on early Greek literature, as did Jaan Puhvel in 1991 in *Homer and Hittite*[1] (whence my title), and myself in several chapters of *How to Kill a Dragon*.[2] In the spring of 1996 I held a seminar on "Greece and Anatolia", devoted to the exploration of some further linguistic and thematic connections between speakers from these geographically

---

[1] Innsbrucker Beiträge zur Sprachwissenschaft. Vorträge und kleinere Schriften 47 (Innsbruck 1991).

[2] *How to Kill a Dragon: Aspects of Indo-European Poetics* (New York 1995).

contiguous regions, who were certainly in contact at various times and in various places in the second millennium B.C. and later. Wendell Clausen in his writings, teaching, and conversation has always been receptive both to the broader notion of cultural influence of Hellenistic Greece on the literature of Rome, and to the finer evidence for that influence in precise philological details. I offer to him here in friendship three brief philological observations arising out of that seminar. These concern (1) the land of *Aššuwa*; (2) the four quadrants of social appurtenance in Hittite Anatolia; and (3) Asianic name and Greek *epiklesis*.

I. The Land of *Aššuwa*

The military activity of the Middle Hittite King Tudhaliyas II in Western Anatolia in the 15th century B.C. is well-documented, in particular his defeat of the "Assuwa coalition" circa 1430 B.C., a group of allies (the Hittite verb is *anda taruppantati* 'were united') of the land of Assuwa (KUR ^{URU}*Aššuwa*) enumerated in KUB 23.11, 12-19, beginning with [KUR ^{URU}Lu-]uq-qa (Lycia) in the southwest and continuing north to KUR ^{URU}U-i-lu-ši-ya KUR ^{URU}Ta-ru-i-ša in the Troad. The correctness of the restoration of *L]uqqa* is supported now by the occurence of the place name in the Hieroglyphic Luvian inscriptions of both YALBURT and SÜDBURG.[3] This group in its geographical spread and apparent military structure and organization strikingly recalls that of the Trojans and their allies (ἐπίκουροι) as described in the Trojan catalogue of Book 2 of the *Iliad*, where the enumeration runs in the reverse direction but with the same boundaries, from the Trojans in the northwest (2.816-18) to the Lycians in the southwest (2.877-78).[4]

---

[3] See now J. David Hawkins, *The Hieroglyphic Inscription of the Sacred Pool Complex at Hattusa (SÜDBURG)*, Studien zu den Boğazköy-Texten, Beiheft 3 (Wiesbaden 1995), and Machteld J. Mellink, "Homer, Lycia, and Lukka," in Jane B. Carter and Sarah P. Morris, edd., *The Ages of Homer. A Tribute to Emily Townsend Vermeule* (Austin 1995) 33-43.

[4] See the discussion and references in "The Language of the Trojans," in Machteld Mellink, ed., *Troy and the Trojan War* (Bryn Mawr 1986); reprinted in Watkins, *Selected Writings*, Innsbrucker Beiträge zur Sprachwissenschaft 80 (Innsbruck 1994) 700-717.

The historicity and immediacy of the Assuwa conflict has been dramatically confirmed by the discovery in Boğazköy in 1991 of a long thrusting sword of Aegean Mycenean design and apparently workmanship, but bearing an Akkadian inscription from Tudhaliyas II:[5]

> i-nu-ma ᵐDu-ut-ḫa-li-ya LUGAL.GAL KUR ᵁᴿᵁA-aš-šu-wa ú-ḫal-liq GÍR$^{HI.A}$ an-nu-tim a-na ᴰIM be-lì-šu ú-še-li
>
> "When Tudḫaliyas the Great King destroyed the land of Aššuwa he dedicated these swords to the Storm-god his lord."

The wording nearly exactly parallels that of the same king's Hittite annals, KUB 21.11, 33: GIM]-*an* KUR ^{URU}*A-aš-šu-wa ḫarninkun* "[Wh]en I destroyed the land of Assuwa." We may assume that this sword was part of the booty from his successful military campaign, and was dedicated to the victor's god.

The Assuwa-Aegean connection presupposed by the presence and provenance of this sword had been earlier documented linguistically on the Greek side, by the presence in the Linear B tablets of persons at Knossos, Pylos and Mycene bearing the name *a-si-wi-jo, Aswijos*. It is reasonable to suggest that the name *Aswijos* as 'man from *Aswa* = *Assuwa*', was originally applied in Greece to refugees from Tudhaliyas' western war of circa 1430 B.C. The divine name *po-ti-ni-ja a-si-wi-ja* attested once at Pylos, the *Potnija Aswija* or 'Aswian Mistress' may be a similar import, a transported goddess.[6]

In Homer Ἄσιος (scanning as Ἄσϝιος) is attested both as a personal name (the Trojan ally Asios, son of Hurtakos, and the Phrygian Asios, brother of Hekabe), and as an adjective of a geographical name,

---

[5] A. Ünal, A. Ertekin, and I. Ediz, "The Hittite sword from Boğazköy-Ḫattuša found in 1991 and its Akkadian Inscription," *Müze/Museum* 4 (1991) 46-52. The same authors in M.J. Mellink, E. Porada, T. Özgüç, edd., *Aspects of Art and Iconography: Anatolia and its Neighbours, Studies in Honor of Nimet Özgüç* (Ankara 1993) 719-30. O. Hansen, "A Mycenean sword from Boğazköy-Hattusa found in 1991," *Ann. Brit. Sch. Ath.* 89 (1994) 213-15. Mirjo Salvini and Lucia Vagnetti, "Una spada di tipo egeo da Boğazköy," *Parola del Passato* 49 (1994) 215-36.

[6] Cf. Watkins (above, n. 2) 151, 290. This account is preferable to connection with a Cretan place name Ἄσος, which will not explain either the phonology or the Homeric scansion.

in the famous simile in *Iliad* 2, likening the host of the Greeks to the great flocks of migratory waterfowl:

> χηνῶν ἢ γεράνων ἢ κύκνων δουλιχοδείρων
> Ἀσϝίωι ἐν λειμῶνι, Καϋστρίο' ἀμφὶ ῥέεθρα.
>
> "geese or cranes or long-necked swans,
> on the meadow of Aswa by the streams of the Kaüstrios."

The birds are pure Indo-European; the morphology and syntax of Ἄσϝιος is both Aeolic and Luvian, as I show elsewhere; and the land of Aswa = Aššuwa in Western Anatolia in the feminine form of its Greek adjective Ἀσία would give its name to an entire continent.

II. "The Four Quadrants of Social Appurtenance"

Our unique source for the history of the Old Hittite kingdom is the introduction to the Proclamation of King Telepinus on lawful succession to the throne, a text which comes at the end of the Old Kingdom, ca. 1500 B.C.[7] The text is a sort of *speculum principum*, and in it we find twice repeated the manifestation of a Hittite and Asianic cultural ideal: a formulaic expression of the unity and cohesion of different social groups under a just ruler.

> DUMU^{MEŠ}-*ŠU* ŠEŠ^{MEŠ}-*ŠU* LÚ^{MEŠ}gaenaš=šešš=a
> LÚ^{MEŠ} ḫaššannaš=šaš *Ù* ERÍN^{MEŠ}-*ŠU* taruppanteš ešer
>
> "His sons (and) brothers, and his relatives by marriage,
> the members of his kindred and his soldiers were united."

The quadripartite formula also is linguistically archaic, as shown by the thematic nom. pl. *gaenaš*(=), from IE **-ōs*, which occurs only once elsewhere in Anatolian, in Palaic *marḫaš* 'gods'. The verb is *tarupp-* 'unite', which we saw in the preceding section was the verb used to describe the Assuwa coalition in Western Anatolia. For the meaning of this verb, compare the Old Hittite *Bauritual* (CTH 414) KUB 29.1 ii 44, of a plaited ritual object:

---

[7] The most recent edition is that of Inge Hoffmann, *Der Erlass Telepinus. Texte der Hethiter* 11. (Heidelberg 1984).

n=at tarup n=at 1^{EN} iya . . .

"Unite it and make it one."

The Telepinus passage describes a quadripartite organization along two parameters: *blood—alliance* and *close—distant.*

close relatives by blood	close relatives by marriage
partisans by kin-fealty	partisans by allegiance

With a nod to Benveniste, we may term these quadrants of social appurtenance:[8] they define the individual's relation to the ruler. We know from another text that each soldier in the Hittite army (ERÍN^{MEŠ}) swore a personal oath of allegiance to the king.[9]

We find in the *Iliad* a remarkable homology to this Hittite societal formula, with the same quadripartite organization along the same two parameters. Perhaps significantly the context is wholly Asianic: words or rebuke spoken by the Lycian Sarpedon to the Trojan Hektor. Sarpedon begins his harangue as follows, *Il.* 5.473-475; the key words are given in boldface.

> Ἕκτορ, πῇ ἱ δή τοι μένος οἴχεται, ὃ πρὶν ἔχεσκες;
> φῆς που ἄτερ **λαῶν** πόλιν ἑξέμεν ἠδ' **ἐπικούρων**
> οἶος, σὺν **γαμβροῖσι κασιγνήτοισί**τε σοῖσι·
>
> "Hektor, where has your battle-rage gone, which you vowed to have? I suppose you said you would hold the city alone without your **own hosts** or **allies**, with just your **sisters' husbands** and your **brothers**."

Kirk in his commentary ad loc., citing Fenik, notes that the two strongest other instances of this "rebuke pattern" both also involve a Lycian ally (Glaukos) rebuking Trojan Hektor. The second of these in book 17 shows clear verbal similarities to the above passage and, more significantly provides the explanation of λαοί 'hosts', *Il.* 17. 144-145:

[8] Emile Benveniste, "Les quatre cercles de l'appartenance sociale," *Le vocabulaire des institutions indo-européennes* 293 (Paris 1969).

[9] Norbert Oettinger, *Die militärischen Eide der Hethiter*, Studien zu den Boğazköy-Texten 22 (Wiesbaden 1976).

φράζεο νῦν ὅπως κε πόλιν καὶ ϝάστυ σαώσῃς
οἶος σὺν **λαοῖς** τοὶ Ϝιλίωι **ἐγγεγάασιν**
"Think now how you can save the city and citadel alone,
with just your **own hosts** who were **born** in Ilios."

Ego's λαοί are thus the 'in-born', partisans by birth as opposed to the ἐπίκουροι, partisans by alliance. They therefore correspond to the Hittite LÚ^{MEŠ}*ḫaššannaš=šaš* rather than to the ERÍN^{MEŠ}-*ŠU*. Sarpedon's model is then:

brothers by birth	brothers by marriage
partisans by birth	partisans by alliance

The role of the root **ĝenh₁-*, the Greek semantic equivalent of Hittite *ḫašš-*, is noteworthy: κασίγνητοι, ἐγγεγάασιν.

Kirk ad 5.471 states that "Since the three most elaborate instances [of the rebuke pattern] all involve Lycian rebukes to Hektor for ignoring the allies, it looks as though this was a fundamental application, therefore that the broader theme of quarreling between Trojans and allies was stronger in the old poetical tradition that *Iliad* suggests." While that might be, I think rather that the concern for the allies is contextually given by the fact that Sarpedon and Glaukos *are* allies. The Hittite ideal is the unity of the four groups of social appurtenance under a wise and just ruler: *taruppanteš ešer*. The theme which is really at stake in the Homeric passages, as appears clearly in Sarpedon's exordium, is an Asianic sentiment of violation of that unity which their culture so highly valued.

III. Asianic Name and Greek Epiklesis

The previous section dealt with the knowledge of a cultural feature that could well have spread in the older pre-1200 period. The direction is clear: an Asianic, specifically Hittite ideal of unity among the groups representing four types of fealty, becomes known to the Greek-speaking world and leaves its trace in a literary topos in the post-1200, Homeric period: spoken by an Asianic Lycian to an Asianic Trojan.

Cultural influence, generated by language contact and some degree of sympathetic bilingualism, may be a two-way street. We have at least one such instance where Greek cultural influence on a segment of Asianic society can be linguistically documented in pre-1200 Hittite sources, and where the Greek knowledge of this phenomenon as an Asianic cultural feature again leaves its trace as a literary topos in the post-1200, Homeric period. This phenomenon is no less remarkable for being well known: it is the "double name" or *epiklesis*, whereby a princely Asianic has both a native name and a semantically transparent Greek name.

We first encounter the phenomenon in the historical name Alaksandus, ruler of Wilusa, whose treaty with the Hittite king Muwatallis (ca. 1300) is preserved in the Hittite archives (CTH 76). No other name for this individual is attested in the Hittite sources, to be sure, but the identification with the Greek name 'Αλέξανδρος is now virtually universally accepted. Sommer's objection to the compound type in the second millennium has been set aside by the Mycenean evidence, the personal name *a-re-ka-sa-da-ra Aleksandrā* (MY V 659, 2). 'Αλέξανδρος "(having the) warding off (of) men" is the "other" name of Paris the Trojan prince and son of Priam. The name Paris (Πάρις) shows clear Asianic affinities, as long recognized.[10]

One must wonder how and why an Anatolian prince like Paris, or one of the four 'kings' of the land of Arzawa like historical Alaksandus came to bear a 'speaking' Greek name. In *How to Kill a Dragon*[11] I speculated that the ruler of Wilusa might have had personal experience of a Greek **apel̯ā* (Doric ἀπέλλα) or *Männerbund*, the "hunter-warrior society of unmarried and propertyless young aristocrats".[12] As such he might have both assumed a Greek "international" name Aleksandros/Aleksandus and taken the Greek divinity *Apel̯ōn* (Apollo) in the form ᴰ]*Āppaliunaš* as his personal god, bringing him

[10] Emmanuel Laroche, "Linguistique asianique," *Minos* 11 (1970 [1972]) 112-135. and Watkins, "The Language of the Trojans," in *Troy and the Trojan War*, ed. Machteld Mellink (Bryn Mawr 1986); reprinted in Watkins, *Selected Writings* (above, n. 4) 700-717.

[11] (above, n. 2) 149.

[12] Kim McCone, "Hund, Wolf und Krieger bei den Indogermanen," *Studien zum Indogermanischen Wortschatz*, ed. W. Meid (Innsbruck 1987) 101-154.

back to Wilusa. ^{D}*Āppaliunaš* is the first of the gods of Wilusa to witness the treaty with king Muwatallis.

While such a scenario is certainly conceivable, we might prefer an explanation of more general application, but still allowing for a period of cultural and language contact, for example intermarriage (cf. Hektor and Andromakhe), or the fosterage of children of the Asianic elite by a corresponding stratum of Greek society. However that may be, the phenomenon of the double name or *epiklesis* is clearly documented among upper-class Trojans in Homer.

Not only do we have Paris-Aleksandros, with the historicity of the Greek name of an Asianic prince of Wilios guaranteed by the altogether remarkable coincidence with Alaksandus of Wilusa, but another in the same family: Hektor's infant son. *Il.* 6.402-3:[13]

> τόν ῥ' Ἕκτωρ καλέεσκε **Σκαμάνδριον**, αὐτὰρ οἱ ἄλλοι
> **'Αστύανακτ'**· οἶος γὰρ ἐρύετο Ἴλιον Ἕκτωρ
>
> "Him Hektor called Skamandrios, but other men
> Astuanaks; for Hektor alone guarded Ilios."

The son's epiklesis is an etymology of his father's name, as commentators since Plato have observed: *Cratyl.* 393 a: ὁ γὰρ ἄναξ καὶ ὁ ἕκτωρ σχεδόν τι ταὐτὸν σημαίνει "For ἄναξ and ἕκτωρ mean just about the same thing."

Virtually the same theme recurs in Andromakhe's first lament for Hektor as his corpse is being dragged off to the ships. *Il.* 22.505-507:

> νῦν δ' ἂν πολλὰ πάθησι, φίλου ἀπὸ πατρὸς ἁμαρτών,
> 'Αστύαναξ, ὃν Τρῶες **ἐπίκλησιν** καλέουσιν·
> **οἶος** γὰρ σφιν **ἔρυσο** πύλας καὶ τείχεα μακρά
>
> "Now he will suffer, having lost his dear father,
> my Astuanaks, whom the Trojans call **by this name**;
> for **you alone guarded** their gates and high walls."

The addressee is of course Hektor.

---

[13] On this figure see Sarah P. Morris, "The sacrifice of Astyanax ," in Carter and Morris, edd. (above, n. 3) 221-245.

The name Skamandrios is derived from the river Skamandros. The divine river, fathered by Zeus, is a Trojan partisan, and very nearly defeats Akhilleus himself in a dramatic battle in *Iliad* 21. This river too has a Greek *epiklesis*: it is couched in the lexical metaphor of 'language of men and language of gods'. Il. 20.74: the great river (μέγας ποταμός):

> ὃν **Ξάνθον** καλέουσι θεοί, ἄνδρες δὲ **Σκάμανδρον**
>
> "Which the gods call **Ksanthos** ('Red'), and men **Skamandros**."

Of the six Homeric examples of this lexical metaphor this is the only one to involve a clear contrast of Asianic name (language of men) and semantically transparent Greek *epiklesis* (language of gods). If the "double name" is an Asianic cultural feature reflecting Greek contact, we may have in the "language of gods / language of men" in Homer a cultural borrowing of diffusion from Asianic to Greek, in other words a Greek cultural feature reflecting Asianic contact. The Old Hittite litanies "To mankind you are X (e.g., ^{D}Tasimettis), but among the gods Y (e.g., Ishtar the queen)" (CTH 733, q.v. for references) are translations from Hattic, and the men's word is a Hattic divine name, while the god's word in an *epiklesis* from another language or pantheon.[14]

The examination hitherto of the pattern of Asianic name and Greek epiklesis of an Asianic aristocrat, first of Paris-Aleksandros and then of Hektor's son Skamandrios-Astuanaks (*Wastuwanaks*) as a surrogate of Hektor brings us to what must be seen as the Greek epiklesis par excellence: that of Trojan Hektor himself. For this we must return to two passages we have already looked at in the preceding section: the "rebukes" of Sarpedon in *Iliad* 5 and Glaukos in *Iliad* 17. For in both the rebuke involves a conscious reference to the synchronic Greek etymology of Hektor's name. In Il. 5.473-475 the etymology is overt (πόλιν+) ἔχω:

---

[14] See Watkins, "Language of Gods and Language of Men: Remarks on Some Indo-European Metalinguistic Traditions," in Jaan Puhvel, ed., *Myth and Law Among the Indo-Europeans* (Berkeley 1970) 1-17; reprinted in Watkins, *Selected Writings* (above, n. 4) 456-472.

Ἕκτορ . . .
φῇς . . . **πόλιν ἑξέμεν** . . .
οἶος . . .

"Hektor ... you said ... you would **hold the city** ... **alone**..."

In *Il.* 17.142-145 the city (πόλις) has been doubled by ϝάστυ, a token of their equivalence, and another verb replaces etymological ἔχω:

Ἕκτορ ...
φράζεο ... ὅππως κε **πόλιν καὶ ϝάστυ σαώσῃς**
οἶος...

"Hektor ... think how you can **save the city and citadel alone**..."

These passages should be juxtaposed to those giving the etymology of Astuanax (Wastuwanaks) as surrogate of Hektor. In *Il.* 6.403 the city is named (Ἴλιος):

Ϝαστύϝανακτ'· οἶος γὰρ **ἐρύετο Ϝίλιον** Ἕκτωρ

"**Wastuwanaks**; for Hektor alone **guarded Wilios**."

In *Il.* 22.506-507 we have a descriptive metonymy of the city:

Ϝαστύϝαναξ ...
οἶος γάρ σφιν **ἔρυσο** πύλας καὶ τείχεα μακρά

"Wastuwanaks ... for **you alone guarded** their gates and high walls."

These are the only passages in the *Iliad* to show the ancient defensive verb ἐρυ- "guard" in this form, the reflex of Mycenean *o-u-ru-to* '*hō wruntoi*'. Otherwise we have a handful of attestations of ῥῡ- reflecting ϝρῡ- or thematized ϝρῡε/ο-, and εἰρῡ- probably reflecting augmented ἐϝρῡ-.[15] The text might once have read *wrūeto* and *wrūso*; the latter is a precious relic of the old athematic paradigm. Other, presumably later forms of this verb in the *Iliad* show shortening of the root vowel as in εἰρύατο, or ῥύεται.

---

[15] The prehistory of this group in Greek is not wholly clear (cf. Chantraine, *DELG*). Forms with augment ἐ-ϝρῡ-(ο-), εἰρῡ-(ο-), appear to have been contaminated with forms from ϝερυ-(ο-) after loss of the initial digamma, whence blended presents like εἰρύαται.

Morphology thus points to the archaism, as well as the thematic centrality, of these four passages. For they are all of a piece: Wastuwanaks is overt in the latter two (344), but πόλιν καὶ ϝάστυ of 2 gives the necessary equation for Hektor's etymology, and the first with πόλιν ἐξέμεν—the last of the 4 to occur in the Iliad—gives the etymology with the root **segh-*, ἐχ-/ἐκ-.

The figure of alliteration should be noted in *Iliad* 6.403 (as restored) **Ϝαστύϝαναξ** ... ϝρύετο **Ϝίλιον**, and 22.506-507 **Ϝαστύϝαναξ** ... ϝρῦσο. The subject of the verb is in each case Hektor, and Ϝαστύϝαναξ bears as name the surrogate epithet for the Trojan hero his father. The old underlying formula finally surfaces in book 24, in Priam's first words to Akhilleus (499-501):

> ὃς δέ μοι οἶος ἔην, **ἔϝρυτο** δὲ **ϝάστυ**καὶ αὐτούς,
> τὸν σύ πρώιην κτεῖνας ἀμυνόμενον περὶ πάτρης
> Ἕκτορα·
>
> "He that alone was left me, who **guarded** the **city** and its people, him you killed just now, as he defended his fatherland, Hektor"

It is another covert etymological explanation of Hektor's name.

One further feature is common to all four passages: the relentless οἶος 'alone', always in the same emphatic verse- or phrase-initial position. These four passages each focus on the meaning of Hektor's name (or his son's), and thereby on his nature: for to be alone is Hektor's tragic flaw. It is, remarkably, an Asianic, not a Greek flaw, as appears from the previous section on the rebukes of Lycian Sarpedon and Glaukos, together with the bitter irony of Andromakhe's address to her slain husband and protector. It is another token of the awareness of Asianic values and sensibilities, resting on cultural contact, which informs the *Iliad.* Akhilleus's solitude is his nobility and his glory; yet once, when brought to his knees by the river Skamandros—a divine surrogate of Hektor—, Akhilleus too had to cry for help from the gods. For Hektor to be alone, οἶος, is no glory but his downfall. He will realize it at his last stand against Akhilleus, when his brother Deiphobos whom he thought beside him vanishes like a will-o'-the-wisp, a divinely wrought illusion.

ON GENDER SWITCHING AS A LITERARY DEVICE IN LATIN POETRY

ROBERT RENEHAN

Magistro Optimo
Litterarum Latinarum
Discipulus Memor et Gratus

An entirely satisfactory account of the genesis of the Indo-European gender system is perhaps no longer attainable; in any event it is not part of my brief in this paper. That the well-attested trichotomy of masculine, feminine, neuter is the result of some combination of two distinct dichotomies, namely male ~ female and animate ~ inanimate, seems plausible enough; it remains but a (partial) theory.[1] What is important for our purposes is to call preliminary attention to certain features of I-E gender as it is manifested in historical Latin, as well as in certain other I-E languages, most relevantly ancient Greek. First, many substantives follow "natural" gender, males being masculine, females feminine; thus *pater*, *vir*, *filius* will obviously be masculine in contrast to such feminines as *mater*, *mulier*, *filia*. This will not take us very far, as will be apparent as soon as, for example, one considers gender usage as applied to *animals*, where the situation is more complex (below). Sense constructions also come into play: *mea Glycerium, quid agis* is flawless idiomatic Latin. Secondly, substantives denoting inanimate objects occur with all three genders: *hasta*, *gladius*, *pilum*. The reason for the gender in a given instance is sometimes explainable, far more often it is not.[2] Thirdly, it sometimes happens that substantives, which are normally *either* mascu-

---

[1] Carl Darling Buck, *Comparative Grammar of Greek and Latin* (Chicago 1933) 170 has some sensible remarks: "Gender is a phenomenon that was already fully developed in the parent speech, its evolution wrapped in the obscurity of a remote past. Apart from certain general probabilities, we must simply take it as we find it." See also Andrew L. Sihler, *New Comparative Grammar of Greek and Latin* (New York and Oxford 1995) 245: "The origin of IE gender has always fascinated scholars, but most ideas on the question are necessarily very speculative."

[2] See J.B. Hofmann and A. Szantyr, *Lateinische Syntax und Stilistik* (Munich 1965) 5-12 (where further bibliography can be found).

line *or* feminine, are used in the other gender.[3] It is this phenomenon which is my concern here. I begin with some documentation. The following is a list, based primarily on my own reading and surely incomplete, of substantives which show a fluctuation between masculine and feminine gender. I give the less common gender in parentheses, thus: "(m.)" or "(f.)." Where there are but one or several exceptions to the normal gender attested in poetry, I also indicate the passage(s). Citations from Latin literature by the ancient grammarians are not indicated here; compare below. If a substantive is not followed by a parenthetic notation of exceptional gender, it means that both genders occur with some, but not necessarily equal, frequency:

accipiter (f., Lucr. 4.1009). *Acheruns* (f., Pl. *Capt.* 999; *Inc. trag.* 77). *adeps*. *aer* (f., Enn. *Ann.* 440 Sk.). *aetas* (m., Pl. *Trin.* 1090?). *alvus* (m., Pl. *Ps.* 823; Acc. *poet.* 1, *trag.* 700; Cinna *poet.* 7(9); Calv. *poet.* 14; Laber. *com.* 157). *amaracus*. *amnis* (f.). *anguis*. *animans*. *anser* (f., Varro *RR* 3.10.3; Hor. *Serm.* 2.8.88?; Colum. 8.14.4). *arcus* (f.). *balanus* (m.). *bubo* (f., Verg. *Aen.* 4.462). *callis* (f.). *calx* = "heel" (m.). *calx* = "pebble" (f.). *canalis* (f.). *canis*. *cardo* (f.). *cimex* (f.). *cinis* (f.). *clunis*. *colus* (m., Cat. 64.311; Prop. 4.1.72, 4.9.48). *corbis*. *cortex* (f., Lucr. 4.50; Verg. *Ecl.* 6.63; Ov. *Met.* 10.512; Mart. 13.86.2, 14.209.1). *crinis* (f., Pl. *Most.* 226; Atta *poet.* 1). *crocus* (f.). *crux* (m., Enn. *Ann.* 359 Sk.). *cupido*. *cupressus* (m., Enn. *Ann.* 223, 511 Sk.). *cytisus*. *damma* (m., Verg. *Ecl.* 8.28, *Georg.* 3.539). *dies*. *ficus* (m.). *finis* (f.). *forceps* (m.). *frons* (m.). *frutex* (f.). *funis* (f. Lucr. 2.1154). *grando* (m. Varro). *grex* (f., Host. *poet.* 2(4); Lucr. 2.662). *grus* (m., Lucil. 168; Hor. *Serm.* 2.8.87). *imbrex* (m.). *lapathus* (m., Lucil. 1235). *lapis* (f., Enn. *Ann.* 567 Sk.; Varro *RR* 3.5.14). *larix* (m.). *latex* (f., Acc. *trag.* 666). *lens* (m., Titin. *Com.* 163). *leo* (f.). *lepus*. *limax* (m.). *linter* (m., Tibull. 2.5.34.). *lodix* (m.). *lupus* (f., Enn. *Ann.* 65 Sk.). *lux* (m., Pl.). *lynx*. *margo*. *messis* (m., Lucil. 707). *metus* (f.,

[3] Fluctuation between masculine and neuter forms and, far less often, between feminine and neuter forms also occurs. I do not consider this phenomenon in the present paper. For it see, e.g., Herbert Zimmermann, "Schwankungen des Nominalgeschlechts im älteren Latein," *Glotta* 13 (1924) 224-41.

Enn. *Ann.* 562 Sk., *Scen.* 407). *mus. myrtus* (m.). *narcissus* (f., Verg. *Ecl.* 5.38?). *natrix* (m., Luc. 9.720). *nox* (m.). *obex. onyx. oscen* (f.). *ovis* (m.). *palumbes. pampinus* (f.). *papyrus. pecus* (m., Enn. *Scen.* 296?). *pedis. penus. perdix* (m.). *phaselus* (m., Catull. 4.1; Ov. *Pont.* 1.10.39). *pollis. polypus* (f., Lucil. 862?). *pulvis* (f., Enn. *Ann.* 315 Sk., *Scen.* 382; Prop. 1.22.6, 2.13.35, 4.9.31). *pumex* (f., Catull. 1.2?). *rubus* (f.). *rudens* (f., Pl. *Rud.* 938). *scobis* (m.). *scrobis* (f., Ov. *Met.* 243; Luc. 8.756). *sentis* (f., *Culex* 55; *Nux* 113). *serpens. silex* (f., Verg. *Ecl.* 1.15, *Aen.* 6.471, 602, 8.233; Ov. *Met.* 9.223, 304, 613). *simia. specus* (f.). *stirps* (m.). *talpa* (m., Verg. *Georg.* 1.183). *termo* (f.). *tibia* (m.). *tigris. torquis. turtur* (f.). *varix. vepres* (f., Lucr. 4.62).

Others will easily make additions to this list, either from the Roman grammarians or from their own reading, but this selection should suffice to demonstrate that gender fluctuation is a common phenomenon in Latin. Curiously, while the usage is undisputed, its *raison d'être* is not. Scholars, ancient and modern, have put forward a variety of explanations; they run the gamut from Aulus Gellius' appeal to euphony alone in a familiar chapter (13.21) to George Goold's pronouncement that "The fact is that ambiguous genders in Latin have no literary significance but constitute a morphological problem."[4] Before discussing any specific passages I state here my own views.

First, some preliminary observations. That some substantives were in actual use with two genders already in the early historical period is certain. That is to say, the poets sometimes had a *choice* of gender. Otto Skutsch, writing of Ennius, could even assert "There are ... so many instances of uncommon gender (sc. in Ennius) ... that they are best explained as being due to a state of the language still somewhat fluid."[5] This perhaps goes too far, if this eminent scholar intended to suggest by the word "fluid" that the Latin of Ennius's time did not yet possess well-established and fixed rules. The "Father of Latin Literature" was writing at a time when Rome was already a major international power and its language was hardly a loose argot. That the Latin of the high golden age

[4] G.P. Goold, "Two notes on Catullus 1" in *LCM* 6.9 (1981) 234.

[5] *The Annals of Q. Ennius* (Oxford 1985) 66-67.

showed some differences merely illustrates the cliché that language constantly changes.[6] Be that as it may, the real issue is *why* Ennius, or another poet, in any given case, elected to use an uncommon gender. That on occasion a poet introduced *suo Marte* a new gender I would not deny. In such cases there must be some linguistic justification for the innovation; not even poets were free to change genders without a by-your-leave. For examples of typical justifications see below.

Second, a methodological principle. Romans could, and did, commit solecisms involving gender, the uneducated doubtless with much frequency. Surviving non-literary documents prove this, if proof were needed. Even the educated had to learn the correct use of gender. It was no superfluous exercise when, say, Flavius Caper composed his (lost, but influential) *De Dubiis Generibus* or, later, Nonius Marcellus included in his compilation *De Compendiosa Doctrina* a book entitled *De Indiscretis Generibus*, to say nothing of the elaborate fifth book of Priscian's *Institutes*, entitled *De Generibus*. But when we come to the literary élite, and in particular to the major poets, we must assume that their employment of gender is always correct and justifiable. It is not impossible that on the odd occasion a poet blundered and used a false gender, but in any given instance it is so unlikely that, if we cannot accept the gender, then the correct procedure is to assume manuscript corruption. A similar practice has always been observed, say, with metrics, and rightly so.

One should not dismiss the occurrence of an unusual gender in a poem as a mere grammatical or morphological curiosity. When poets introduce the rare and uncommon, be it a gender, a word, phrase or grammatical construction, they do so of set purpose. The Romans could hear the difference and we should make an effort to do so too. When Dylan Thomas began his famous villanelle with "Do not go gentle into that good night", the substitution of the adjective "gentle" for the usual and expected adverb "gently" makes all the difference. Such is the stuff of poetry. It is the same with a Roman poet who knew his craft. Naturally, generalities of this sort do not tell the whole tale. It is often

---

[6] Consider: if no Latin earlier than the first century B.C. had survived, there would still be more than enough examples of gender fluctuation to permit one to talk of a fluid state of the language, were one so inclined. It would be a misguided inclination.

possible to be more specific. Thus an uncommon gender may produce an archaic coloration or be a learned Grecism. Sound doubtless often played a part in the choice although, if truth be told, we can only rarely attain any certainty on this score. It is well and good for Aulus Gellius (13.21.12, after Probus) to cite Verg. *Aen.* 2.554, *haec finis Priami fatorum* and observe "*si mutes 'haec' et 'hic finis' dicas, durum atque absonum erit, respuentque aures, quod mutaveris*", but the fact remains that, unless Vergil had used the uncommon feminine gender with *finis* in the first place, no one would ever have commented on the sound of the verse. The feminine *haec finis* catches our attention; that is primary. Servius, who is silent on the sound of the expression, remarks "*HAEC FINIS ut 'dies', si tempus longum significat, generis feminini est.*" Note the appeal to analogy. Gellius in the same chapter (13.21.21) calls attention to *funis* used as a feminine in Lucretius (2.1154): *Lucretius aeque auribus inserviens 'funem' feminino genere appellavit in hisce versibus: 'haut, ut opinor, enim mortalia saecla superne | aurea de caelo demisit funis in arva'; cum dicere usitatius manente numero posset: 'aureus e caelo demisit funis in arva'*. *Funis* is normally of masculine gender. Quintilian 1.6.6, *'funem' masculinum esse 'funiculus' ostendit*, suggests that there was a certain doubt about its gender, that is to say, that there existed some passage(s) where *funis* was feminine.[7] Lucretius, *loc. cit.*, is one such and the *OLD* cites Vitruvius 10.2.4 as another, wrongly. (The text is very uncertain and Vitruvius regularly has *funis* as a masculine.) Gellius justifies Lucretius's departure from the norm with the two words *auribus inserviens*, "as a service to our ears." This will not do. One could scarcely substitute an ungrammatical genitive for a dative, a subjunctive for an indicative, a participle for an infinitive, on the grounds of euphony. No more can one shuffle genders without a good reason. Lucretius's reason cannot be determined with certainty, but a reason there was. Ernout and Meillet state the situation succinctly: "*funis*, *-is* m. (mais tend à passer au féminin comme les autres substantifs en *-is*: *finis*, etc., sans doute d'après *restis*; Lucrèce écrit *aurea* ... *funis*, 2.1154, dans un passage où il songe à l'homérique

---

[7] Quintilian is actually more explicit; just above (1.6.5) he writes: *Comparatio in nominibus aut genus deprehendit aut declinationem; genus, ut si quaeratur, 'funis' masculinum sit an femininum eqs.*

σειρὴν χρυσείην, *Il.* 8.19, etc.)... "[8] Here are two possible explanations, not necessarily mutually exclusive, 1) by analogy with the feminine synonym *restis* and 2) a Grecism in imitation of a familiar Homeric expression. I doubt whether the second explanation *alone* would suffice,[9] but it could reinforce a natural tendency. Note that another synonym of *restis*, namely *rudens*, normally a masculine, is feminine in Pl. *Rud.* 938. Whether Lucretius was the first to make *funis* feminine or had earlier parallels, now lost, we cannot know. The reason why Lucretius did not use the metrically equivalent (and feminine!) *restis* seems clear: Lucretius, Vergil, Horace, and Ovid all use *funis*; none of them uses *restis* which is found chiefly in prose and the comic poets (also Juvenal and Martial). *Funis* was the more formal word and the striking feminine gender helped to create a genuinely poetic phrase.

That the employment of a specific gender in poetry can be a literary device which imparts a poetic tone to the language is nicely illustrated by a single word, *tigris*.[10] Ernout-Meillet (see above, n. 8) s.v. *tigris* state "La prose emploie le nom au masculin, la poésie au féminin..." The old Lewis and Short *Latin Dictionary* already had noted s.v. *tigris* "...in prose *masc.*, but *fem.* Plin. 8.10; 8.66; in the poets always *fem...*" For Pliny add 6.73 and 8.65, but the rule is of general validity; observe that *a literary genre can determine gender*. This is a question of grammatical gender, not sex.[11] The man who woos Chloe in Hor. *Carm.* 1.23 ("*Vitas inuleo me similis, Chloe*") proclaims (9-12):

---

[8] A. Ernout and A. Meillet, *Dictionnaire Étymologique de la Langue Latine,* ed. 4 (Paris 1967) 262.

[9] The case is different with the fluctuating genders of such Greek and Latin doublets as ἀήρ / *aer* which involve the same word.

[10] For the history of the word see A.S. Pease's note to Verg. *Aen.* 4.367.

[11] See on this my remarks in "Of Mice and Men in Aristotle," *CP* 89 (1994) 247 n.4. There is considerable variation of gender in the case of some "animal" nouns, and none at all in the case of others. O. Skutsch on Enn. *Ann.* 332: "*canis* (*es*), especially when referring to the hound, is often feminine: *ThLL* CANIS 253.11f." See also his remarks at *Ann.* 542. Greek κύων reveals a similar practice; W.S. Barrett on Eur. *Hipp.* 18: "κύων of a hound is ordinarily fem. (X. *Kyneg.* passim) unless used specifically of a male..." See also A.W. Bulloch on Call. *H.* 5.114. Illuminating is Barrett's observation at E. *Hipp.* 231: " ... Tragedy (and lyric; epic vacillates) normally has ἵπποι and πῶλοι fem. when used of a team of horses ... this ... is a question not of sex but of gender (cf. esp. S. *El.*, where Orestes' team are fem. in the pl. (703f., 734f., 737f.) but its members masc. in the sing. (721f.,744))."

atqui non ego te tigris ut aspera
Gaetulusve leo frangere persequor:
tandem desine matrem
tempestiua sequi uiro.

The suitor is hardly denying that he is chasing her down "like a lady tiger and a gentleman lion"; *tigris* is feminine, because the poets made *tigris* of feminine gender whatever the context. Clearly that sounded more poetic and the educated Roman would detect the difference from ordinary prose usage. The ancient grammarians frequently call attention to such unusual genders in Latin poetry, a fact which tends to confirm this assertion. I confine myself (for reasons of space) to a few examples. Verg. *Aen.* 4.462 writes *sola ... bubo*; the noun is normally masculine. Servius ad loc.: "*'Sola' contra genus posuit; Lucanus* (5.396) *'et laetae iurantur aves bubone sinistro', item Ovidius* (*Met.* 5.550) *'infandus bubo.' et hoc est in usu. sed Vergilius mutavit, referens ad avem. plerumque enim genus relicta specialitate a generali sumimus, ut si dicas 'bona turdus' referendo ad avem.*" [12] Quintilian 9.3.6: "*Fiunt ergo et circa genus figurae in nominibus; nam et 'oculis capti talpae' et 'timidi dammae' dicuntur a Vergilio; sed subest ratio, quia sexus uterque altero significatur, tamque mares esse talpas dammasque quam feminas certum est.*" The passages in question are, respectively, *Ecl.* 8.28 *cum canibus timidi venient ad pocula dammae*; *Georg.* 3.539 *timidi dammae cervique fugaces*; *Georg.* 1.183 *aut oculis capti fodere cubilia talpae.* Servius at *Georg.* 3.539 annotates: "*TIMIDI DAMMAE. mutavit genus, ut vitaret homoeoteleuton: Horatius* [*Carm.* 1.2.11-12] *'et superiecto pavidae natarunt | aequore dammae...'*" There are similar observations at *Ecl.* 8.28. Quintilian's remarks to the effect that *damma* and *talpa* (as many other nouns denoting animal species) may be used of either sex are true enough and the suggestion of Servius (or his source) that Vergil wished to avoid the sound that *timidae dammae* would have produced may be valid, but the paramount point is that cultured Romans heard in *timidi dammae* and *oculis capti talpae* recherché and *poetic* phrases. Consider now Hor. *Carm.* 2.13.39-40:

nec curat Orion leones
aut timidos agitare lyncas.

[12] Compare also Nonius Marcellus 194.2-6.

Lynx is normally feminine. The masculine here was sufficiently unusual to attract the attention of the grammarians. Priscian *Inst.* 6.26 (*GL* I.218.3-8): "*Horatius in II carminum: 'Nec curat Orion leones | aut timidos agitare lyncas.' (et notandum etiam de genere, quod masculino genere protulit Horatius, cum Vergilius feminino in III georgicon* [= 3.264]: *'Quid lynces Bacchi variae?').*" Nisbet and Hubbard on Hor., loc. cit., remark "... *lynx* is normally feminine ... but the vowel-sounds of *timidas agitare lyncas* might have seemed to lack variety (in spite of the changes of quantity)." This is couched in cautious enough language; ultimately, of course, we can never know whether Horace found *timidas ... lyncas* a displeasing collocation. What we can know is that the phrase *timidos ... lyncas* is exquisite, in the Latin sense of that term. The Pseudo-Acron scholium here is obscure and significant: "*AUT TIMIDOS AGITARE LINCAS: Lincas dammas* (v.l. *Lincas idest dammas), ut Vergilius* [*Georg.* 3.264]: *'Quid linces Bacchi variae?'*" It is absurd to equate the *lynx* and the *damma* and difficult to believe that any scholar would have done so. With the help of the passages from Servius and Priscian cited above one can reconstruct the earlier comment on which our extant comment is based. 1) The *Georgics* passage was cited to illustrate the usual feminine gender of *lynx* (compare Priscian). 2) The unusual gender of *lynx* in Horace *was compared with* the unusual masculine gender of *damma* in Vergil; the fact that both substantives had the same epithet (*timidus*) presumably helped to suggest the comparison. In the Pseudo-Acron scholium unintelligent abridgement has produced nonsense. Nevertheless we can still extract a significant piece of information from it, namely that, just as *timidi dammae*, so also *timidos lyncas* had a rare (read: poetic) ring to it.

It is easy for us to miss the delicate overtones of seemingly ordinary diction and it is a serious mistake to ignore the precious comments of the Roman grammarians. For example. D Serv. at Verg. *Aen.* 2.355: "*sane apud veteres 'lupus' promiscuum erat, ut Ennius* [= *Ann.* 65 Sk.] *'lupus femina feta repente'.*" Festus, p.364.1-9 Lindsay: "*'Recto fronte ceteros si novit' Cato in dissertatione consulatus* [= fr.25]. *Antiquae id consuetudinis fuit ut cum ait Ennius quoque 'a stirpe supremo'* (= *Ann.* 166 Sk.) *et ... 'lupus feta'* [= *Ann.* 65 Sk.] *et 'nulla metus'* [= *Scen.* 407]. *Etiam in commentariis sacrorum pontificalium frequenter est hic ovis et haec agnus ac porcus.*[13] *quae non ut vitia, sed ut antiquam consuetudinem testantia, debemus accipere.*" Compare also Quintilian 1.6.12. Otto Skutsch, with his customary learning, comments at Enn. *Ann.* 65 "*lupus femina*: epicene names of animals can be made into feminines by adding *femina*. This in classical Latin is still common with names such as *bos* and *sus* which cannot be converted more economically. *-os* nouns prefer to develop an *-a* feminine such as *equa, capra, ursa, porca,* (*porcus femina* in a ritual context Cato *agr.* 134.1; Cic. *leg.* 2.57), *agna* (but *agnus femina* in ritual language *Lex Numae ap.* Paul. Fest. 222; Gell. 4.33..." Phrases such as *lupus femina* and *porcus femina* may still have been in ordinary use in Ennius's and Cato's day, but by the high classical period they had become archaic and were heard as such. Their continued use in religious or legal formulae does not gainsay this because of the well-known conservative nature of such language. (Compare the still current "Our Father which art in heaven" with its archaic use of the relative pronoun and archaic form of the verb.)

There are times when the introduction of a different gender is a matter of meaning as well as of diction; *res* and *verba* coincide. Hor. *Serm.* 2.8.85-88, from a description of a pretentious dinner:

> deinde secuti
> mazonomo pueri magno discerpta ferentes
> membra gruis sparsi sale multo, non sine farre,
> pinguibus et ficis pastum iecur anseris albae

Here grammatical gender does involve biological sex. *Grus* (v.87) is normally feminine and *anser* (v.88) masculine, whatever the sex;[14] not

---

[13] The point is that *ovis* is normally feminine, *agnus* and *porcus* normally masculine. To make this clearer it has been proposed to emend to *hic <et haec> ovis et <hic et> haec agnus <et hic et h>a<e>c porcus.* For such "*hic et haec*" formulae see, to give but one instance, Priscian *Inst.* 5.42 (*GL* II.169, 6 ff.), but Festus's text is intelligible without change.

[14] There are, as often, a couple of departures from the norm; *grus* is masculine ap. Laberius *Com.* 47 and *anser* feminine ap. Varro *RR* 3.10.3 and Columella 8.14.4. (The language of this passage suggests a direct borrowing from Varro, loc. cit.). The full expressions would be *grus mas* and *anser femina* (even in the classical period: neither is a second declension noun ending in *-us*). Priscian *Inst.* 6.81 (*GL* II.265.16-21) pronounces "*hic et haec ... grus*" and cites Verg. *Aen.* 10.265 *Strymoniae ... grues* for the feminine, our passage for the masculine.

so here. The commentators explain the point; thus Kiessling-Heinze: "Dass es Kranichmännchen waren (*grus* ist der Regel nach Femininum, Maskulinum noch bei Laberius 47 R.), sowie dass die Leber (88) von einer weissen Gans und keinem Gänserich stammte, und warum das so angeordnet sei, hat Nasidienus redselig seinen Gästen mitgeteilt: das sind die *causae et naturae rerum* 92." E.P. Morris:[15] "*gruis*: here masc., though commonly fem., as *anser*, commonly masc., is here made fem. The gender of such words is grammatical and somewhat shifting, but apparently the unusual gender is chosen to indicate sex, *as though the epicure could tell the sex by the taste* (emphasis added)." There is a similar passage in Hor. *Serm.* 2.4.44: *fecundae leporis sapiens sectabitur armos*. The *OLD* lists *lepus* as masculine only and that is in fact the normal gender of the word.[16] Here, *exempli gratia*, is E.P. Morris' note ad loc.: "*fecundae*: *prolific*, in general; but the use of the feminine appears to be intentional. The emphasis is upon *armos*; the true epicure will select for his guests the forelegs of the female hare. Cf. *Sat.* 2.8.89." The poet has thus employed uncommon gender to very good effect. The distinction in both instances was a fine one and might easily be missed. In fact it was, for the most part, and already in antiquity. In *Serm.* 2.4.44 it is the epithet *fecundae* which indicates the gender of *leporis* and at *Serm.* 2.8.89 *albae* proves that *anseris* is feminine. In the one passage the MSS (and Donatus at Ter. *Eun.* 426) have *fecundi*, in the other they have *albi*. The true readings were preserved in the famous lost codex Blandinius vetustissimus of Jacob Cruquius. Richard Bentley in his edition of Horace acutely argued the case for the feminines;[17] most editors have followed him.

To come to some nouns denoting inanimate objects or abstractions. *Pulvis* is normally masculine; Ennius uses it as a feminine several times. *Scen.* 382 *crassa pulvis oritur, omnem pervolat caeli fretum. Ann.* 315

---

[15] *Horace. The Satires,* with Introduction and Notes by Edward P. Morris (New York 1909).

[16] Quintilian 1.6.12 describes *lepus* as "*epicoenon*" and so also Priscian *Inst.* 3.29 (*GL* II.104.19); compare Varro *RR* 3.12.4: *Quis item nescit, paucos si lepores, mares ac feminas, intromiserit eqs.* Priscian *Inst.* 5.42 (*GL* II.169.13-15) includes *lepus* in a list of "*hic et haec*" substantives; here some MSS have a variant *lupus*.

[17] P. Lejay's note to *Serm.* 2.4.44 is also well worth reading, see *Oeuvres d'Horace ... par F. Plessis et P. Lejay. Satires publiées par Paul Lejay* (Paris 1911).

Sk. *pulvis fulva volat*. A third possible example is *Ann.* 264 Sk. *iamque fere pulvis ad caelum vasta videtur*, where Skutsch accepts Ernout's correction *caeli* for *caelum*, which would then be governed by *vasta* understood as a neuter plural. (See Skutsch's discussion.) For our purposes the reading here matters little as there are at least two certain occurrences of a feminine *pulvis* in Ennius. (All three passages, incidentally, are preserved by Nonius, the two passages from the *Annales* cited specifically as examples of the feminine gender.) Whether *pulvis* f. is an innovation of Ennius himself or rather a vestige from a period when the gender of *pulvis* was still not fixed we cannot say.[18] No matter. What is significant is that, apart from Ennius, this very common word is always masculine in classical Latin—with the exception of Propertius who uses *pulvis* seven times, four times as a masculine (1.17.23; 1.19.6, 22; 4.2.40) and thrice as a feminine (1.22.6; 2.13.35; 4.9.31).[19] The editors of Propertius cite no variants. How contemporaries of Ennius heard *pulvis* f. we cannot know for sure. On the other hand the extant classical evidence is quite adequate to show that the feminine gender would have been unusual, and unprosaic, in Propertius's day. The later grammatical tradition has preserved traces of this. Charisius, *GL* I.89.22-23 (= p.113.5-7 Barwick): *Pulvis masculini generis est, quamvis Propertius* (2.13.35) *dixerit 'qui nunc iacet horrida pulvis'*. Anon. *De Dubiis Nominibus*, *GL* V.558.5-6: *Pulvis generis masculini; sed Propertius dicit 'horrida pulvis'*. Why Propertius chose to use the feminine gender in a particular passage is matter for speculation, perhaps not entirely idle. The first occurrence of the feminine is in the concluding poem of the *Monobiblos*, 1.22.6:

(sic mihi praecipue, pulvis Etrusca, dolor,
tu proiecta mei perpessa es membra propinqui,
tu nullo miseri contegis ossa solo)

Here *pulvis* is poetically put for *terra*; "*terram Etruscam alloquitur*", as Enk rightly observes ad loc. Words for "earth" (γῆ, χθών, *terra*,

[18] Contrast Hofmann-Szantyr p.11 (a Grecism after κόνις) with Skutsch in his edition of the *Annals*, p.67 ("... best explained as being due to a state of the language still somewhat fluid"); see also his comments on *Ann.* 315 and my remarks above, p. 216.

[19] In Cic. *Q. fr.* 3.1.4 *eo loco pulvis non glarea iniecta est*, the feminine *iniecta* is presumably due to attraction to the nearer noun.

tellus) are usually feminine; the notion of earth as mother is more than ancient.[20] Analogy may be one factor in determining the gender. The last two elegies of Book I (21 and 22) are closely related. I.21 laments the death of a relative and I.22 refers to the same death. Wilamowitz accurately pronounced "Sind doch die beiden Schlussgedichte von Properz I schlecht und recht Epigramme."[21] *Elegy* I.22 ("*Qualis et unde genus, qui sint mihi, Tulle, Penates eqs*.") reveals a definite style. Wilamowitz again: "Das ist ja gerade eine Form des Grabepigramms."[22] Here is another feminine *pulvis* from Propertius, 2.13.35:

> et duo sint versus: Qui nunc iacet horrida pulvis,
> unius hic quondam servus amoris erat.

That is clearly "Grabepigramm" style. *Pulvis*, used of human remains, is often not clearly distinguished from the *earth* which receives it. *Gen*. 3.19 is as good an illustration as any of the reason for this: *in sudore vultus tui vesceris pane donec revertaris in terram de qua sumptus es | quia pulvis es et in pulverem reverteris.* Whether or not such notions influenced the gender of *pulvis*, two things are clear. First, grave epigrams, real or literary, affect a conservative and archaic tone. Second, in Propertius' day *pulvis* in the feminine gender was old-fashioned. In sum, the poet uses the gender in these two poems as a conscious archaism.[23] The effect is pleasing. It is of course the exception when one can even speculate with such specificity; in most cases we must be content with the recognition that by the use of an uncommon gender a poet can fashion an uncommon, and often elegant, phrase.

To give one example, *cupido*, when not used of the deity Cupid, is usually a feminine substantive. Hor. *Carm.* 2.16.15-16: *nec levis somnos timor aut cupido | sordidus aufert.* The masculine gender surprised Porphyrio: *cupidinem hic non puto pro Amore* (correxi: *amore*)

---

[20] Compare Plato's striking comment, *Menex.* 238A: οὐ γὰρ γῆ γυναῖκα μεμίμηται κυήσει καὶ γεννήσει, ἀλλὰ γυνὴ γῆν.

[21] *Sappho and Simonides. Untersuchungen über Griechischen Lyriker* (Berlin 1913) 297.

[22] loc. cit. (above n. 21).

[23] It is perhaps worth noting that in modern Italian *polvere* is feminine; the history of the word from classical Latin to Italian through vulgar Latin could be instructive.

posuisse, sed pro cupiditate. Quare miror masculino eum genere cupidinem dixisse, cum hac significatione haec cupido dicatur. Kiessling-Heinze ad loc. state the facts: "*cupido* ist bei Horaz stets, auch als Abstraktum, Mask.; das ist, wie die alte Personifizierung (und Plaut. *Amph.* 840) zeigt, das ursprüngliche; seit Sallust und Lukrez wird es, nach Analogie von *libido* u.a., sonst meist als Femin. gebraucht." For *cupido* m. compare further Hor. *Carm.* 3.16.39; 3.24.51; *Serm.* 1.1.61; *Epist.* 1.1.33; Ov. *Met.* 8.74; 9.374; 10.636; Sil. 4.99. What is of most interest here is that Horace, clearly departing from contemporary usage, *always* employs *cupido* as a masculine. But once one has called attention to the fact that *cupido* m. is an uncommon gender, deliberately chosen and possibly with an old-fashioned patina on it, there is little to add.

Perhaps the most familiar problem of gender in Latin literature is that of *pumex* in Catullus 1.2:

> Cui dono lepidum novum libellum
> arido modo pumice expolitum?

The difficulty is that Servius at Verg. *Aen.* 12.587 (*inclusas ut cum latebroso in pumice pastor*) states the following: *'in pumice' autem iste masculino genere posuit, et hunc sequimur, licet Catullus dixerit feminino.*[24] Most editors follow Servius and print *arida ... pumice*; the noun is not attested elsewhere as a feminine. A. Klotz[25] and T.P. Wiseman[26] may be mentioned as strong champions of *arida*; G.P. Goold[27] on more than one occasion has condemned *arida* in language of emphatic challenge. I doubt that there will ever be unanimity on this question, but more perhaps remains to be said.

First, the easy issue, that of euphony. It can settle nothing. Forcellini in his great *Lexicon* began his entry thus: "*PUMEX m. perraro f.*

---

[24] D Serv. adds after *sequimur* the words *nam et Plautus ita dixit*. See *Aul.* 297 and *Pers.* 41.

[25] A. Klotz, "Zu Catull," *Rh. Mus.* 80 (1931) 342-46. Klotz has interesting things to say about uncommon genders in Latin poetry in general; I think, however, that he overplays the Greek card.

[26] T.P. Wiseman, *Clio's Cosmetics. Three Studies in Greco-Roman Literature* (Leicester 1979) 166-71. Wiseman has a very full bibliography to which I refer the reader rather than repeat.

[27] (1) G.P. Goold, "Servius and the Helen Episode," *HSCP* 74 (1970) 134-39; (2) Id., "Two Notes on Catullus 1," *LCM* 6.9 (1981) 233-35.

... *Feminino genere extulit Catullus* ... *non tam quod ita ferret usus, quam quia male sonabat 'arido modo'*..." Catullus may indeed have wished to avoid the repeated *-do* sound (especially after *dono* in v.1), but there are no absolute rules,[28] and *lepidum novum libellum* | ... *expolitum* did not bother the poet. At the other end of the spectrum from Forcellini is H. Dettmer.[29] She writes that *arido modo* "surely is what Catullus wrote." Here is her proposal (the "novelty" of which she recognizes): "By having the '*do*' of *ari*-DO *mo*-DO in v.2 echo the '*do*' of DO-*no* in v.1 [Catullus] is reinforcing the central idea in these introductory verses of the poet as gift-giver." In other words we are supposed to hear echoes of *dono* (or *do*) in the final syllables of both *arido* and *modo*; that the vowel quantities of the first and second *-do* differ is not mentioned. This proposal is a specimen of contemporary fantasy and may be ignored. Whether *arido modo* or *arida modo*—or neither—would have offended Catullus's ear we cannot say with any confidence.

Next, the tradition direct and indirect. Both the archetype of our Catullus MSS and those authorities who cite the passage[30] have *arido*; the only exception is Servius. This is of less significance than it may appear. Since *pumex* is regularly masculine, *arido* would almost inevitably eject an original *arida*. Such a change is the easiest sort of trivialization and it would not have been surprising if no trace at all of an original *arida* had survived. Most major corruptions occur very early in a MS tradition; as copies proliferate, collation of MSS serves as a corrective. Assume that *arido* was an early corruption, a very reasonable assumption, and its presence in Terentianus Maurus, say, or Isidore is easily explained. What is less easily explained is *arida* in Servius, because there is no obvious reason for inventing such a statement, going, as it does, against the normal gender of *pumex*. It has the ring of

---

[28] For such *homoeoteleuta* see Appendix IV in Norden's edition of *Aeneid VI* (Leipzig 1916) 405-407. He rightly distinguishes between "Theorie" and "Praxis"; for Catullus see p. 407.

[29] Helena Dettmer, "Two Catullan Notes: 56.5 and arido vs. arida in 1.2," *LCM* 11.8 (1986) 131.

[30] For references see Klotz, pp. 345-46; Goold (1), p. 138; (2) p. 233.

authenticity. In other words, *arida* here is a classic case of *lectio difficilior*.[31]

Arida, then, deserves to be taken seriously. But can one explain the irregular gender? It is often explained as a Grecism which takes its gender from the corresponding Greek noun, κίσηρις, ἡ. So for instance Klotz and Wiseman. This seems very artificial, especially in the absence of an appropriate Greek model for the passage.[32] At most, the gender of κίσηρις could have been a contributing factor to the choice of a gender for *pumex* which already found its justification in the nature of the Latin language itself. Substantives ending in *-ex* (whether the *e* be long or short) are mostly masculine, but some are feminine and, more significantly, some fluctuate between masculine and feminine, for example *cortex*, *frutex*, *latex*, *obex*, *silex*. The generic words for "stone" in Greek and Latin, λίθος and *lapis*, primarily masculine, also occur as feminines. This is frequent with λίθος, especially when a particular kind of stone is meant (LSJ s.v. II), and Ennius (*Ann.* 567 Sk.), Varro (*RR* 3.5.14) and other writers use *lapis* as a feminine. A number of specific "stone" words in Latin are feminine, such as *rupes*, *cautes*, and, most pertinently, some fluctuate in gender: *calx* and *silex*. This last in fact provides *pumex* as apt an analogy as one could wish, Clausen on Verg. *Ecl.* 1.15: "*silice in nuda*: feminine here as elsewhere in Verg. where its gender can be determined (*Aen.* 6.471, 602, 8.233, where see Eden), but masculine in Lucretius, and both masculine and feminine in Ovid." Detractors of feminine *pumex* seem to forget that cases of a substantive extant in a single passage with exceptional gender, *hapax eiremena* of gender as it were, do occur. See for example *funis*, *latex*, *messis*, *vepres*; of these *latex* in particular is a close morphological parallel. If Catullus wished to write *arida* ... *pumice*, whether from a desire to pro-

[31] I write this fully aware that one must always take what Servius says with the greatest caution. Here Goold argues forcefully that Servius or his source misread a (hypothetically reconstructed) lexicon, the eye skipping from one entry (*pumex*) to another (*pulvis*) so that the statement *licet Catullus dixerit feminino* would refer to *cinis*, a word which Catullus does use in the feminine, rather than to *pumex*. See his discussions (above, n. 27) for details. This is ingenious, and possible (cf. R. Kaster, "Servius and Idonei Auctores," *AJP* 99 (1978) 199, n. 48), but it can scarcely be considered certain. It must remain a clever guess; if it is wrong, we are left with an explicit attestation of a feminine *pumex* in Catullus, whatever the ultimate source of it.

[32] Klotz (above, n. 25) 343-45 takes such an approach to uncommon Latin gender to excess.

duce a poetic phrase or from considerations of euphony (or both), he would not have exceeded the boundaries of the grammatically permissible by making *pumex* a feminine. He shows himself sensitive to uncommon genders elsewhere, for example *cinis* (f., 68.90; 101.4); *colus* (m., 64.311); *finis* (f., 64.217).[33] A striking illustration of this, hitherto unnoticed so far as I know, occurs in 4.1:

> Phaselus ille, quem videtis, hospites,
> ait fuisse navium celerrimus

The syntax of v.2 is a Grecism as has long been recognized. *Phaselus* in v.1 is in origin a Greek word, φάσηλος. It has two meanings, a kind of "bean" and a "skiff" or "yacht", so-called from its shape. The second meaning survives chiefly in Latin, but see Appian *BC* 5.95: ... δέκα φασήλοις τριηριτικοῖς, ἐπιμίκτοις ἔκ τε φορτίδων νεῶν καὶ μακρῶν. φάσηλος in Greek is of masculine gender. *In Latin poetry, where the gender can be determined*, phaselus *is usually feminine*. This is true of both meanings. For "bean" see Columella 10.377 (a hexameter). For "boat" see Ov. *Am.* 2.10.9; Sen. *Her. O.* 695; Luc. 5.518; Mart. 10.30.13. Ps.-Acron on Hor. *Carm.* 3.2.29 misquotes Verg. *Georg.* 4.289 (*et circum pictis vehitur sua rura phaselis*); he has *picta ... phaselo* (*pictis ... phaselis* is a v.l.). For the masc. see Ov. *Pont.* 1.10.39.[34] In short, when Catullus wrote *phaselus ille* in the fourth poem of the collection, a poem with considerable Greek background, he deliberately abandoned the commoner gender to introduce a Grecism;[35] the first two words of the poem set the tone.

---

[33] Lucretius reveals a similar practice: *accipiter* (f., 4.1009); *cinis* (f., 4.926); *cortex* (f., 4.50); *finis* (f., 1.107, 551, 561); *funis* (f., 2.1154); *grex* (f. 2.662); *vepres* (f., 4.62).

[34] Cic. *ad Att.* 14.16.1 writes "*conscendens ... in phaselum epicopum has dedi litteras ...*" *Epicopum* is a Greek adjective (ἐπίκωπος) meaning "equipped with oars." Elsewhere in Latin it seems to occur only at *ad Att.* 5.11.4 "*... habebam et aliquid* ἐπικώπων", whence Tyrrell and Purser conjectured ἐπίκωπον for *epicopum* at *ad Att.* 14.16.1 (fort. *epicopon*?). Whatever the spelling, Cicero was clearly using Greek nautical language and presumably took *phaselum* as a masculine. (Note, however, that, if ἐπίκωπον—or *epicopon*—be accepted, the adjective is of *two* terminations and, strictly speaking, the gender of the noun is uncertain.)

[35] Hor. *Carm.* 3.2.25-26 affects a Grecism by a different means, a Greek termination: *fragilemque mecum | solvat phaselon.*

I conclude with a passage where, as so often, decision is difficult, Verg. *Ecl.* 5.38-39:

> pro molli viola, pro purpureo narcisso
> carduus et spinis surgit paliurus acutis

Clausen ad loc. explains the situation succinctly: "*purpureo narcisso*: 'duo homoeoteleuta' (D Serv.) ...The grammarian Diomedes seems to have read *purpurea narcisso*, *GLK* i.453.34: '*primus modus soloecismi fit per immutationem generum nominis, cum dicimus 'atra silex'* (*Aen.* 6.602) *aut 'amarae corticis'* (*Ecl.* 6.62-63) *aut 'purpurea narcisso', cum* ater silex *debeat dici et* amari corticis *et* purpureus narcissus'. Ribbeck adopted *purpurea narcisso*, and Hofmann-Szantyr 11 list *narcissus* along with *cortex* and *silex* as feminine in Verg. Although a rarity in Greek, the feminine occurs in Theocr. 1.133 and Meleager 46.2 G.-P." Here we have, essentially, the same situation as in Catullus 1.2. Is it not curious that the one lection (*arida*) has found so much favor and the other (*purpurea*) so little? There are similarities. In both cases the unusual gender is attested by only one authority (Servius, Diomedes) and in both one must decide whether this is to be explained as a fortuitous vestige of authentic erudition or as a mere fluke. But there are differences. νάρκισσος *is* found as a feminine—and where we would most wish to find it, in Theocritus. One can more readily explain a feminine *narcissus* as directly modelled on ἡ νάρκισσος than justify a feminine *pumex* on the analogy of a completely different word, ἡ κίσηρις. Note further that Ennius used *cupressus*, normally feminine, as a masculine at least twice, *Ann.* 224 (where see Skutsch) and 511. Yet, as remarked, *arida* has found much more favor than *purpurea*. Not that everyone has rejected *purpurea*. Clausen mentions Ribbeck and Hofmann-Szantyr. Klotz for another also came out firmly in support of *purpurea*.[36] The feminine is not so easily disposed of as Sir Arthur Hirtzel, in the apparatus criticus to his OCT text of Vergil (1900), supposed: "purpurea *testatur Diomedes* (*cui immutatum genus soloecismus recte videtur*) *legit Ribbeck*." Can he really have been unaware that Diomedes cites *purpurea narcisso* in company with two other "solecisms" that are printed in every text of Vergil? *Purpurea* strikes me as a genuine *lectio difficilior* which, while hardly certain, has some chance of being correct.

---

[36] (above, n. 25) 344.

If so, the feminine *narcissus* is a deliberate Hellenistic, or rather Theocritean, conceit. It is noteworthy that Wendell Clausen in the Introduction to his splendid edition of the *Eclogues* singles out this very passage, *Ecl.* 5.36-39, as being, in its employment of botanical diction, uniquely comparable to Theocritus.[37]

[37] *Virgil Eclogues.* Edited with an Introduction and Commentary by Wendell Clausen (Oxford 1994) xxix-xxx: "… There is only one passage quite like [Theocr. 5.128-31] in the *Eclogues*…"

DE RE PVBLICA AND *DE RERVM NATVRA*

JAMES E.G. ZETZEL

"Lucreti poemata ut scribis ita sunt, multis luminibus ingeni, multae tamen artis. sed cum ueneris." Cicero's brief and enigmatic comment in a letter to his brother Quintus written in February 54 (*Q. fr.* 2.10.3) is the sole contemporary reference to Lucretius. Presumably, despite the plural *poemata*, it concerns the only work of Lucretius of which we know, the extant *De rerum natura*, but even that is not necessary, and much as we would like to know what Cicero planned to say to Quintus when they met, we do not. His description of the combination of *ingenium* and *ars* is high praise of Lucretius' poetic style, but it is frustratingly compressed, and provides no indication of how Cicero read *De rerum natura* and what he thought of its content or structure—if, indeed, he was giving his impression of the whole poem and not just excerpts of it.[1] And yet, the repeated allusions to *De rerum natura* in Cicero's philosophical writings make it clear that he knew it well; it is also significant that the earliest allusions to Lucretius occur in the work which Cicero began to compose only a few months after the letter to Quintus, *De re publica*.[2] It is the interpretation of those allusions that concerns me here; I will argue that Cicero's response to Lucretius is above all—as the letter to Quintus would suggest—a literary one, and that it needs to be seen in the context of the intellectual debate over the proper way to employ the

---

[1] For the problems of interpretation surrounding this letter, cf. Shackleton Bailey *ad loc*. The following works are referred to by the author's name alone: E. Andreoni, "Sul contrasto ideologico fra il *De re publica* di Cicerone e il poema di Lucrezio (la genesi della società civile)," in *Studi di poesia latina in onore di Antonio Traglia* (Rome 1979) 1: 281-321; J. Fontaine, "Le *Songe de Scipion* premier Anti-Lucrèce?," *Mélanges d'archéologie et d'histoire offerts à André Piganiol* (Paris 1966) 3: 1711-29; E. Heck, *Die Bezeugung von Ciceros Schrift De re publica*, Spudasmata 4 (Hildesheim 1966); T. Maslowski, "The Chronology of Cicero's anti-Epicureanism," *Eos* 72 (1974) 55-78; J.L. Penwill, "Image, ideology and action in Cicero and Lucretius," in A.J. Boyle, ed., *Roman Literature and Ideology* (Bendigo 1995) 68-91; G.C. Pucci, "Echi lucreziani in Cicerone," *SIFC* 38 (1966) 70-132; J.E.G. Zetzel, ed., *Cicero De re publica: Selections* (Cambridge 1995). I am grateful to Susanna Zetzel for considerable improvements to this paper.

[2] For a thorough survey of Cicero's allusions to Lucretius with earlier bibliography, cf. Pucci. On the chronology of the composition of *De re publica*, cf. Zetzel 3-4; the testimonia are collected by Heck 17-21.

new-found appreciation of Greek culture in the construction of new forms of Roman literature. Wendell Clausen has illuminated a major aspect of this debate in his studies of Roman Callimacheanism; this paper attempts to extend his analysis of the New Poets to the new philosophical literature of the last years of the Roman Republic.

Various types of allusion to Lucretius have been detected in *De re publica*, not all of them equally convincing.[3] There are, in the preface and elsewhere, clear references to and attacks on Epicureanism, that need not refer specifically to Lucretius; whether or not they do will be discussed below. Everything in *De re publica* that can legitimately be connected with *De rerum natura* itself is found in the *Somnium Scipionis* which concluded Cicero's dialogue; and here again, degrees of plausibility vary. The mere presence of lexical items or unusual usages common to the *Somnium* and *De rerum natura* says very little: most of them (e.g. the use of *templum* in its archaic sense to mean a region of the sky or the poeticism *Grai* for *Graeci*) can be explained as the result of the archaic and poetic color of the *Somnium*: such language may recall Ennius rather than Lucretius.[4] Even the rare use of *anfractus* to refer to the orbit of the sun may be a technical term taken from a source no longer extant.[5] There are, in fact, only two passages that supply unmistakable verbal reminiscences of *De rerum natura*, and in both cases more than

[3] Andreoni argues (with far more detail than Maslowski) that the argument of *De re publica* is directed against Epicureanism, and goes so far as to suggest that the debate on justice in Book 3 is a covert attack on Epicureanism, with Carneades as a displaced Epicurus; the argument is very unconvincing. She also argues that the account of the origins of society in the introduction to Book 3 and the definition of the *res publica* at 1.39 are intended specifically as responses to the account of civilization in Lucretius Book 5. That they are opposed goes without saying, but the ideological divide goes back to the early Peripatos and Epicurus himself, and I am not convinced by any of the links Andreoni sees between Cicero and Epicureanism, far less between Cicero and Lucretius in specific. The reference to *natura nouerca* (Andreoni 294-96) at *Rep*. 3.1 is striking, and is the first instance of the image in Latin, but it corresponds to nothing in Lucretius (cf. Heck 150-51), and the similarities in the anthropologies of the two are for the most part commonplaces. The famous definition of the *res publica* at 1.39 is not specifically anti-Epicurean (cf. Zetzel *ad loc*.), nor is it easy to see why *soliuagus* in that passage must allude (Andreoni 316-17) to *uolgiuago* at Lucretius 5.932.

[4] For *templum* cf. 6.15.2, 17.1; for *Grai*, cf. 6.16.1. Pucci 72-73 is rightly skeptical.

[5] Accepted as a Lucretian echo by Pucci 73 together with a few others not susceptible of proof.

verbal reminiscence is involved. When Scipio describes himself as *a pueritia uestigiis ingressus patris et tuis* (6.26.1), it is hard not to think of Lucretius' reference to his own adherence to Epicurus, "cuius ego ingressus uestigia" (5.55):[6] not only is the language similar, but Scipio's declaration of his adherence to his ancestors and the *mos maiorum* is in pointed contrast to Lucretius' allegiance to a Greek philosopher whom Cicero despises. When Cicero describes the ultimate abode of dedicated statesmen with the phrase *certum esse in caelo definitum locum, ubi beati aeuo sempiterno fruantur* (6.13.1), it recalls Lucretius' description of the eternal calm of the Epicurean gods, *omnis enim per se diuum natura necessest | immortali aeuo summa cum pace fruatur* (1.44-45).[7] Again, verbal parallels lend strength to the contrasting beliefs: Cicero's statesmen have immortal souls with immortal blessings, while in Lucretius such bliss is reserved for the non-doing Epicurean divinities. Both these allusions refer to prefatory and programmatic statements; both criticize Lucretius' Epicureanism. The *Somnium* as a whole is very much, as Jacques Fontaine saw, an "anti-Lucrèce" in philosophical terms: it presents a Platonic universe and afterlife, with immortal souls; it describes a spherical cosmos centered on the earth; it envisages a cosmic concern for the affairs of humans; and it provides an eschatological justification for political activity. All these ideas are completely incompatible with Epicureanism; and given the opposition of the two texts, it is reasonable to believe that when, in the final sentence of the *Somnium*, Cicero condemns to age-long wandering the souls of those who *impulsu ... libidinum uoluptatibus oboedientium deorum et hominum iura uiolauerunt* (6.29.2), he is attacking the Epicurean, and perhaps specifically Lucretian, goal of *uoluptas*.

The interpretation of these allusions to *De rerum natura* in the *Somnium*, however, is problematic: are they to be seen purely as criticism of the Epicureanism of Lucretius, or are they in some way concerned with more literary issues? What is striking about the verbal echoes examined above is not that Cicero rejects Lucretius' philosophy, but that he does

---

[6] Cf. Pucci 73, Maslowski 75. The parallel is closer than to Lucr. 3.3-4 *inque tuis nunc | ficta pedum pono pressis uestigia signis*. The construction at Lucr. 5.55 (*ingredior* + accusative) is less frequent than Cicero's use of the dative.

[7] Cf. Pucci 73-74, Maslowski 74-75, citing Lucretius' repetition of the lines at 2.646-47. I accept the authenticity of 1.44-49, but the allusion does not depend on it.

so by employing the same kind of allusive technique that characterizes the Alexandrian poets and their Roman followers. To express his objections to *De rerum natura* as philosophy, Cicero displays his knowledge of *De rerum natura* as text.[8]

That is also true, in a far more complicated fashion, of one final reference to Lucretius in the *Somnium*, one that does not involve a verbal echo. In introducing his dream, Scipio describes a late evening spent in conversation with the aged Numidian king Masinissa, in which Masinissa discoursed at length about his association with the elder Africanus, the conqueror of Hannibal. After that, Scipio says, he went to bed and fell into the deep sleep in which the vision of Africanus appeared to him; the narration of the *Somnium* itself follows. The sentence in which Scipio introduces his dream is interrupted by a very long parenthesis explaining it. The vision of Africanus, he supposes, came to him because our waking thoughts and conversations produce dreams about the same subjects: *fit enim fere ut cogitationes sermonesque nostri pariant aliquid in somno tale, quale de Homero scribit Ennius, de quo uidelicet saepissime uigilans solebat cogitare et loqui* (6.10.3).[9]

Ennius reported his dream in the opening of the first book of the *Annales*; he said that Homer appeared to him, explaining the Pythagorean theory of the transmigration of souls and his own psychic identity with Ennius himself (and an intervening peacock).[10] Ennius' dream was reported in turn by Lucretius, in Book 1 of *De rerum natura*: in summarizing the various theories of the nature of the soul, he referred to Ennius for the idea that the soul passes from humans into other

---

[8] Although Fontaine's argument that the influence of Lucretius pressed Cicero to become a moral reformer and "le directeur de conscience des 'optimates' " (1728) in the *Somnium* is unconvincing, I am greatly indebted to his discussion of the larger correspondences between the two works. He rightly identifies (1725) Cicero's rejection of excessive desire for *gloria* in the *Somnium* as closely related to Lucretius 3.59-63, but it is also implicit in Laelius' speech on justice at 3.40. For the later development of the attack on *gloria* in *De officiis*, cf. A.A. Long, "Cicero's politics in *De officiis*," in A. Laks and M. Schofield, edd., *Justice and Generosity* (Cambridge 1995) 224-33.

[9] Cf. Pucci 73-75.

[10] For the interpretation of Ennius' dream, cf. O. Skutsch, *The Annals of Quintus Ennius* (Oxford 1985) 147-50 citing earlier bibliography, and the commentary on fragments ii-x of Book 1; see also now S. Goldberg, *Epic in Republican Rome* (Oxford 1995) 89.

creatures; and after praising Ennius for his poetry, he described the dream (1.120-126):

> Etsi praeterea tamen esse Acherusia templa
> Ennius aeternis exponit uersibus edens,
> quo neque permaneant animae neque corpora nostra
> sed quaedam simulacra modis pallentia miris.
> unde sibi exortam semper florentis Homeri
> commemorat speciem lacrimas ecfundere salsas
> coepisse et rerum naturam expandere dictis.

Lucretius obviously disagrees with the Pythagorean theory expounded by Ennius, and he obliquely criticizes the dream itself: if the soul passes from one creature into another, then there is no place for an underworld; and if Homer's soul is now in fact Ennius' own soul, then what appeared to Ennius in his dream was not Homer's soul, but merely some *simulacrum* or *species*. In other words, what Ennius saw was just an Epicurean *eidolon* without underlying reality, and hence its report of the transmigration of souls can have no real authority: the dream is self-refuting.[11] Cicero, in referring to Ennius' dream, says nothing about its content; he merely points out the plausibility of his having dreamt about Homer—and in so doing, he may in fact echo Lucretius' own statement in Book 4 that dreams correspond to our waking concerns.[12] At the same time, however, while Cicero never says so explicitly, the subject of Scipio's dream (the soul and the afterlife) is precisely parallel to Ennius' own dream—and perhaps just as irrelevant to the kind of waking thoughts that provoked it: what Masinissa said about Scipio will have had as little to do with the nature of the soul as Ennius' own diurnal study of Homer did with the Pythagorean peacock. Furthermore, it is worth noting that just as Lucretius describes Ennius' dream as concerning *rerum naturam*—the subject of Lucretius' poem, not Ennius'—so too Scipio's dream echoes, and forms a suitable coda to, the ideals of public service expressed by Cicero in the preface to *De re publica*. One cannot be absolutely certain, but it seems highly probable that Cicero's

[11] For the problem, cf. Bailey on Lucretius 1.120; for a full interpretation of Lucretius' allusion to Ennius, cf. E. J. Kenney, "Doctus Lucretius," *Mnem.* 23 (1970) 373-80. See also Penwill 79-80.

[12] Cf. Lucretius 4.962-70. The same idea appears earlier in a fragment of Accius; see Bailey *ad loc.*

reference to Ennius' dream at the opening of the *Somnium* is intended to recall at once the passage of the *Annales* and Lucretius' use of it in *De rerum natura*.

Once again, moreover, the allusion to Lucretius involves both literary and philosophical considerations. In the first place, although neither the form of *De re publica* nor the content of the *Somnium* is the same as its Ennian equivalent, they correspond much more closely than they do to the intervening Lucretian text: both Ennius and Cicero believe in an afterlife of the soul, and (Ennius at the beginning of his work, Cicero at the close of his) they use that belief to justify the larger goals of the work itself, in one case the writing of heroic epic, in the other active participation in public life. What is more, the subject of both Cicero and Ennius is Rome itself. Lucretius, in contrast, attempts to co-opt Ennian epic to a philosophical purpose and a devaluation of public service quite alien both to Ennius himself and to Cicero. Secondly, there is a more explicit response to Epicureanism in Cicero's construction of the *Somnium*: by offering a rationalistic explanation of Scipio's dream, making it the product of his waking concerns rather than a supernatural revelation of the afterlife, Cicero transformed his Platonic model, the Myth of Er which concludes the *Republic*, in the light of Epicurean criticisms of Plato—criticisms which, it appears, Scipio expressly mentions within the dramatic dialogue of *De re publica*.[13]

According to Macrobius (*Comm.* 1.2.1-4 = *Rep.* 6.7), Epicurus' pupil Colotes had attacked the Myth of Er not merely on philosophical grounds, but because the idea of a man's dying and coming back to life after twelve days to report his post-mortem experiences was ludicrously improbable. Cicero, acccording to Augustine (*C.D.* 22.28 = *Rep.* 6.4), explicitly stated that the implausibility of Plato's tale showed that he was not entirely serious, *lusisse*, in having the revived corpse tell a tale that so closely matched Plato's own ideas. In *De re publica*, Cicero both in his own voice and through his characters distances himself from Plato's political theories;[14] in the introduction to the *Somnium*, he apparently went further and had Scipio refer both to the Myth of Er itself and to the criticisms levelled against it by Colotes, in order to show (according to

---

[13] For this, cf. Zetzel 14, 223-4.

[14] See below, p. 246.

Favonius Eulogius) "that the things reported about the immortality of the soul and about the heavens were not the fictions of dreaming philosophers nor the unbelievable tales laughed at by the Epicureans, but were the speculations of wise men (*prudentium coniecturas*)."[15] In replacing a revelation the source of which was the object of Epicurean scorn by the plausible fiction of a statesman's dream, Cicero both eliminated the mystical dogmatism which neither he (as a skeptic) nor the Epicureans (as materialists) could comfortably accept as genuine revelation, and simultaneously gave to his account of the afterlife the *auctoritas* that should be accorded to the rest of the ideas that he attributes throughout *De re publica* to the beliefs of one of the greatest Roman statesmen of the preceding century.

In constructing the *Somnium*, therefore, Cicero dealt with Lucretius and Epicureanism on several levels. Through allusion and through the eschatology of the dream, he rebutted Lucretius' exposition of materialist psychology and by extension the entire argument of *De rerum natura*. Through the careful setting and the reply to Colotes, he gave his dream a verisimilitude the lack of which the Epicurean had rightly criticized in the Myth of Er. Through the reference to Ennius' dream, he replied to Lucretius' distortion and criticism of Ennius and placed the *Somnium* squarely within the grand tradition of Roman literature. What seems most significant in his treatment of Epicurus and Lucretius in the *Somnium*, however, is not the fact of his reply to the philosophy but the literary quality of the exchange. Cicero draws attention to the careful and artistic construction of his own work; he replies to Colotes' rationalistic observations about the implausibility of Plato's fictional Er; and he elevates Ennius over Lucretius just as, it might be suggested, in his political analysis he elevates the early customs of Rome over recent and disastrous innovations. In employing such an oblique and allusive technique to place himself in the Roman literary tradition, not only is Cicero responding to Lucretius' allusion to Ennius, he is perhaps also suggesting that the relationship of his own work to its Platonic model is parallel

[15] Favonius 1.5 = *Rep.* 6.3, accepting Ziegler's insertion of *nec* before *somniantium*. This is rejected by Heck 64 on the grounds that the dismissal of 'dreaming philosophers' is not possible in the introduction to the Dream of Scipio; but Cicero's point is to distinguish between implausible dreams and Scipio's philosophically based vision. Ziegler prints the quotation given above as Cicero's actual words, but that is unlikely: cf. Heck.

to that of Ennius to Homer—a relationship that Ennius had himself announced through the oblique and allusive use of Callimachus' famous dream in the *Aetia*.

To concentrate on the transformations of Ennius' dream in Lucretius and Cicero, however, gives a somewhat one-sided perspective: it characterizes the relationship between *De re publica* and *De rerum natura* as a purely literary debate, as if—in the manner of the erudite polemics of the Alexandrians—it were intended merely to provoke the admiration of a learned audience. But that is precisely what is not true of *De re publica*, and probably not of *De rerum natura* either: both texts claim to be protreptic, one towards active political participation, the other towards belief in the tenets of Epicureanism. Cicero had a well-documented and consistent aversion to Epicureanism, and he attacks it fairly explicitly in the introduction to *De re publica* as well as in the *Somnium* itself. As noted above, the final sentence of the *Somnium*, attacking those who have surrendered themselves to *uoluptas*, is both a rebuttal of Epicureanism and an echo of his own introduction to *De re publica*. It is a mistake, however, go to the other extreme and understand the introduction, and by extension the entire dialogue, as a single-minded refutation of Epicureanism and its implications for political life.[16] Cicero's general goal, to expound and promote an ideology of political participation and public service, has not one opponent but two. One is Epicureanism, a philosophy that believed in pleasure as the highest good, and which declared that one should not take part in public life unless necessary; the other is philosophy in general, and in particular the belief in the primacy of the philosophical life over the active life. And of these two, the latter is clearly dominant.

Disentangling the arguments of the introduction to *De re publica* is not easy: aside from a few fragments, the opening (slightly more than half of the whole introduction) is lost.[17] Working backward from the final paragraph, however, the basic themes are clear. In the first place, as noted above, Cicero describes it as protreptic: *Haec plurimis a me*

---

[16] So Maslowski, esp. 75-78.

[17] For various reconstructions of the preface, see E. Bréguet, ed., *Cicéron: la république* (Paris 1980) 1: 21-25, with bibliography; also K. Büchner, *M. Tullius Cicero de re publica* (Heidelberg 1984) 69-76 and Zetzel 95.

uerbis dicta sunt ob eam causam, quod his libris erat instituta et suscepta mihi de re publica disputatio; quae ne frustra haberetur, dubitationem ad rem publicam adeundi in primis debui tollere (1.12.1). The preceding argument in favor of taking part in public life was divided into a positive section—making the case for political activity—which is largely lost and ends at 1.3, and a negative section refuting those who discourage it (1.4-11). The positive arguments derive from both obligation and—in a very extended sense—utility: we are obligated to serve the state because of our debt to the *patria* as an *antiquior parens quam is qui creauit* (fr. 1a Ziegler); it is in our interest to serve the state for two reasons, because it is only through the activity of its citizens that a state can survive and be successful (fr. 1f, 1.1)—something that in turn assures the well-being of the citizens themselves—and because it is only through activity, of which political activity is the highest form, that individual *uirtus* can be perfected: *uirtus in usu sui tota posita est; usus autem eius est maximus ciuitatis gubernatio* (1.2.1). The refutation of arguments against public service is very similar: to those who cite the dangers of politics and the ingratitude of the people, using both traditional examples of Athenian fickleness and Roman instances culminating in Cicero's own exile (1.4-7),[18] he responds by reiterating the quasi-filial service owed to the *patria* which brought us up (1.8); to the objection that the philosopher (*sapiens*) should not participate in public life because of the need to struggle *cum impuris atque immanibus aduersariis* (1.9.2), he replies that the best argument for participation is to preserve the state from the rule of *improbi* (1.9.3); and against those who say that the *sapiens* should not participate in public life unless compelled by necessity (1.10.1) he adduces his own career: how could he have defeated Catiline if he had not been consul, and how could he have become consul without having spent his whole life in the public service? (1.10.2-3) More generally, he offers the practical argument that it is foolish to claim to rule in a crisis without knowing how to do so in normal times; and if it is admitted that the *sapiens* will engage in public affairs if necessary, then that admission itself entails acquiring the knowledge and experience to make such an engagement both possible and useful (1.11).

[18] Maslowski 63-65 argues that the unnamed opponents here are specifically Epicureans, but in context they are a vague category of admirers of Cicero, probably imaginary; cf. Zetzel on 1.4.1.

Some of the arguments made in the introduction are very clearly directed against the Epicureans in particular. When Cicero cites the *exceptio* that the *sapiens* will not take part in public life *extra quam si eum tempus et necessitas coegerit* (1.10.1), he clearly has Epicurus in mind,[19] and that is true in several earlier passages as well. In the phrase *a qua isti auocant* cited by Arusianus Messius (fr. 1d Ziegler = Arus. 457.14K), the relative must refer to *res publica* or *patria* and the derogatory *isti* to the Epicureans.[20] In the first paragraph of the extant text, Cicero first refers to Epicurean criticisms of the elder Cato for his lifelong participation in public affairs,[21] and then contrasts the natural necessity of virtue and public participation with Epicurean love of *uoluptas* and *otium*: *unum hoc definio, tantam esse necessitatem uirtutis generi hominum a natura tantumque amorem ad communem salutem defendendam datum, ut ea uis omnia blandimenta uoluptatis otique uicerit* (1.1.5). The use of *uoluptas* makes the object of Cicero's attack quite evident, and it is likely that most of the pejorative uses of *otium* in the introduction are also directed against Epicurean ideals.[22]

Nor are these the only references to Epicureanism. The list of philosophic ideas that in fact derive from lawgivers is qualified by *quod quidem recte honesteque dicatur* (1.2.2)—an oblique attack on Epicurean utilitarianism and legal relativism.[23] At the end of the exposition of his positive arguments in favor of civic participation, moreover, he

---

[19] Cf. Seneca, *De otio* 3.2: "Epicurus ait: non accedet ad rem publicam sapiens, nisi si quid interuenerit."

[20] Cf. Heck 219. *isti* in the preface can refer either to Epicureans (as here and at 1.1.3, cited next n.) or to philosophers in general; on the problem, cf. Andreoni 283 n.11 with bibliography.

[21] *Rep*. 1.1.3: "sed homo demens ut isti putant, cum cogeret eum necessitas nulla, in his undis et tempestatibus ad summam senectutem maluit iactari, quam in illa tranquillitate atque otio iucundissime uiuere." Here *necessitas* presumably refers to the Epicurean *exceptio*, the derogatory *isti* refers to the Epicureans as in fr. 1d, the contrast between calm and storm is characteristic Epicurean language, and the use of *otium* is also a derogatory reference to Epicureanism (see next n.). On this passage, see Andreoni 284 n.13.

[22] On Cicero's use of *otium* in the introduction, see Zetzel on 1.1.2. At 1.8, "tutum perfugium ... et tranquillum ad quietem locum" is also presumably directed against Epicurean attitudes.

[23] For a recent discussion of Epicurean ideas of law, cf. A. Alberti, "The Epicurean theory of law and justice," in Laks and Schofield, edd. (above, n. 8) 161-90.

offers a ringing defense of his own beliefs, specifically directed against the Epicureans: *et quoniam maxime rapimur ad opes augendas generis humani, studemusque nostris consiliis et laboribus tutiorem et opulentiorem uitam hominum reddere, et ad hanc uoluptatem ipsius naturae stimulis incitamur, teneamus eum cursum qui semper fuit optimi cuiusque, neque ea signa audiamus quae receptui canunt, ut eos etiam reuocent qui iam processerint* (1.3.5). Here Cicero produces his own form of *uoluptas* to counter the Epicureans, and his own idea of nature diametrically opposed to theirs: civic responsibility and community feeling are both natural and valuable.[24] At the very end of the preface, after rejecting in his final negative argument the *exceptio* of the Epicureans, Cicero gives a clear statement of the beliefs which animate his vision of political life: *neque enim est ulla res in qua propius ad deorum numen uirtus accedat humana, quam ciuitates aut condere nouas aut conseruare iam conditas* (1.12.4). The importance of *uirtus* in the context of political activity is combined with his sense of the divine order which inspires and informs it; it is this conjunction which is celebrated and indeed immortalized in the *Somnium.*

It is obvious, at the very least, that Cicero's political, ethical, and indeed cosmological views are antithetical to those of Epicureanism, and that there is a very good reason for him to offer polemical comments specifically addressed to the Epicureans. Cicero views civic responsibility and the very existence of states as something natural, based on humans' innate sense of community and on shared values of law and justice; he views virtue as the goal of human existence; he accepts a quasi-Stoic view of natural law and cosmic order combined with a Platonic belief in the immortality of the soul.[25] For an Epicurean, the soul is material and mortal; the origin of the state is purely utilitarian, and justice is not a metaphysical constant but changes with time and circumstance; the wise man will avoid taking part in public life unless compelled to do so; and, of course, the goal of human existence is not virtue, but happiness derived from static pleasure. It is no wonder that, in the debate on justice

---

[24] Cf. Zetzel *ad loc.*

[25] On Cicero's variations between Stoic and Platonic language (particularly on the question of whether the supreme divinity is the author of, or identical with, natural law), see now J.-L. Ferrary, "Statesman and law in Cicero's political philosophy," in Laks and Schofield, edd. (above, n. 8) 67-68.

which occupies the bulk of Book 3 of *De re publica*—unfortunately now very fragmentary—the Epicurean view that moral behavior is based solely on the fear of punishment is regarded in Philus' speech against the possibility of justice as being closest to the truth, while in Laelius' defense of justice the Epicureans occupy the lowest possible position, confusing human beings with animals.[26] The entire vision of the cosmos presented in the *Somnium* is diametrically opposed to Epicurean beliefs: not only is the soul immortal, but the divinity who rules the geocentric cosmos takes a direct and active interest in the affairs of mankind and honors with immortal life those whose actions, primarily in public affairs, most closely approximate the divine justice that animates the universe itself.

And yet, on closer inspection, there is little reason to believe that the dominant impulse behind the composition of *De re publica* was the refutation of Epicureanism, far less the refutation of *De rerum natura*, as has been suggested. Cicero was hostile to Epicureanism before he read Lucretius, and the strength of his feeling in *Pro Sestio* and *In Pisonem* should not be underestimated, even though it may have been provoked more by his hatred for the Epicurean Piso than by the doctrine itself.[27] In *De oratore*, indeed, he treats Epicureanism more with benign contempt than with active hostility, but in the context of rhetoric there was little reason to devote much energy to Epicurean beliefs: they were not orators, they knew it, and Cicero could easily dismiss them as irrelevant to his argument. In discussing the value of various philosophical schools to the aspiring orator in *De oratore* Book 3, he deals with Epicureanism very summarily, and rightly so: it was completely useless, and had no interest in rhetoric.[28]

In *De re publica*, on the other hand, there was good reason to pay attention to Epicureanism, as the one school that totally and completely rejected political involvement and public duty; indeed, for the purposes of his argument in the preface, one might suggest that if Epicureanism had not existed, then Cicero would have been compelled to invent it.

[26] For a reconstruction of the debate on justice, cf. J.E.G. Zetzel, "Natural law and poetic justice," *CP* 91 (1996) 297-319.

[27] Maslowski 68-71. The most important passages are *Sest.* 23 and 138.

[28] *De orat.* 3.63; cf. Maslowski 72-73.

Whether for the sake of his argument or because of his sympathy for the philosophical position of Antiochus of Ascalon, Cicero was inclined to blur the distinctions among the ethical theories of the early Academy, the Peripatos, and the Stoics; hence the only philosophical position that was completely and genuinely antithetical to his own was that of Epicurus. If only in order to clarify *e contrario* his own theories of political life, Cicero needed the handy straw man supplied by Epicureanism.[29] Although Cicero was obviously impressed by Lucretius as a poet, it seems unreasonable to believe that Cicero saw *De rerum natura* "as a threat to the very existence of the state."[30] What he saw, one might suggest, was a great poet embracing a philosophy that took to an extreme the general failing of all philosophers: thinking that (to adapt the aphorism of a later writer) it was enough to interpret the world without taking part in it.

It is, then, the conflict between political theory and political action, between the speculative and the active lives, that is the principal focus of the introduction. *De re publica*, from the outset, is emphatically not merely a work of political theory, designed for the armchair philosopher: in the lost portion of the introduction, Cicero quoted Lucilius (an unfortunately corrupt fragment) to define his audience as not the most learned, but merely the average intelligent reader; the elder Pliny, in citing this passage, calls Cicero's attitude *publica ... eruditorum reiectio* (*N.H. praef.* 7 = fr. 1c). And although Cicero announced that he was following Plato as a model,[31] he was also firm in his polite contempt for mere theorists: *profecto omnis istorum disputatio, quamquam uberrimos fontes uirtutis et scientiae continet, tamen conlata cum eorum actis perfectisque rebus uereor ne non tantum uideatur utilitatis adtulisse negotiis*

[29] Fontaine's argument (see above, n.8) that the highly moralistic criticism of Roman politics in Lucretius underlies Cicero's attempt in *De re publica* to give a more ethical basis to public life is appealing, but ultimately unconvincing: a similarly ethical view of oratory pervades *De oratore*, which was written before Cicero had read *De rerum natura*.

[30] Maslowski 76, cf. J.D. Minyard, *Lucretius and the Late Republic* (Leiden 1985) 75-76. Although they overestimate Cicero's fear, they are right to emphasize that it was the quality of Lucretius' poetry that made it significant to Cicero (even if I do not believe that it seemed dangerous to him). For a useful analysis of Lucretius' political ideas with copious bibliography, cf. D.P. Fowler, "Lucretius and Politics," in M. Griffin and J. Barnes, edd., *Philosophia Togata* (Oxford 1989) 120-50; see also Fontaine 1715-16.

[31] Pliny, *N.H. praef.* 22 = fr. 1b: *de re publica Platonis se comitem profitetur.*

hominum, quantam oblectationem otiis (Lact. *Inst.* 3.16.5 = fr. 1e).[32] A similar attitude is evident in his discussion of active *uirtus*: as quoted above, he describes its greatest use as the governance of states, and continues ... *et earum ipsarum rerum quas isti in angulis personant reapse non oratione perfectio* (1.2.1).[33] In both passages, *isti* is philosophers in general, not Epicureans in particular, and there is a contrast between accomplishment, *perfectio*, and empty talk, no matter how intelligent. He goes on to cite the Platonic philosopher Xenophanes, who said, when asked what in fact he taught his pupils, that he taught them to do willingly what the laws compelled them to do (1.3.1); nothing that a philosopher can write is preferable to the well-ordered state created by the lawgiver (1.3.3-4). So too at the end of the introduction: after describing his desire to move people to take part in public life, and after allowing that philosophers, even those who play no active political role, do perform some service in writing about political philosophy, he reaffirms his belief that the true *sapientes* are men like the Seven Sages, *quos ... omnes paene uideo in media re publica esse uersatos* (1.12.3).

In his admiration for the Seven Sages as practical statesmen and philosophers *avant la lettre*, Cicero repeats what he had argued at the beginning of the excursus on philosophy at *De oratore* 3.56 and follows the views of the Peripatetic Dicaearchus.[34] In fact, the contrast between the active and the contemplative lives was a source of dissension among the early Peripatetics, and the debate on the subject between Theophrastus and Dicaearchus was well known to Cicero and pervades the preface to *De re publica*;[35] it is also parallel to another aspect of *De oratore*, the contrast between the value of technical training (*ars*) and practical experience (*usus*) for the orator.[36] More important, it is central to both the composition and the dramatic setting of *De re publica* itself: there is an obvious tension in the preface and in the dialogue between what the

[32] The fragment clearly belongs to *De re publica* rather than the *Hortensius*; for its location, cf. Bréguet *ad loc.* against Heck 76-77.

[33] For the Platonic allusion here (and in the parallel passage at *De oratore* 1.57) cf. Zetzel *ad loc.* and id., "Looking Backward," *Pegasus (Journal of the Exeter University Classics Society)* 37 (1994) 25-29 .

[34] Cf. Zetzel on 1.12.3.

[35] Cf. Zetzel on 1.2.1.

[36] Cf. (e.g.) *De orat.* 1.5, 1.107-121.

book is—a philosophical discussion of constitutions and citizenship—and what it advocates—action rather than writing. Cicero explains his book by describing it as a practical contribution to the public good based on the combination of his considerable experience with his wide philosophical reading; it is clear that he would rather have been an active participant in the affairs of state, but under the political circumstances that was impossible. The conversation which he creates in *De re publica* among Scipio Aemilianus and his friends is similarly the embodiment of informed engagement and practical philosophy at once. Their lives and their words represent the synthesis of the Peripatetic contrast between the active and the speculative lives: practical politics is ennobled by philosophical training, and philosophical speculation is based on a lifetime of practical experience.[37]

It is for this reason that the attack on Epicureanism in the introduction and elsewhere in the dialogue is not in itself of overriding importance. Cicero assumes the truth of Aristotelian ideas of active *uirtus*, and of the primacy of civic life; Epicurean attitudes, while useful as a focus of attack, were so absurd that their falseness could simply be assumed.[38] The role of philosophy—at least of properly moral philosophy in the Platonic-Aristotelian tradition—was more problematic, and is central to the entire construction of *De re publica*: philosophical analysis of government provides necessary distance and insight, but to act is better than to write. For that reason, although Cicero announces his debt to Plato, translates from the *Republic*, and adapts the Myth of Er, he frequently contrasts his own practical discussion with the abstract and impractical theories of Plato, and attacks Plato's imaginary state as *praeclaram ... sed a uita hominum abhorrentem et a moribus* (2.21.3).[39] It is in that context that Cicero's response to Lucretius needs to be seen, and that context which explains the literary quality of his allusions to *De re-*

[37] For this interpretation, cf. Zetzel 15-17.

[38] Note, for instance, Cicero's incredulity when the lawyer Trebatius Testa becomes an Epicurean in 53: cf. *Fam.* 7.12. There were indeed a number of prominent Epicureans in Rome at this time, including Cicero's friend Atticus who generally observed the injunction against political activity. The most prominent, however, was Julius Caesar, who clearly did not. On conversions to Epicureanism and Epicurean political activity—mostly in the 40s—cf. A.D. Momigliano, *Secondo contributo alla storia degli studi classici* (Rome 1960) 379-88.

[39] Cf. also 2.3.2, 2.51.1.

rum natura: it is not so much the philosophy itself that bothered him—no sensible Roman, in Cicero's view, could accept that—but the book that contained it. For Cicero, Lucretius' poem was a misuse of the Ennian style in the service of a vile philosophy, a brilliant piece of writing the form of which was debased by its content.

One further consideration may support this reading. Cicero began to write *De re publica* as a dialogue in nine books; after various second thoughts about its dramatic setting in October of 54, he reshaped it into six books. The structure of the dialogue as Cicero revised it (and as we have it) is remarkably similar to that of *De rerum natura*: each work has six books; in each, the books are in pairs. In *De rerum natura*, the three pairs deal respectively with atomic motion, the constitution and nature of the human *animus*, and the larger structures of the world and cosmos, moving from the smallest constituents of nature to the largest.[40] In *De re publica*, they proceed from the administrative and constitutional order of the state, to the institutions (law and education) which shape it, to the role of the individual citizen and statesman. And while in one sense the progression in *De re publica* is from large to small, from the state down to the human beings who constitute it, there is also a progression in the opposite direction, from the administrative workings of government to the natural law which governs states and individuals alike to the cosmic rewards that await the true statesman.[41] If the allusion to Ennius' dream is a response to Lucretius' allusion to the same passage, then the entire organization of *De re publica* is meant to evoke the similar organization of *De rerum natura*, and the sublime Platonic vision of the cosmos in the *Somnium* answers the horrific vision of the plague with which *De rerum natura* concludes.

Recent scholarship has rightly emphasized the connections between *De rerum natura* and the epic tradition;[42] not only do its cosmic concerns align it with the grand scope of hexameter epic, but its style (together with the specific reference in the dream) specifically recalls Ennius. So

---

[40] Cf., for example, Bailey 1: 31-32.

[41] For the structure of *De re publica*, cf. Zetzel 16-17, 24; Ferrary (above, n.25) 49-51. There are clearly other important structural patterns in both works, but the division into pairs of books is dominant in both.

[42] See particularly M.R. Gale, *Myth and Poetry in Lucretius* (Cambridge 1994) 99-128; also Penwill 76-77.

too does Cicero, and not only in the archaic solemnity of the *Somnium*. *De re publica* is of course in prose; but through numerous allusions and through its very subject matter, Cicero places himself squarely in the Ennian tradition. The tightly organized structure of *De re publica* recalls *De rerum natura*, and its style and content—even though Cicero's view of Rome's history is far more tragic than that of Ennius[43]—draw attention to the model which they have in common. In *De re publica*, however, form matches content: it displays a unified cosmos in which human effort and political order, however flawed in practice, are justified and rewarded by the order of the universe. The *Somnium* looks back to and explains both the argument in favor of political activity of Cicero's introduction and the opening discussion in Book 1 about the importance of astronomy: abstract learning is given a genuine and profound role in the understanding of public life. At the same time, moreover, it alludes to the opening of Lucretius' poem, drawing attention to the wide divergence of content cloaked in very similar forms.

In acknowledging the genuine and important philosophical differences between Cicero and Lucretius, it is important not to forget the literary framework in which they are displayed. One of the most striking characteristics of the literature of the 50s BCE is the concentrated attempt by Roman authors to assimilate to their own needs and traditions the Greek intellectual heritage which, during the first half of the first century, the Roman elite had come to see as its own, as a result in part of conquest, travel, and the arrival of Greek intellectuals in Rome itself.[44] Wendell Clausen has demonstrated the important role of Parthenius in introducing Roman poets to Callimachean poetics; and many of the same tensions between public and private, between the Ennian inheritance and the new learning, are evident in philosophical writing as well.[45] Cicero adopts the form of Platonic dialogue while dismissing Plato's impractical ideas as useless for Roman government; he adopts the structure of *De rerum natura* in order more obliquely to criticize Lucretius' mis-

[43] Cf. Zetzel (above, n. 33) 29-32.

[44] On the intensification of the Greek cultural presence in Rome in this period, cf. E. Rawson, *Intellectual Life in the late Roman Republic* (Baltimore 1985) 1-17 (and passim).

[45] For the similar importance of Ennius even in Catullus, cf. J.E.G. Zetzel, "Catullus, Ennius and the Poetics of Allusion," *ICS* 8 (1983) 251-66.

take in adapting the manner of Ennius to propound a Greek philosophy antithetical to the public and epic world of the *Annales*, a philosophy at least as uncomprehending of Roman realities as Plato's *Republic*. Cicero deliberately uses the shape of *De rerum natura* and its own Ennian allusions to shape his reversal of Lucretius' hierarchy of Greek and Roman, of public and private. Platonic ideas are subordinated to Roman achievement; Roman values are ennobled with the methods of Greek philosophy. Cicero's deepest concern in *De re publica* was with the well-being of the Roman state; but like Catullus and like Lucretius himself, he was also attentive to the problem of how best to shape a national literature that could match that of Greece. The failure of Lucretius, in Cicero's view, was not merely his acceptance of the wrong philosophy; it was the failure to use Greek culture to understand, and to improve, Rome.

BECOMING 'CICERO'

ROBERT A. KASTER

> Kitsch ... must derive from the basic images people have engraved in their memories: the ungrateful daughter, the neglected father, children running on the grasss... Kitsch causes two tears to flow in quick succession. The first tear says: How nice to see children running on the grass! The second tear says: How nice to be moved, together with all mankind, by children running on the grass! It is the second tear that makes kitsch kitsch... Kitsch is the aesthetic ideal of all politicians and all political parties and movements... Kitsch is a folding screen set up to curtain off death.
>
> Milan Kundera. *The Unbearable Lightness of Being*

> Much later, Caesar came upon one of his grandsons with a book by Cicero in his hands. When the stricken boy tried to hide it in his cloak, the emperor took it from him and read a large part of it, standing just as he was. Then, handing it back, he said: "An eloquent man, my child, an eloquent man, and a patriot."
>
> Plutarch. *Life of Cicero*

My title, "Becoming 'CICERO'," has a double sense, alluding to two interrelated processes. There is, first, the process of the rhetorical schools of Rome and the Romanized provinces, by which students fit themselves to the model of Cicero and sought, in a limited way, to "become" the orator, or at least an orator like him.[1] Second, and the subject that will be at the center of discussion below, there is the process by which Cicero himself became the cultural icon 'CICERO'—all upper case, because the icon is larger than life, and in inverted commas, because the icon bears only a partial and inconstant resemblance to the

[1] On the use of Cicero in the schools of the early empire see esp. Michael Winterbottom, "Cicero and the Silver Age," in *Éloquence et rhétorique chez Cicéron*. Entretiens sur l'antiquité classique, vol. 28 (Geneva 1981) 237-74, with further references. More generally helpful is Will Richter, "Das Cicerobild der römischen Kaiserzeit," in Gerhard Radke, ed., *Cicero: Ein Mensch seiner Zeit* (Berlin 1968) 161-97.

human being whom it represents. How did this process begin? How did Cicero enter the canon?

The question is worth asking, because the outcome was not obviously predictable from the final decade of Cicero's life, when the factions he supported in the state were on the defensive or in retreat and when his personal reputation for eloquence was under attack. A fully rounded answer to the question would open a broad window onto the cultural life of Rome from the triumviral period through the end of the next century, embracing (for example) the Cicero of the historians, the relations between the historians and the rhetorical schools, and the contrast between the rhetorical schools recalled by a Seneca and the very different sensibility of a Quintilian. All this of course goes beyond the scope of a modest paper like the present one, which examines only one corner of the field: the rhetorical schools of the Augustan age and the Cicero constructed by the *scholastici* who frequented them.

We can start, then, with a question asked by an obscure Roman declaimer named Capito some time in the reign of Augustus (Sen. *Contr.* 7. 2. 6): "Ciceronem quisquam occidere potuit qui audiit?"—"Could anyone who heard Cicero speak bear to kill him?" The question, when I first encountered it, stopped me in my tracks because of its sheer and compounded fatuousness: it is fatuous because it is asked with reference to Cicero's killers, who certainly *had* heard him speak; and it is fatuous because it is asked by a man who almost certainly could *not* himself have heard Cicero speak, for reasons of simple chronology. Since the question as a question is quite pointless—taken literally, it very plainly can only be answered "yes"—it seems that the true point of the question must lie elsewhere: it must be taken to serve as a kind of implied contrafactual condition, by which the speaker distinguishes *himself* from the killers—"*I* certainly couldn't have brought myself to kill Cicero," he means to say, "had *I* been in *their* shoes." Capito thus uses the question as a bit of claptrap to place himself on the side of the angels, and on the side of his audience.

Yet an aggressive member of the audience—the sort of heckler, perhaps, who sometimes haunted the declaimers' *auditoria*—might be tempted to ask: "But Capito, why *not* kill Cicero?" What, that is, are the values underlying Capito's sentimental question? Where does it locate the redeeming worth of Cicero's life? The implied answer is clear enough, according to the terms of the question in its context, and it clearly enough reduces Cicero, by equating the value of his life with the aesthetic value of his speech: one is expected to forbear from killing Cicero because he was a damn fine orator. Here Capito's question begins to connect directly with the topic of this essay, the place that Cicero came to occupy in the canon. Cicero's claim to life—his claim to "immortality," as we shall see—rests on his speech, and on the good judgment of the connoisseurs of speech: in asking his question Capito was in effect saying, not just "*I* wouldn't have killed him," but "*I* wouldn't have killed him, because *I* am capable of recognizing immortal genius when I hear it." And this even though *he* has *not* heard it, even though (as it happens) *he* does not much imitate Cicero's speech in his own. All the more one begins to suspect that in asking his question Capito is appropriating a particular current conception of 'CICERO', so that he can wear it like a badge. Or to put it a bit differently: one begins to suspect that we are in the presence of classic kitsch, of self-referential sentiment that serves to reduce experience.

Now the canonization of any classic is inevitably a reductive process, constrained by a classic's very function: to help us form the orderly patterns that are useful for the business of living. Classics are the "exemplary visions [that we invoke] to make authoritative sense of the world, sorting our experience of the particular and the momentary into categories and hierarchies of value … [which are] assumed to be permanent and universal."[2] Consequently, any author or text that is complex and interesting enough in the first place to be thought worthy of playing this authoritative cultural role must be subtly changed, shaped, tidied, and simplified, as one feature is exaggerated here, another pared away there, to stress the traits that are found useful for controlling the uncertainties and instabilities of the moment. For most of the Latin classics the beginnings of this process of simplification and appropriation are for all practical purposes lost from sight: this is true even of those authors who long have had the greatest cultural weight—a Vergil or Horace or Tacitus. In the case of Cicero, however, we have a chance at

[2] Christopher A. Stray and Robert A. Kaster, "Reinterpreting the Classics," *Annals of Scholarship* 10.1 (1993) 1.

least to glimpse the process at work in the generation or two after his death, as his epigoni created the 'CICERO' that suited them in the declamations remembered by Seneca.

Three texts provide the main testimony: all elaborate fictions surrounding Cicero's death. The first is *Controversiae* 7.2, "Popillius Ciceronis Interfector": as the statement of the theme tells us, "Cicero [had] defended Popillius on a charge of parricide; he was acquitted. When Cicero was proscribed, Popillius was sent by Antony to kill him, and he brought back his head to Antony. [Popillius] is accused of misconduct."[3] Beyond the fact that a certain Popillius apparently did kill Cicero, all of this is imaginary: Popillius had not been defended by Cicero on a charge of parricide, nor would he have been liable to a civil action *de moribus* even if all else were true.[4] The theme must already have been in common use during the first two decades of Augustus' reign, since Seneca excerpts the treatment of at least one declaimer who did not survive the end of the first century B.C.E.[5]

The other two texts are both *suasoriae*—exercises in which the speaker does not accuse or defend but seeks to give advice—and both concern Cicero's relation to Antony.[6] In *Suasoria* 6, "Cicero deliberates whether to beg Antony's pardon": the declaimer must urge Cicero one way or the other, and as Seneca reports, "almost no one ventured to exhort Cicero to beg pardon of Antony: they had too high an opinion of Cicero's spirit" (§12). Like *Controversiae* 7.2, this piece has a fictional premise; it too can certainly be dated no later than the first half of Augustus' reign, on the same ground. Seneca also does us the favor here of quoting liberally from treatments of Cicero's death by several contemporary historians, including Livy and Asinius Pollio: as we shall see, these quotations at least suggest that the sensibility at work in the

[3] Translations of Seneca are taken from Michael Winterbottom's Loeb edition (Cambridge, Mass., and London 1974) with occasional minor modifications

[4] Stanley F. Bonner, *Roman Declamation in the Late Republic and Early Empire* (Liverpool 1949) 124.

[5] Porcius Latro (*Contr.* 7. 2. 1, 8-10, 14), whose death is dated to 4 B.C.E. in Jerome's *Chronicle*; on his chronology see further my edition of Suet. *Gramm.* (Oxford 1995) 329-30.

[6] See also Quint. 3. 8. 46, Martial 5. 69.

declamations was not confined to the schools of rhetoric, and that the canonization of Cicero cut across the literary culture of the period.

Finally, *Suasoria* 7 takes Cicero's imaginary dealings with Antony one step further: now he no longer merely deliberates whether to beg Antony's pardon, he deliberates whether to accept Antony's offer of pardon in return for the destruction of his own writings. According to Seneca (*Suas*. 6. 14 f.), this variation on the theme was directly inspired by Asinius Pollio, who alleged that Cicero had promised to deliver speeches denying his criticisms of Antony, were he allowed to live. The allegation was made in the published version of Pollio's speech defending one of the Aelii Lamiae: the date of this trial, and so the *terminus post quem* it would provide, are unknown;[7] and there is some reason to doubt that Seneca is correct in identifying the source of inspiration in the first place, since the similarity is not terribly close.[8] Be that as it may, the *suasoria* surely belongs to the same Augustan milieu as the other two texts.

Now two general features of these texts are especially curious and suggestive, and deserve some preliminary notice here. First, there is the fact that all three texts center on Cicero's death. This emphasis is in fact doubly odd. It is odd because there were surely other episodes in Cicero's long public career that could have served as interesting, challenging raw material for such exercises (one thinks, for example, of the political risks incurred by the defense of Sex. Roscius of Ameria, or the choices faced at the outbreak of civil war in 49 B.C.E.)—to say nothing of the limitless opportunities for creating fictional or semi-fictional

---

[7] Susan Treggiari, "Cicero, Horace, and Mutual Friends," *Phoenix* 27 (1973) 249-51, identifies the Aelius Lamia in question as Cicero's friend L. Aelius Lamia (pr. 42?) and conjectures that his trial occurred in December 43—not a very helpful *terminus* for our purposes in any case.

[8] It must be remembered that Seneca can be unreliable in such matters: cf., e.g., his defective sketch of declamation's history (*Contr*. 1 pr. 12 with Janet Fairweather, *Seneca the Elder* [Cambridge 1981] 104 ff.), or his claim that all teachers of *pulcherrimae disciplinae* before Rubellius Blandus were freedmen (*Contr*. 2 pr. 5 with Suet. *Gramm*. ed. Kaster, p. 292), or his claim that Asinius Pollio was the first Roman to recite his own compositions to an invited audience (*Contr*. 4 pr. 2 with Alexander Dalzell, "Asinius Pollio and the Early History of Public Recitation at Rome," *Hermathena* 86 [1955] 20-28).

scenarios in which 'CICERO' could be made to figure.[9] And it is odd because it is so very different from the case of Demosthenes, the other great orator whose life became fodder for declaimers: when Demosthenes is enlisted as a figure in declamatory themes—as he is dozens of times—it is his career as a forensic orator and statesman that provides the (real or fictional) context, not his death from poison.[10] This double oddity may suggest that something about Cicero's death itself is at some level crucial to the kind of appropriation that is occurring: I will return to this suggestion at the conclusion of this paper.

Then there is the second curious feature: it is only in these themes that the speakers give much evidence of actually having *read* Cicero. That is to say, the speech of the declaimers—in Seneca and elsewhere—is not notably Ciceronian in style, in the way that (say) the hexameter poetry of Statius or Silius Italicus is indelibly Vergilian.[11] Moreover, even where the declaimers give evidence of having read Cicero, through allusion or quotation, the range of texts exploited is quite narrow:

ad Brut.: *Suas.* 6. 1 (alludes to 8. 2?)

ad Fam.: *Suas.* 6. 4 and 6.8 (alludes to 10.1.1)

Cat. 1: *Suas.* 6. 3 (from §2)

Cat. 4: *Contr.* 7. 2. 10 (alludes to §3), *Suas.* 6. 12 (alludes to §3)

Verr.: *Contr.* 7. 2. 1 (alludes to 2 *Verr.* 5. 118), *Suas.* 6. 3 (alludes to 2 *Verr.* 5. 161)

Phil. 1: *Suas.* 6. 4 (alludes to §38)

---

[9] Two late antique handbooks in fact imagine such themes, drawn from earlier stages in Cicero's career: Sulp. Vict. *RLM* ed. Halm p. 314. 16 ("deliberat Cicero an post consulatum eat in provinciam"), Iul. Vict. *RLM* ed. Halm p. 380. 24 ("damnato Milone exclamavit M. Tullius Romam sedem bonis viris esse non posse et reus est laesae rei publicae"); but there is no trace of such themes in the sources of the earlier empire. In this respect the contrast with Demosthenes (next n.) is especially striking.

[10] The themes in which Demosthenes figures, often in rivalry with Aeschines or in opposition to Philipp, are catalogued by Richard Kohl, *De scholasticarum declamationum argumentis ex historia petitis* (Paderborn 1915) 66-82 (nos. 259-328), cf. 63-65 (nos. 245, 248-52, 256-57).

[11] See Winterbottom (above, n.1) 251-54, on both the declaimers recalled by Seneca and the *Declamationes minores* and *maiores* ascribed to Quintilian.

Phil. 2: *Contr.* 7. 2. 5 (quotes §64), 7. 2. 10 (alludes to §119), 7. 6. 13 (alludes to §5?), *Suas.* 6. 3 (alludes to §63 f.), 6. 4 (alludes to §64), 6. 5 (from §67), 6. 7 (alludes to §77), 6. 12 (alludes to §119), 7. 1 (alludes to §24)

post red. in sen.: *Contr.* 7. 2. 5 (alludes to §39?)

Lig.: *Suas.* 6. 13 (general allusion)

Marc.: *Suas.* 6. 4 (alludes to §25?)

Mil.: reading implied by Cestius' "response," *Contr.* 3 pr. 16 (see also Quint. 10. 5. 20); *Suas.* 6. 2 (from §§92, 105), 7. 3 (from §101)

Rosc. Amer.: *Contr.* 7. 2. 3 (quotes §72), *Suas.* 7. 2 (general allusion)

Rab. Perd.: 9. 2. 10 (alludes to §11 f. ?)

loc. ignot.: *Suas.* 7. 5[12]

As this compilation shows, the speeches that are quoted or alluded to more than once make for a very short reading list indeed—*pro Milone, pro Sex. Roscio Amerino,* one of the *Verrines*, the second *Philippic*, the fourth *Catilinarian*—and there are virtually no such echoes beyond *Controversiae* 7. 2 and *Suasoriae* 6-7. I take this to mean that insofar as Cicero the orator and stylist matters to what is going on in these declamations—and that aspect of Cicero is in fact quite central—the *idea* of Cicero the orator and stylist matters more to the declaimers than his oratorical style itself, just as Cicero dead or on the point of death matters more than a Cicero conceived in mid-career as advocate and politician. For what is "going on" is that Cicero is being heroized: he is becoming a culture hero, an icon more important as an abstract representation than the historical reality of the man and the sensible reality of his words. How does this happen, what is the icon made to represent, and why?

To sketch an answer we can look at four aspects of the phenomenon, under the headings "amplification," "competition," "sentimentalization," and "simplification." I should stress that I do not

[12] "cara est cuiquam salus quam aut dare aut eripere potest Antonius?": not found in extant texts, but cf. *Phil.* 2. 5, 60.

suggest that these are *stages* in some sequential process, such that one stage would have to reach completion before the next could begin: I do not know how one would go about demonstrating such a thing even if one believed it; and in fact I am inclined to suspect that all four aspects of the process were going on simultaneously, in ways intricately related and now untraceable. I offer these aspects here as little more than convenient analytical categories, or as gross and visible symptoms of the underlying cultural metabolism.

By "amplification," first, I mean the pronounced tendency to speak of Cicero in terms that draw him larger than life—that place him, in fact, beyond all human proportion. So for example we hear Seneca himself at the very start of his work, explaining that civil war and the general political unrest of the 40s B.C.E. kept him at home in Cordoba as an adolescent and prevented him from pursuing his education at Rome (*Contr.* 1 pr. 11): "otherwise," he says, "I might have got to know that genius (*ingenium*), the only possession of Rome to rival her empire." A certain kitschiness and sentimentality hang about that conceit—a counterfeit nostalgia for something that one never knew—which recalls Capito's question, "Could anyone who heard Cicero speak bear to kill him?" But it is of course the sheer extravagance of the conceit that is truly breathtaking, as Cicero's individual *ingenium* is equated with the value of the vast expanse of empire. The stress on *ingenium* is something to which we will return. For the moment, it is enough to note that this initial amplification of Cicero by Seneca himself is matched subsequently by a no less extravagant conceit favored by the declaimers, who frequently use metaphors of sacredness and immortality in a similar kind of amplification.

For example, in *Suasoria* 6. 4 the speaker assures Cicero, as he considers whether to beg Antony for life, that "if [you have regard to] the memory of your works, you will live for ever," *semper victurus es*. In *Suasoria* 7, which concerns the supposed offer of clemency in return for the destruction of his writings, there is a string of arguments based on the premise of immortality: "Antony has realized that so long as the products of Cicero's eloquence survive, Cicero cannot die" (§2); "that the body, which is frail and fleeting, should be saved, is the genius, which is eternal, to perish?" (§7); "he reprieves you from death, at the cost of the death of the only part of you that is immortal," *quod in te solum inmortale est* (§8). Or again, in *Suasoria* 6, "Memory, undying guardian of human works, through which great men attain to eternal life, will hand you down to all future generations, sacrosanct," *sacratum* (§5): here the language of immortality is complemented by the language of sacredness, which in turn is found again in other passages. So at *Suasoria* 7.3 a quotation from the *pro Milone* is termed *sacra illa vox tua*, "that sacred utterance of yours." And at *Suasoria* 6.19—a passage from the history of Cremutius Cordus that summons up the scene in the Forum as Cicero's remains are exposed on the Rostra—we are told that "what most set men weeping and wailing was the sight of his right hand, tied by the side of his head: the hand that had been the servant of that god-like eloquence," *divinae eloquentiae ministra.*

Such language of course later becomes a commonplace in speaking of Cicero: Quintilian, for example, applies it to his great model several times, placing him thereby in the same company as Plato.[13] Its applications to Cicero by the declaimers in Seneca's text are among the earliest instances of the phenomenon (only a fragment of Cornelius Nepos stands any chance of being earlier);[14] and such language is used of Cicero *alone* by the declaimers in Seneca's text. Plainly he is being spoken of in language appropriate to a hero—a figure specially marked and set apart (the essence of being *sacer*), enduring and of special worth, a figure looked to as an embodiment of some crucial aspect of right order. Equally plainly, if one asks what aspect of right order he embodies to merit such praise, it is the exercise of transcendant *ingenium*, as manifested in his eloquence. It is therefore worth pointing out that when Cicero, in his own voice, uses such language, it is almost always with regard to *political* matters, to denote statesmen and affairs of state that he regards as being of transcendant value: it does not occur at all, for example, in the survey of earlier orators in the *Brutus*.[15] A shift seems to

---

[13] Quint. 1. 6. 18 *quae M. Tullius in Oratore diuine ut omnia exequitur*, 2. 16. 7 *diuina M. Tulli eloquentia*, 4. 1. 70 *diuina illa pro Ligario ironia*, 4. 3. 13 *ille diuinus orator*, 11. 1. 62 *diuine Cicero seruauit*; cf. 10.1.81 (on Plato), 10.1.83 (on Theophrastus, playing on his name after Cic. *Or.* 62).

[14] Winterbottom (above, n. 1) 242 n. 4.

[15] See Cic. *ad Fam.* 2. 15. 1, 5. 12. 1, 10. 9. 2, 10. 14. 2, 10. 28. 1, 11. 6a. 1, 12. 14. 4, 13. 16. 4; *ad Att.* 1. 19. 6, 7. 1. 4, 14. 14. 3, 15. 1. 5; *ad QFr.* 1. 1. 7, 1.

have occurred between Cicero himself and the men who are appropriating him as their cultural icon, a shift from the sphere of public action to the concerns of individual talent.

Turning from "amplification" to "competition" we see first that the figure of a Cicero thus "amplified" serves as a kind of cultural yardstick—not surprisingly, since that is a central function of a canon. Thus Seneca comments on the account of Cicero's death offered by the cross-grained Asinius Pollio (*Suas*. 6. 25): "I can swear to you that there is nothing in his history more eloquent than the passage I have cited, so that I think Pollio on that occasion did not praise Cicero [something Pollio in fact nowhere did] but competed with him." When Seneca here invokes the idea of rivalry—*certasse cum Cicerone videatur*—he reminds us of the service that a canon performs in the conduct of certain kinds of competition: men seeking to establish a pecking order of cultural status through (in this case) the exercise of eloquence use Cicero as the standard against which they measure themselves and others—and, often, as a stick with which they can thrash each other more or less publicly.

The phenomenon is nicely illustrated by an anecdote that Seneca relates in the preface to his third book of *Controversiae* (§§14-16). Seneca here claims to be reporting the comments of Cassius Severus, one of the most noted forensic orators of Augustan Rome, who is trying to explain why he does not much engage in or like declamation.[16] After pointing to several considerations—the fact that no one has equal talent in all areas, that his own talent for practical oratory is not suited to the artificiality of declamation—Severus concludes:

1.3 3; *ad Brut.* 5. 1, 23. 4, 7; *Arch.* 16, *Balb.* 40, *Flacc.* 30, *Har. Resp.* 2, *Leg. Agr.* 2.90, *Leg. Man.* 10, 33, 36, *Marc.* 2, 26, *Mil.* 21, 91, 99, *Mur.* 61, 75, *Phil.* 2. 33, 39, 3. 3, 14, 4. 3, 4, 5. 23, 9. 10, 10. 7, 11, 13. 19, 43, 44, 14. 33, 38, *Planc.* 77, *Post Red. ad Pop.* 1, 7, 15, *Post Red. in Sen.* 1, 3, 25, 28, *Prov. Cons.* 32, 38, *Rab. Post.* 42, *Sest.* 50, 85, 86 (contrast: *Post Red. ad Pop.* 5 *tam divino atque incredibili genere dicendi*; *Post Red. in Sen.* 1 *tam divinum atque incredibile genus orationis*). Note that in *De oratore* Cicero causes his interlocutors to apply the language of "divinity" to eloquence (e.g., 1. 40 *Ser. Galbam ... divinum hominem in dicendo*); but when he uses such language of Crassus, speaking in his own voice (3. 4-6), it is with reference to the *res publica*.

[16] On Cassius Severus see Michael Winterbottom, "Quintilian and the *Vir bonus*," *JRS* 54 (1964) 90-97.

> I should take more pains in my defence ... if I did not know that Asinius Pollio, Messalla Corvinus and Passienus (now our leading orator) are rated as declaimers below Cestius or Latro.[17] Do you think this is the fault of the speakers—or of their hearers? *They* are not worse speakers; the audience is judging by worse standards... They prefer their Cestius to the eloquent men I have just mentioned—and they would prefer him to Cicero if they did not fear a stoning. In fact, they *do* prefer him to Cicero, in the one way open to them: they learn Cestius' declamations by heart while not reading Cicero's speeches—except the ones to which Cestius has written replies.
>
> I recall going into his school when he was going to recite a speech against Milo. Cestius, with his usual admiration for his own works, said: "If I were a gladiator, I should be Fusius. If I were a mime, I should be Bathyllus. If I were a racehorse, I should be Melissio." I couldn't contain my rage. I shouted: "If you were a sewer, you'd be the Great Sewer." Universal roars of laughter. The school-men looked at me to discover who this bull-necked lout was. Cestius, who had taken on himself to reply to Cicero, could find nothing to reply to *me*...

The episode is shaped by conflicting appropriations of the classic. Despite their differences, Cestius and Severus clearly both treat Cicero as CICERO, the upper-case icon, and it is their contrasting uses of this icon that help make their differences visible. For Cestius, Cicero is the object of explicit rivalry, the cultural father who must be displaced. When Cestius recites an *In Milonem* to "respond" to Cicero's *Pro Milone,* he in effect seeks to elevate himself to Cicero's level:[18] the implied syllogism apparently runs "If Cicero is larger than life, and if I reply to him successfully, then I too am larger than life." (Also apparently implied is the minor premise "Since this is my school, I can define what constitutes a 'succesful' reply.") Conversely, in telling this story Severus stresses the supreme standing of Cicero—one who (like him-

---

[17] On Cestius Pius and Porcius Latro, then two of the most distinguished professors of rhetoric at Rome, see Suet. *Gramm.* ed. Kaster, pp. 327-31.

[18] Quint. 10. 5. 20 provides testimony independent of the hostile anecdote.

self) was *not* a mere *scholasticus*—all with the aim of venting his indignant contempt for the *scholastici* and putting them in their place. Can you believe it (he says in effect), those morons actually prefer Cestius to Cicero—they ought to be stoned! Can you believe it (he says in so many words), Cestius presumed to reply to Cicero, but could not even reply to me: Severus does not have to add "What a pipsqueak!" for us to hear that judgment. Here the classic is deployed as part of the tactics in a cultural skirmish: Cestius tries to elevate himself above the level of the common *scholasticus*, and Severus tries to cut Cestius and the other *scholastici* down to size. In this competitive context, the different appropriations of the classic have little to do with the classic itself—Cicero himself is little more than a foil in all this—and a great deal to do with the different agenda of those doing the appropriating.

There is, however, one other use to which Cicero is put that has the effect of blunting competition—not by criticizing competition as such, but by stressing the general cultural trough into which *all* latter-day Lilliputians have fallen. Here Cicero serves as a benchmark to measure the distance between past greatness and the present. So Seneca advises his sons (*Contr*. 1 pr. 6 f.):

> … an imitator never comes up to the level of his model. This is the way it is; the copy always falls short of the reality. Moreover, you can by this means judge how sharply intellectual standards are falling every day, how far some grudge on nature's part has sent eloquence into a decline. Everything that Roman oratory has to set alongside or even above the haughty Greeks reached its peak in Cicero's day: all the geniuses who have brought brilliance to our subject were born then. Since, things have got worse every day. Perhaps this is due to the luxury of the age… perhaps, as this glorious art became less prized, competitiveness transferred itself wholly to sordid businesses that bring great prestige and profit; perhaps it is just Fate, whose grim law is universal and everlasting—things that get to the top sink back to the bottom, faster than they rose.

Note that the reasons given for the decline—the bogeymen of luxury, trivial pursuit of base distinctions, and vague "fate"—are the merest clichés: taken together, they amount to a respectable Roman way of giving a shrug of the shoulders; they are certainly not explicitly political, attributing decline of eloquence to the loss of *libertas*. That would be a subject more difficult to confront than the strawman of luxury.

Relegating greatness to the past is a way to deflect attention from the present, where things can be taken to be as they are because—well, that is just the way they are. And as this kind of distance is established, sentimentality and kitsch rush in to fill the gap between prior greatness and present unease. The classic is enveloped in the glow of nostalgia for a past which those waxing nostalgic could not have experienced for themselves; and this false nostalgia generates large gestures of easy and unearned sentiment.

This sentimentalization, or "kitsch-ification," of Cicero is of course exemplified by the question with which we began, "Could anyone who heard Cicero speak bear to kill him?," which transforms the nasty reality of Cicero's murder into a cozy and self-regarding gesture, as the speaker simultaneously winks at his audience and pats himself on the back. It is exemplified, too, in another passage we have already seen, in which Cremutius Cordus goes over the top in evoking the grisly spectacle on the Rostra: "what most set men weeping and wailing was the sight of his right hand, tied by the side of his head: the hand that had been the servant of that god-like eloquence" (*Suas*. 6. 19). It is exemplified when the declaimer Marcellus Aeserninus, seeking a *color*, casts Antony and Cicero as figures in a historical romance (*Contr*. 7. 2. 10):

> Antony was wondering: "What punishment shall I invent for Cicero? Shall I have him killed? But he has long since fortified his mind against fear of that. He knows [and here Antony is made to allude to *Cat.* 4. 3 and *Phil*. 2. 119] that "death is not premature for an ex-consul or wretched for a wise man." Let us have something new, that he neither expects nor fears. [Here one imagines a dramatic pause.] He does not resent offering his neck for an enemy to sever—but he will resent offering it to a client. Let someone call Popillius, so that Cicero may learn how much profit there was for him in defending the guilty."

And this sentimentalization is exemplified in the melodramatic script that the historian Aufudius Bassus composed for Cicero's last moments (*Suas*. 6. 18): "Cicero drew aside the curtain a little, and seeing the armed men said: '*I* am stopping here; approach, soldier, and if you can do *this* properly, cut off my head.' Then, as the soldier trembled and hesitated: 'What if I had been the first victim to whom you came?'" The script of course is capped with an epigram worthy of—a declaimer.

This aestheticizing of Cicero's murder depends for its effect upon a sharing of easy sentiment, which is in turn made possible only by a simplification of Cicero, a reduction of him to something everyone can agree on, instead of the multifaceted and uneven and awkward person that he was. And what is it that can everyone agree on? That Cicero was the embodiment of verbal *ingenium*, that he *was* his oratory, nothing less and nothing more, in so far as the oratory conveyed the *ingenium* that was his essence. Or as Seneca observed, in commenting on *Suasoria* 7 (§10): "I know of no-one who declaimed the other side in this *suasoria* [the side, that is, that would urge Cicero to burn his writings to save his life]; everybody worried about Cicero's books—no one about Cicero himself."[19]

After reading the declamations, one sees Seneca's point—though one is also tempted to ask what other concern *scholastici* could be expected to have. Certainly the strongest single thread that runs through these declamations is the schoolmen's understanding of Cicero's *ingenium* in exclusively literary terms—his writings and style—and the complete identification of his writings and style with his life. Consider the following, found in *Suasoria* 7 alone:

7.1: Q. Haterius equates Cicero's past life with his writings: "[Antonius] says he gives you your life — having found a way to suppress even the fact that you had lived";

7.2: Cestius Pius terms Cicero's writings his *melior pars*;

7.8: Argentarius says "He reprieves you from death—at the cost of the death of the only part of you that is immortal. What

[19] Seneca goes on to suggest that Cicero, unlike the declaimers, would at least have thought the other side worth considering (*...cum adeo illa pars non sit mala ut Cicero, si haec condicio lata ei fuisset, deliberaturus non fuerit*).

sort of bargain is that? ... He does not want *you* to live—he wants you to outlive your genius... Can you tolerate Cicero burying what is best in him before he is buried himself?";

7.8: the elder Arellius Fuscus offers an even stronger statement of the "classic" Cicero—identifying the "real" Cicero with his writings, deriving his personal "immortality" from those writings, and imagining the eternal effect of those writings—before assuring Cicero that his life is the "cheapest part" of him, *vilissima pars:* "the *real* Cicero"—*verus Cicero*—is constituted by the writings, the Cicero "who Antonius thinks cannot be proscribed save *by* Cicero."

From the perspective of the the first century's end, Quintilian remarked that "'Cicero' is now considered the name, not of a human being, but of eloquence," *non hominis sed eloquentiae* (1. 10. 112). The transformation in fact seems to have been largely complete 100 years earlier: already in the declamations recalled by Seneca, Cicero the *homo* has all but been replaced by Cicero the icon of *eloquentia.* That Cicero should be thus reduced to his *ingenium*, and that this essential *ingenium* should further be identified solely as verbal *ingenium*, is not surprising: for verbal *ingenium* is of course the chief attribute of the declaimers themselves. Probably not for the first time, and certainly not for the last, the classic takes on the properties of a mirror: the celebrant casts his gaze upon the icon reverently, and sees himself.

One may wonder, however—and this is a point I suggest only hesitantly, by way of conclusion—whether this mirroring also helps to explain one curious feature of these exercises mentioned above. Unlike his counterpart Demosthenes, who enters the school exercises as a vigorous orator in mid-career, the Cicero of the declamations is a Cicero on the point of death. He is not the effective advocate or sometimes effective statesman of life; he is a brilliant yet utterly impotent orator, an orator unable to work his will upon a political world in which speech has been shoved to the margins. Did the declaimers—brilliant yet utterly impotent orators, orators unable to work their will upon a political world in which speech had been shoved to the margins—did the declaimers gropingly, unintentionally perhaps, but revealingly choose as their icon a Cicero who represented their own condition more faithfully than they might care to say?[*]

[*] It is a great pleasure to offer this paper to Wendell Clausen, whose teaching and scholarship have put me, like so many others, greatly in his debt. For helpful responses to an earlier version of this paper I want to thank the faculty and students at the University of Virginia; and for more good criticism than this version reflects, Sander M. Goldberg, Ralph Johnson, Peter White, and James E. G. Zetzel.

REPETITION AND EMPIRE IN THE *AB VRBE CONDITA*

CHRISTINA S. KRAUS

"The meaning of a text goes beyond its author not sometimes but always ... one understands differently when one understands at all."

H.-G. Gadamer, *Truth and Method*

Beyond perhaps all other genres, history needs its readers. Whether we regard Livy's *monumentum* as a treasury or a jumbled heap, it demands our attention, insists that we compare, contrast, interpret, and, above all, judge.[1] Only the consensus of readers, following the historian's clues, can identify a given character or event as a good or bad *exemplum*, a subject of imitation or avoidance; the sum total of such identifications establishes the narrative as an authoritative version of the past. This process relies on the reader's recognition of familiar story patterns: conventional, repeating elements of narrative and theme whose deployment "avoids unmeaningful particularities and produces an image that has meaning for all."[2] Like the tradition-dominated world which it

[1] These acts are required of readers of any genre (see in general S. Fish, *Is There a Text in This Class? The Authority of Interpretive Communities* [Cambridge, Mass. 1980]); for the particular situation of history see, e.g., H. White, *The Content of the Form* (Baltimore 1987) 11-20 and cf. Livy, *Praef.* 9-10 for his reader's job. W. Liebeschuetz, *JRS* 85 (1995) 315, reviewing Levene, is overly sceptical about the extent to which Livy's readers are to participate in the project. The following works will be cited by author's last name: J.D. Chaplin, *Livy's Use of Exempla and the Lessons of the Past* (Diss. Princeton 1993), C.S. Kraus, *A Commentary on Livy Ab Vrbe Condita Book VI* (Cambridge 1994) and (=1994b) "No Second Troy: *Topoi* and Refoundation in Livy Book V," *TAPA* 124 (1994) 267-89, D.S. Levene, *Religion in Livy* (Leiden 1993), T.J. Luce, *Livy, the Composition of his History* (Princeton 1977), D. Quint, *Epic and Empire* (Princeton 1993), S. Rimmon-Kenan, "The Paradoxical Status of Repetition," *Poetics Today* 1 (1980) 151-59. Standard commentaries on ancient authors (e.g., Ogilvie on Livy 1-5) will be cited by author's last name alone.

[2] T.F. Carney, "Formal Elements in Livy," *PACA* 2 (1959) 6. See D.F. Kennedy, "'Augustan' and 'Anti-Augustan': Reflections on Terms of Reference" (in A. Powell, ed., *Roman Poetry and Propaganda in the Age of Augustus* [Bristol 1992]) *passim*, esp. 36: "some 'stability' is necessary for even the most minimally organised society to function. This is the place of convention, the encoding of a specific relationship of power and social structure within an expression or act so that it is

reflects, Livy's text abounds with these patterns: son emulating father, ritual observance repeated over the years, story echoing story.

We have become accustomed to seeing these repetitions as the textual manifestations of the Romans' ideological construction of their past, which similarly comprised repeating chains of events. Together with the fullness and redundancy of expression characteristic of Livian narrative, repeating elements form a narrative *cosmos* that re-presents the Roman *imperium*. Livian readers have analyzed the effect of *topoi* and other conventional elements, the link between narrative *exempla* and Roman ideals, and the way in which repeated, formal elements create a kind of generic *mos maiorum*.[3] These approaches share a treatment of the *Ab Vrbe Condita* as its own intertext, the text "which the reader must know in order to understand a work of literature in terms of its overall significance": the Livian reader's "compulsory response" is to identify and interpret the relation of his history not only with one or more other outside texts, but primarily with the *Ab Vrbe Condita* itself.[4] Over and over again Livy quotes Livy explicitly, or implicitly refers to earlier Livian stories, themes, and images: allusion and intertextuality proliferate within the bounds of a single text.[5]

---

taken for granted, affirmed, and reproduced"; he is building on the work of P. Bourdieu, *Language and Symbolic Power* (Oxford 1991) and others.

[3] For the terms see P. Hardie, *Cosmos and Imperium* (Oxford 1986); for formal language as a reinforcement of empire see F. Hickson, *Roman Prayer Language: Livy and the* Aeneid *of Vergil* (Stuttgart 1993) 145; on the stability of genres in general see F. Cairns, *Generic Composition in Greek and Roman Poetry* (Edinburgh 1972) and G.B. Conte, *The Rhetoric of Imitation* (Ithaca 1986). For history's exemplary repeated stories see e.g., L. Catin, *En Lisant Tite-Live* (Paris 1944), Chaplin (with extensive bibliography), C.S. Kraus, "*Initium Turbandi Omnia a Femina Ortum Est*: Fabia Minor and the Election of 367 B.C.," *Phoenix* 45 (1991) 314-25, and A.J. Woodman, "Self-imitation and the Substance of History: Tacitus, *Annals* 1.61-5 and *Histories* 2.70, 5.14-15," in A.J. Woodman and D. West, edd., *Creative Imitation & Latin Literature* (Cambridge 1974) 143-55.

[4] The quotations are from M. Riffaterre, "Compulsory Reader Response: The Intertextual Drive" in M. Worton and J. Still, edd., *Intertextuality: Theories and Practices* (Manchester 1990) 56; see also the editors' introduction (1-44). There are of course outside texts at play in Livy's narrative as well, especially Sallust; but most are now largely lost, and I am interested here only in *internal* allusions.

[5] See W. Moskalew, *Formular Language and Poetic Design in the Aeneid* (Leiden 1982) for a similar process in Vergil, whose repetitions set up a network of allusion that works forward and then, when we reread the epic, backward; the effect can be either to reinforce or to ironize a given image, action, or utterance (see also

Any literary allusion is a kind of quotation, and intertextuality is no exception. But quotation is a double-edged sword. By forcing the reader's attention back to another, prior text it can question the quoting author's authority, integrity, originality; but it can also encourage synthesis, inviting us to incorporate the quoting and the quoted text, to make connections that enrich both.[6] Most studies of Livy's abundant "self-allusion" have concentrated on its constructive, architectonic function, seeing it as the historian's way of founding and increasing his textual *Vrbs*, which reflects the real *urbs* and *imperium* of Rome. But there is another kind of self-allusion in Livy, a destructive intertextuality: the doublets and repeated incidents that have prompted readers to level against him the charge of "bad form." The centrifugal effect of quotation has been dominant in reactions to these doublets, whose study has been the single most productive tool for dismantling the *Ab Vrbe Condita* into its component parts (i.e., Livy's sources).[7] Some (most notably Luce) do try to put the history back together again; but particularly where doublets are concerned, Livy has all too often appeared to be little more than the sum of his parts. Concomitantly, the historian's authority, both literal and figurative, has been challenged, sometimes denied altogether.[8] I would like to take a different tack: to propose a reading of these doublets that asks what they can teach us about the text and about the world it

next n.). A distinction is often made between intertextuality and allusion, but I find it more helpful to think of a continuum with Moskalew's specific verbal allusions on one end and Riffaterre's pervasive intertext on the other.

[6] P. Hammond, "The Play of Quotation and Commonplace in *King Lear*," in L. Hunter, ed. *Toward a Definition of Topos* (London 1991) esp. 79-80; on the tension between the centripetal and centrifugal impulses of intertextual readings see Worton-Still (above, n. 4) 11.

[7] For a recent example of Livian *Quellenforschung* see A.-M. Adam, ed., *Tite-Live: Histoire Romaine XXXIX* (Paris 1994) xxvi-li. It is generally agreed that these doublets arise when Livy switches from one to another of his many narrative sources; in making the disparate strands into a single text Livy does not notice that he is putting down the same piece of information twice, either because the first time was many pages earlier or else because he has taken a break between sources or has simply not been paying attention (these latter comprise what one might call the "coffee break" theory of Livian composition).

[8] The tendency to dismiss Livy is not nearly as widespread as it was, but these parts of the text are still hard to look at. For similar problems in another author see G. Williams in *Ramus* 20 (1991) 196-200 on the doublets in the *Fasti*: by finding a reading that accounts for doublets as part of an authorial strategy, Williams tries to refute the charges that Ovid was careless or unrevised.

creates. It is a reading that is less interested in the production of the *Ab Vrbe Condita* than in its reception, and in the crisis of interpretation and authority generated by this "bad" repetition.

I. Good Repetition

It is perhaps largely owing to their formal resemblance to Livy's exemplary scenes, *topoi*, and other conventional elements that his doublets have been so troublesome. I begin, therefore, with a brief look at two prominent types of "good" repetition: places where a physical *locus* duplicates a significant place in earlier history, and "mirror scenes," in which the repetition is linked more to character and situation than to location. These are primarily repetitions of type rather than of event or of language, though sometimes a later episode will precisely copy an earlier one and, indeed, be narrated in similar or identical language (the *deuotiones* of the Decii Mures are a notable example). The narrator, either alone or with a character, often directly identifies the event as an exemplary repetition.

A *locus* that has already seen a significant action may serve as a mnemonic trigger for a repetition of that action.[9] In 380 BCE (6.28.5-29) the Praenestini stage a battle at the Allia, where the Gauls routed the Romans a decade earlier (5.37-8). In response, the Romans refuse the omen of the place, seeing in it instead a chance to erase the memory of the Allia. In so doing they choose two new paradigms, sites of their victory over the Gauls: *locum insignem memoria cladis irritaturum se potius ad delendam memoriam dedecoris quam ut timorem faciat, ne qua terra sit nefasta uictoriae suae: quin ipsi sibi Galli si offerantur illo loco, se ita pugnaturos* <u>*ut Romae*</u> *pugnauerint in repetenda patria,* <u>*ut postero die ad Gabios*</u>*, tunc cum effecerint ne quis hostis qui moenia intrasset nuntium secundae aduersaeque fortunae domum perferret* (6.28.8-9). Sixty years later the Romans avoid the destiny of Caudium, which re-

[9] There is a close overlap here between real and rhetorical *loci*: see Kraus 1994b and M.K. Jaeger, "*Custodia Fidelis Memoriae*: Livy's Story of M. Manlius Capitolinus," *Latomus* 52 (1993) 350-63. For the Allia, Caudium, and Cannae as *exempla* see Chaplin 110, 63-95; for such *loci* in other authors see, e.g., J. Henderson, "Lucan/The Word at War," *Ramus* 16 (1987) 126. There are famous examples of this place-doubling outside Livy, e.g., Pharsalus and Philippi (as at Luc. 1.694) and the Tacitean forests; cf. Woodman (above, n. 3).

curs several times quite soon after its first, paradigmatic, appearance as a *locus* of defeat (9.2-6). At 9.31, Caudine topography nearly traps a Roman army, and the text looks set to repeat itself: "Caudium" is not mentioned but verbal allusions point clearly to the earlier narrative.[10] The beleaguered Romans extricate themselves with tactics familiar from the second Allia, including the topographical chauvinism (*sed quem esse iam uirtuti Romanae inexpugnabilem locum?*) and the choice of alternative past sites as models for the present (*Fregellana arx Soranaque et ubicumque iniquo successum erat loco memorabantur*, 9.31.13). Shortly thereafter, this time explicitly mindful of Caudium, the Romans avoid potential ambush: *eam* [sc. *siluam Ciminiam*] *intrare haud fere quisquam praeter ducem ipsum audebat; aliis omnibus cladis Caudinae nondum memoria aboleuerat* (9.36.1). These passages underscore a recurring tension between memory and forgetting: to move ahead the Romans have to forget, to *delete* their memory; but only preserving the knowledge of what happened before allows them to change.

At Nola, a repeated battle site in the second Punic war, the situation is more elaborate. At the time of the first battle it is not yet a significant place, so in their *hortationes* the enemy commanders choose another *locus*, Cannae, as an argument. It appears as a sort of narrative seed in the first skirmish for Nola: a townsman, who had been rescued by Hannibal after the Romans left him for dead at Cannae, wants to betray his city to Carthage; but by promising him rewards for his bravery Marcellus entices him to remain loyal (23.15.7-15). Two possible meanings for "Cannae" are here suggested, first by the narrative voice, second by Marcellus: it is a place of death, especially for Romans (23.15.8 *ad Cannas in aceruo caesorum corporum*) or a place of glory (23.15.12-13[11]). The next year Hannibal boasts that he will obscure the memory of

[10] Cf. 9.31.7 *saltum auium*, 31.9 *in uetere disciplina militiae iam sine praecepto ullius sua sponte struebatur acies* ~ 9.2.12-14, esp. §13 *pro se quisque nec hortante ullo nec imperante ad muniendum uersi castra*, 31.13 (quoted in text) ~ 9.3.2 *omnia aequa et plana erunt Romano in perfidum Samnitem pugnanti*. Caudium's potential vulnerability is already signalled in the aftermath of the first disaster: *saltus Caudinos non ubique Samnitibus fore* (9.7.5).

[11] *multos sibi referre ... qui uir esset ille quaeque et quotiens pericula pro salute ac dignitate populi Romani adisset utique Cannensi proelio non prius pugna abstiterit quam prope exsanguis ... sit oppressus.* Marcellus makes Bantius sound like a modern version of Cato's paradigmatically brave (but insufficiently rewarded) tribune (Cato fr. 83P *eumque inter mortuos defatigatum uulneribus atque quod sanguen eius*

Cannae by a *maior et clarior uictoria* (23.43.4), a claim shortly repeated by Hanno (23.43.12). Marcellus counters this with an Allian ploy, choosing a different *locus* to inspire his soldiers: Nola itself, where they recently defeated Hannibal (23.45.1). Marcellus does not ignore Cannae but boldly uses it as a metaphor for *Carthaginian* defeat (23.45.2-4 *qui pugnent, marcere Campania luxuria ... Capuam Hannibali Cannas fuisse*). And at Cannae's next (and final) appearance in this episode, it is evident that Marcellus has successfully redefined its exemplary meaning: it serves Hannibal as a bitter reproach for his troops' degeneration since that victorious winter (23.45.6, 8).

In these episodes Livy shows both his external readers and his characters that there are two ways of reading history, either by seeing it as something one is doomed to repeat (as the Praenestini and Hannibal believe) or as something that one can escape by repeating with a difference. Only the latter leads to progress, illustrating one of the paradoxes of repetition, that "constructive repetition emphasizes difference, destructive repetition emphasizes sameness (i.e., to repeat successfully is not to repeat)."[12]

The same paradox operates in the kind of exemplary repetition with which we are perhaps most familiar, that involving people. Some of the best known occur within individual families, such as the Manlii.[13] In accepting the challenge of the Gallic champion, Manlius Torquatus refers to the heroism of his great-uncle who threw the Gauls from the Tarpeian rock during the sack of Rome. Manlius' imitation of that feat will establish nothing less than his lineage: *si tu permittis, uolo ego illi beluae ostendere ... me ex ea famila ortum quae Gallorum agmen ex rupe Tar-*

defluxerat cognouere ... at tribuno militum parua laus pro factis relicta, qui ... rem seruauerat).

[12] Rimmon-Kenan 153; see also S. Bann, *The Inventions of History* (Manchester 1990) 73-75 and cf. 24.8.20 *lacus Trasumennus et Cannae tristia ad recordationem exempla sed ad praecauenda similia documento sunt*. This is an area in which psychoanalytical models of post-traumatic repetition have influenced narratological models; see P. Brooks, *Reading for the Plot* (New York 1985), P. Hardie, *The Epic Successors of Virgil* (Cambridge 1993), and Quint. On the necessity for innovation see further below, n. 46.

[13] On their single combats see S.P. Oakley, "Single Combat in the Roman Republic," *CQ* 35 (1985) 392-410, and A.M. Feldherr, *Spectacle and Society in Livy's* History (Diss. Berkeley 1991) 192-234. *Exempla* and exemplary scenes are, of course, not restricted to the first decade; see Chaplin *passim*.

peia deiecit (7.10.3). His deed is soon challenged and appropriated by Valerius Corvus, whose victory over a Gaul in similar circumstances is held up by his commander as a subject worthy of imitation: *M. erat Valerius tribunus militum adulescens, qui haud indigniorem eo decore se quam T. Manlium ratus, prius sciscitatus consulis uoluntatem, in medium armatus processit ... ostentansque insignem spoliis tribunum, 'hunc imitare, miles' aiebat* [sc. *consul*] (7.26.2, 7). The singular *miles* with which the consul particularizes his exhortation focusses the issue: each individual soldier can emulate this individual hero, as the Manlian tradition becomes a type-scene to which any brave and obedient soldier can aspire.[14] Within the family, however, the disturbance caused by Manlius Capitolinus' appearance at 7.10.3 as a model (he was, after all, executed on charges of treason: 6.11-20) is shown by the next generation's failure to carry on the exemplary tradition: Torquatus' son gets it wrong by fighting without his commander's permission.[15] As a consequence, while Torquatus' and Corvus' exploits won them each a cognomen, the execution of Manlius the son by his father implicates the latter in a new, undesirable aetiology: his severe discipline, known ever after as "Manlian," will itself provide an *exemplum*, salutary but distressing (8.7.17, 22). Again, we come up against paradox. Manlius dies because he failed to repeat *exactly*, i.e., failed to ask his commander's permission. In a sense, he is like the Praenestini, and does not understand which elements of an exemplary act must be repeated and which must, or need, not be. His inability to make that judgment catapults him into the category of never- rather than always-to-be-imitated; like Manlius Capitolinus, however, he will remain forever as a (sinister) model.

---

[14] Valerius' rivalry with Manlius is the first step in the exemplary chain, *aemulatio* in the same generation (the older soldier will recognize the younger's merit: 7.26.11-12). The patrician Valerius' competitive challenge to inherited Manlian heroism is later recalled in his very Marian self-description: *facta mea non dicta uos ... sequi uolo nec disciplinam modo sed exemplum etiam a me petere. non factionibus nec per coitiones usitatas nobilibus sed hac dextra mihi tres consulatus summamque laudem peperi* (7.32.12).

[15] The narrative follows the son's lead in repeating his father's act: cf. 8.7.7 *ut nostro duorum iam hinc euentu cernatur* ~ 7.9.8 *ut noster duorum euentus ostendat* (both in the challenge) and 8.7.13 *'ut me omnes ... pater, tuo sanguine ortum uere ferrent, prouocatus equestria haec spolia capta ex hoste caeso porto'* ~ 7.10.3 (quoted in text).

The capture of Samnite spoils by successive Papirii and the *deuotiones* of the Decii at Veseris and Sentinum offer further notable instances of father-son imitation.[16] The traditional display of the two sets of spoils allows people to compare them—a public demonstration of the son's *aemulatio*, which is further demonstrated by his completion of the temple begun by the elder Papirius.[17] In Livy's narrative of the second Decius' *deuotio*, imitation extends to the discourse as well: not only does the son appeal to family tradition but the narrative itself refers the reader back to the first *deuotio* for details, creating a sort of textual paternity (10.28.13, 15 *datum hoc nostro generi est ut luendis periculis publicis piacula simus ... deuotus inde <u>eadem precatione eodemque habitu quo pater</u> P. Decius ad Veserim bello Latino se iusserat deuoueri*). Furthermore, the first *deuotio* is intertwined with Torquatus' *triste exemplum*: though the dream advising a *deuotio* precedes Manlius' single combat and execution (8.6.10), the decision as to which consul will dedicate himself is not in fact made until *after* it. The shocking juxtaposition with Manlius' extermination of his own descendant highlights the fruitful outcome in the second generation (of which Livy's readers are already well aware) of the elder Decius' act.[18]

In these repeating pairs or triads we see enacted the double challenge of living up to ancestral glory and of confirming one's own merit, thereby proving membership in a *gens* and becoming a model for later generations of that same *gens*.[19] Here too, however, as in the *locus*-related examples, the task is to avoid mindless exact repetition: characters

---

[16] Since it is not in Livy's extant text, I do not include the third generation Decius, whose *deuotio* at Ausculum in 279 is perhaps pure invention (and as such a nice example of "life" imitating art); arguing for its authenticity are O. Skutsch (*The Annals of Quintus Ennius* [Oxford 1985] 353-55) and T.J. Cornell (*JRS* 76 [1986] 248-49).

[17] Esp. 10.46.4 *inspectata spolia Samnitium et decore ac pulchritudine paternis spoliis, quae nota frequenti publicorum ornatu locorum erant, comparabantur* (cf. 9.40.16) and 10.46.7 *ab dictatore patre uotam filius consul dedicauit exornauitque hostium spoliis* with Weissenborn-Müller's n. on the practice. For father-son rivalry / emulation in contemporary sculptural groups, see C.B. Rose, *Dynastic Commemoration and Imperial Portraiture in the Julio-Claudian Period* (Cambridge 1997).

[18] Levene 222-23; on the second devotion as an "abridged" version of the first, see Levene 236.

[19] The challenge is a familiar one: see, e.g., the Scipio epitaphs, now collected in E. Courtney, ed., *Musa Lapidaria* (Atlanta 1995) nos. 9-13.

and reader alike must be able to see both the kinship and the difference, even as the Romans can compare the two sets of *spolia* dedicated by the two Papirii. That difference and the likeness together generate recognition and a kind of educational profit, an exemplary self-consciousness, as it were.[20] Like the spectators of the spoils, the reader is intimately involved in making these stories work.

II. Bad Repetition

For every one of these productive doublings in the *Ab Vrbe Condita* there is, it seems, a doublet.[21] Some appear to be simple mistakes: Livy has reported the same thing twice, producing two stories that as a consequence of their new independence tend to have slightly different details.[22] Others appear to be stories about different people, places, or times, but are in fact probably multiple narratives of the same original event. Since our awareness of these depends on our sensitivity to the probability of things happening more than once, there is not always a

[20] Nearly exact repetition can occasionally be profitable for the reader if not for the characters: such is the case with the repeating Claudii throughout the first decade, as seen, e.g., in the doubled conflict between two consular Claudii, father and son, and two Laetorii (2.27.4-7 ~ 2.56.5-16; see Ogilvie *ad* 2.56.5 and 6). T.P. Wiseman's synchronic analysis (*Clio's Cosmetics* [Leicester 1979] 57-103) shows clearly the lack of change and learning in most Claudian portraits, though both he and A. Vasaly ("Personality and Power: Livy's Depiction of the Appii Claudii in the First Pentad," *TAPA* 117 [1987] 203-226) demonstrate the complexity and sophistication in Livy's deployment of these stock characters from whom the reader can learn how not to behave in a Republic. On the difficulties in speaking of "exact" repetition, see n. 22.

[21] The term commonly used for "inadvertent" or "erroneous" repetitions. Though doublets abound in his history, Livy is sensitive to the problem they pose; cf. 6.42.5-6 (a doublet firmly avoided), 44.14.5-13 ~ 45.3.3-8 (presented as a variant notice: see Luce 117 n.6) and especially 29.35.2 *duos eodem nomine Carthaginiensium duces duobus equestribus proeliis interfectos non omnes auctores sunt, ueriti, credo, ne falleret bis relata eadem res.*

[22] One might object that as a result of these differences we are no longer dealing with true repetition. But this is one of its paradoxes: "difference is introduced through the very fact of repetition, the accumulation of significance it entails, and the change effected by the different context in which it is placed" (Rimmon-Kenan 152-53).

scholarly consensus about which are genuine repetitions and which cases of coincidence.[23]

Livy's doublets range from simple to complex. The simple ones, with which I begin, fall into certain natural categories, involving people (especially when on journeys or embassies), battles, constitutional and ritual elements. (Each of these is also a conventional element in historiographical narrative, an important point to which I shall return in section III below.) One of the most notorious is the repeated *ouatio* of M. Fulvius Nobilior, a "clear example of careless composition."[24] The notices are very close together indeed, one forming a sort of coda after the report of the victory at Thermopylae (36.21.10-11), the other toward the end of the book, in a section of *res internae* separating phases of the wars with Antiochus (36.39.1-2). Similarly, Lucretius leaves for Brundisium twice (42.35.3 ~ 42.48.5), while Aelius departs for Gaul at 32.9.5 and 32.26.1-3. Embassies, too, are particularly susceptible to doubling, e.g., the *legatio* of Claudius and Decimius (42.19.7-8, 26.7-9) is repeated by that of Claudius, Postumius, and Iunius (42.45), and there is a double Macedonian embassy to Rome (42.36.1-7 ~ 42.48.1-4).[25] Just as easily multiplied are battles, especially in the early books,

---

[23] After all, history *does* repeat itself: for some fairly bizarre modern examples, see A.J. Woodman, "From Hannibal to Hitler: The Literature of War," *University of Leeds Review* 26 (1983) 119-20. I have chosen examples that are generally accepted and could easily be multiplied. Although it is not always certain that Livy would have acknowledged these as doublets—and though he may have inherited them—I include them here because they have an effect *on the reader* similar to that produced by the simple errors. (That effect does vary with the reader's background: so a specialist in the history of early Rome may be more attuned to the implausibility of the endless battles with the Volsci than a reader with more exclusively literary preoccupations; for a beginning on a text's "horizon of expectations" see Fish [above, n.1], H. R. Jauss, *Toward an Aesthetic of Reception* [Minneapolis 1982] 22-28 and R.R. Nauta, "Historicizing Reading," in I.J.F. de Jong and J.P. Sullivan, edd., *Modern Critical Theory and Classical Literature* [Leiden 1994] 207-21.)

[24] Luce 80. A *supplicatio* with *hostiae maiores* precedes each report of Nobilior's return (36.21.9 ~ 36.38.7), an unrelated repetition that may have contributed to Livy's error (which would thus be a sort of narratological dittography).

[25] Books 42-45 are particularly full of doublets: Luce 123-35.

where the same neighboring tribes attack Roman territory over and over again.[26]

The sizeable group of repeated constitutional and ritual elements includes duplicated laws (e.g., *prouocatio* at 2.8.2 ~ 3.55.4-5 ~ 10.9.3-6 and on plebescites at 3.55.3 ~ 8.12.14), prodigies (e.g., 39.46.5 ~ 39.56.6), vows of temples (e.g., the temple to Jupiter Stator, 1.12.6 ~ 10.36.11[27]), and decrees of games (e.g., the *Ludi Apollinares*, which are twice decreed *in perpetuum*: 26.23.3 ~ 27.23.5-7[28]). Readers have seen patterns in these repetitions, suggesting, for instance, that they can be read as framing devices;[29] whether these frames are deliberate on Livy's part or a success in the reader's view, the latter's need to identify such patterns reflects a wish to impose a form on the very places in the text which most cry out for interpretation and regularizing: places, that is, where narrative coherence has broken down and the historian's authority is seriously challenged.

More interesting are doublets such as the deaths of Pleminius, a Roman garrison leader who terrorized the Locrians and looted their shrines. At his first demise (204 BCE, 29.22.7-10), a gloss incorporated into the text from a very early point in the MS tradition reports that Clodius Licinus narrated his death in the year of Scipio's second consulate and the *ludi uotiui* then performed (i.e., 194).[30] When Livy himself reaches 194 he reports the death again, in "end-of-year" material (34.44.6-8). There is a cross-reference in this second telling (34.44.6),

[26] Livy doubts the probability of some of these skirmishes, doubts echoed by modern historians; see Kraus 28 with n. 121 and add T.J. Cornell, *The Beginnings of Rome* (London 1995) 304-7, 318-24 (not sceptical).

[27] The Republican temple was built (in theory) on the site of Romulus' *fanum* (if any): see Ogilvie *ad* 1.12-13.5, L. Richardson, Jr., *A New Topographical Dictionary of Ancient Rome* (Baltimore 1992) 225.

[28] See Weissenborn-Müller *ad locc.* for possible reasons for the duplication.

[29] The prodigies in book 39 may frame the year (so Levene 97-98); since Aelius' departures are set on either side of Quinctius' activities in Greece, the second resembles a conventional return-via-repetition to a narrative after a digression (T. Suits, "The Structure of Livy's Thirty-Second book," *Philologus* 118 [1974] 258-59).

[30] S.P. Oakley, "Livy and Clodius Licinus," *CQ* 42 (1992) 547-51 shows conclusively that §10 is not by Livy; but it is found already in the fifth century *codex Puteanus* and as such forms part of the Livian *reading* tradition. As Oakley notes (551), its excision does nothing to mitigate the duplicated death.

which sends the reader back not to book 29 but to Galba's vow of the *ludi* in 200 (31.9.6-10). Levene is surely right to suggest that there is a point to Livy reminding us of Pleminius in Scipio's second consulate, since his criminal behavior took place in Scipio's first; but as the obtrusive set of references to other parts of the *Ab Vrbe Condita* link not those magistracies but a set of unrelated *ludi*, I would add that they exacerbate the reader's sense of dislocation, while failing to enhance the possibly edifying connection between Scipio's consulates.[31] Pleminius's deaths do exhibit a distinct narrative self-consciousness, however, a meta-narrativity concentrated in the footnotes to other parts of the story.

This self-consciousness is also found in many of the more complex doublets, such as the capture of Croton, which is reported both in 216 (a brief mention at 23.30.5-8) and in 215 (24.1-3). The second, expansive narrative is full of "double-talk," especially in the description of Croton itself, in which Livy elaborates the divisions between inhabited and uninhabited sections, between outer city and citadel, between sea and land approaches (24.3.1-3, 7-8). Not only is the town divided, it is also doubled: within it is the equally wealthy *templum* of Juno, more famous even than the famous city, in which are mixed wild and cultivated (*silua* surrounding *pastua*), nature and art (flocks *sine pastore* and a *columna aurea*). There is doubleness in the narrative form—the story of the *templum* is ringed by the description of the town—and theme as well: in this story of treason both town and enemy have divided loyalties (indeed, there are *two* enemies, the Brutii and the Carthaginians).

Similarly heightened meta-narrativity can be found in some of the engagements which, like the second Allia, are fought at or near the same place, usually within a short space of time. However, they differ from the Allia battles in that, though Livy plays with the similarity between them, neither he nor his characters see them as an original + copy/reversal. Two sets in book 6 will illustrate: first, a pair of engagements at Satricum (6.7-8 ~ 22-24), which Livy links and contrasts by verbal echoes, opposing particularly the behavior of the soldiers (*cunctabundi* at 7.2, over-eager at 23.3-4) and Camillus' response to

[31] Levene 83. The form-seeking reader might note that the phrase *ludi uotiui* occurs at 29.22.10 as well, thus linking the three passages; but the links, like the cross-references, underscore Livy's carelessness by emphasizing that there are *two* reports of Pleminius' death (see also below, n. 40).

his own advanced age (vigorous despite it in the first episode, initially willing to take the sidelines in the second). Quotation of the first battle in the narrative of the second tempts the reader to consider the two as a pair, though Camillus' contradictory and confusing actions, in particular, make that pairing seem pointless. Even more closely connected are two battles at Sutrium (6.3-4 ~ 9), in whose narratives the doubleness is both external and internal: after capturing the town the Etruscans are themselves captured by the Romans, repeated action flagged with *iterum igitur eodem die Sutrium capitur* (6.3.6); in Livy's narrative of the second skirmish, three years later, a remarkable accumulation of parallel expressions enhances the story of a town physically divided.[32] These narratological "flags," like those in the Croton episode, invite us to notice rather than gloss over the doubling of the battle; they are "formal witticisms" drawing attention to Livy's *ars*.[33]

Finally, these complex, self-conscious doublets can occur in what appear to be thematically determined clusters.[34] One such nexus occurs in the narrative neighborhood of Lake Regillus, the battle that marked a critical stage in the relations between Rome and the Latins. Traditionally it ended with the appearance of the archetypal Latin dual, Castor and

[32] Cf. 6.9.7 *partem oppidi iam captam ... ex parte altera ... oppidanos uim hostium ab se arcentes*. For the repeated elements in Sutrium (1) see Kraus 101; for the double-speak in Sutrium (2) see *ead.* 141 and cf. 143 on its pendant piece, a battle at Nepet in which political division throws into relief the physical and linguistic divisions at Sutrium.

[33] The term is E. Bartman's ("*Decor et Duplicatio*: Pendants in Roman Sculptural Display," *AJA* 92 [1988] 224). The more conscious we are of the *ars*, the more fragile the narrative illusion and (potentially) the author's control; cf. H. Gotoff, *Cicero's Elegant Style* (Urbana 1979) 55-56, 72-73 on a writer whose authority—like Livy's—depends to a great extent on his ability to produce a transparent picture of the world without tipping his rhetorical hand. The grey area between the Sutrium scenes and the second Allia type is shown by the doubling of a battle at Herdonia, each between Hannibal and a Roman commander named Fulvius (25.21 ~ 27.1; on the doubling see J. Briscoe in *The Cambridge Ancient History*^{2} Vol. VIII [Cambridge 1989] 54 n. 52). Though they are probably reports of the same event Livy treats them as he does his "exemplary" battles; cf. 27.1.9 (Hannibal encourages his men with the very argument that the Praenestini use at 6.28): *ipse Cn. Fului similitudinem nominis—quia Cn. Fuluium praetorem bienno ante in iisdem deuicerat locis—increpans, similem euentum pugnae fore adfirmabat. neque ea spes uana fuit.*

[34] It is a nice question whether in these passages Livy is manipulating material or the reader imposing a thematic order so as to make sense of what would otherwise seem random content.

Pollux; though omitting the theophany, Livy coyly alludes to it by reporting the vow of a temple to Castor (2.20.12).[35] In addition to the external wars, there are terrible political divisions in this period, hinted at early on by the dedication of the *aedes Mercuri*, a merchants' cult on the Aventine (2.21.7), an act of plebeian self-assertion later repeated as the cause of the first duel between Claudius and Laetorius (2.27.4-7: above, n. 20). Closely following that duel is the first *secessio plebis* (2.31.7-33.3), which leads to the tribunate, the plebeian "magistracy" to head their "second city" within Rome; the creation of the tribunate is itself later doubled, when after the second duel between Claudius and Laetorius the Romans revise the manner in which the tribunes were elected.[36] In the meantime, the new tribunate sparks the violent deeds of the violent patrician Marcius Coriolanus, a struggle in which the magnitude of civil strife turns a *ciuis* into a *hostis* who eventually marches on Rome, divisiveness replacing what should be indivisible (2.34.7-40.12).[37] Within this larger context Livy sets 2.21, a chapter bridging the gap between 499 and 495 BCE. Reflecting the doublings and strife in the text around it, this single paragraph of "archival" details contains no fewer than four actual or near-doublets.[38] Some readers may find the disturbance minimized by the fact that this is virtual "end-of-year" material, which by its very nature is non-narrative and a conventional place for variants; on the other hand, in this chapter a basic architectural unit of Livy's monument is rocked by the high proportion of doublets, three out of four of which go unchallenged by the historian.

[35] For the Dioscuri at Lake Regillus see Ogilvie *ad* 2.20.12; for other Livian allusiveness see Kraus *ad* 6.42.12.

[36] For this somewhat mysterious event see Ogilvie *ad* 2.58.1; for the second city of the *plebs* cf., e.g., 6.40.5 *plebi tamquam aliam incolenti urbem* and see Kraus *ad* 6.11.7, adding Cornell (above, n. 26) 258, n. 64.

[37] Civil war is of course particularly susceptible to twins/doublings: from Eteocles and Polyneices on, the one who should be the same is suddenly the other, enemy replacing friend.

[38] They are: the foundation of the Saturnalia (§2 ~ 22.1.20, actually a reorganization of the festival, not its institution: Ogilvie *ad* 2.21.2 and H.H. Scullard, *Festivals and Ceremonies of the Roman Republic* [London 1981] 206); the battle itself (§3 *hoc demum anno ad Regillum lacum pugnatum apud quosdam inuenio*); a "second" colony to Signia (§7, but the first was probably a fiction: Ogilvie *ad* 1.56.3); and the dedication of the *aedes Mercuri* (above). In the course of the chapter Livy confides that this is a particularly difficult period to sort out (§4 *tanti errores <res> implicant temporum*)—another flag.

Doublets can also cluster around a person, as is notoriously true of Scipio Africanus.[39] His death, which may or may not have followed hard on his disgrace and voluntary removal from Rome, is reported twice, first at 38.53.8 (187 BCE). Though Livy gives him an obituary (53.9-11) and resumes his narrative of Scipio's case with the words *morte Africani* (54.1), he tells us shortly thereafter that there is no consensus on the date of Africanus' death or the location of his tomb (56.1-3). Four years later he twice reports a variant dating synchronizing Scipio's death with those of Philopoemen and Hannibal (39.50.10, 52.1). Though rejecting 183 as a death-date for Africanus (39.52.1-6), Livy nevertheless uses it to introduce a *syncrisis* among the three leaders: *trium clarissimorum suae cuiusque gentis uirorum non tempore magis congruente comparabilis mors uidetur esse quam quod nemo eorum satis dignum splendore uitae exitum fuit. iam primum omnes non in patrio solo mortui nec sepulti sunt... Scipio etsi non exul neque damnatus ... uoluntarium non sibimet ipse solum sed etiam funeri suo exilium indixit* (39.52.7-9).[40] In "formal" (textual) terms, Scipio dies twice.

Nor are the physical remains in this case trustworthy. Scipio is first buried at Liternum (38.53.8), information to which Livy will later refer as if it were unproblematic; indeed, the historian himself claims to have

---

[39] For the latest discussion of what might have really happened and when in this most confused of all Livian episodes see Adam (above, n. 7) lii-lxiii and *ad locc.* and Walsh *ad* 38.56.1, 39.52.1 and 9. The Scipio narratives can be seen as simple variants rather than as doublets: I include them here, however, because both deaths are functional narrative units despite Livy's rejecting the later as the "true" demise. Nor does he explicitly choose between the two tombs: each is vouched for by signs (the proof that Scipio is inside is the statue outside), and though autopsy and Book 45 (quoted below, n. 41) support Liternum, the great public, family tomb is at Rome. For a discussion of the Scipionic narrative and Livy's *persona* asa historian, see M.K. Jaeger, *Livy's Written Rome* (Ann Arbor 1997).

[40] The closing words *non sibimet ... indixit* pose a riddle whose answer lies in the first narrative of Scipio's death (38.53.8-9 *silentium deinde de Africano fuit. uitam Literni egit sine desiderio urbis; morientem rure eo ipso loco sepeliri se iussisse ferunt monumentumque ibi aedificari, ne funus sibi in ingrata patria fieret*). The subtle cross-reference connects the notices, paradoxically giving the first a kind of authority by answering the puzzle and at the same time undercutting that authority by accentuating the chronological confusion (in case you had forgotten, Scipio is *already dead*). See also above, n. 31, and on the synchronism, see Walbank *ad* Pol. 23.12.

seen Scipio's tomb and statue there.[41] He may also, however, be buried in the family tomb at Rome, which Livy sites precisely outside the *porta Capena*, and which has its own statues: *utroubique monumenta ostenduntur et statuae; nam et Liternae monumentum monumentoque statua superimposita fuit, quam tempestate disiectam nuper uidimus ipsi, et Romae extra porta Capenam in Scipionum monumento tres statuae sunt, quarum duae P. et L. Scipionum dicuntur esse, tertia poetae Q. Ennii* (38.56.3-4). Tombs ought not to be unstable: we should be able to say *hic situs est* and mean it. But Scipio was so famous that, like Homer or Aeneas (cf. Dion. Hal. 1.54.1), his siting was contested; he had two *monumenta*, which are duly represented in Livy's own, doubled *monumentum* to him. Yet the Livian Scipio's doubleness comes less from his fame than from his essential ambiguity. Over and over the historian shows how hard it is to evaluate Africanus: was he a charismatic manipulator, or the best of all Romans? a crook, or a great servant of the state? one who deserved to end his days in glory, or in hiding? Livy's treatment of his demise and burial reflects this essential doubleness, leaving the reader still uncertain as to whether Scipio was entirely a Good Thing.[42]

There are, then, two broad categories of "bad" repetition in the *Ab Vrbe Condita*: a simpler type that seems to pass unnoticed by Livy but disconcerts readers who care about consistency in history (including things like the non-revertibility of the past and the impossibility of bilocation); and a more complex type that is noticed, even emphasized by the

---

[41] For the Liternum tomb cf. last n. and 45.38.7, where Servilius uses Africanus as an *exemplum*: *Literni domicilium et sedem fuisse domitoris Africae, Literni sepulcrum ostendi erubescamus* (for *ostendere* of evidence that proves a point cf. 38.56.3, quoted above). Modern readers are confronted by a further Scipionic doubling that reflects the two images of him created by Livy's narrative. F. Coarelli has reconstructed the façade of the Roman *sepulcrum Scipionum* as a triple-arched structure with the statues inside the arches; Richardson, disliking what he perceives in this reconstruction as a strong resemblance to a *scaenae frons*, "would prefer to have the program of statuary crown the attic of the tomb, as it might a triumphal arch" (above, n. 27, 360 with references). We are thus offered two possible reconstructions of the tomb, one resembling a stage set with figures (evocative of the potential slippage in Scipio's behavior throughout his lifetime between genuine and assumed), the other, a triumphal arch, model of Roman power, solidity, and endurance.

[42] On Scipio's ambiguity see, e.g., Levene 61-2, 73-4 and n. 120, Walsh's "Introduction" to Book 38; other great and ambivalent figures are Marius and Julius Caesar (Kraus 9 and n. 37).

historian and hence becomes more problematic for his readers. Though sharing some traits with productive, exemplary repetition, "bad" repetition does not show characters learning from the past, and hence lacks moral utility—a utility which, after all, is based on the principle that the past can provide salutary paradigms of good and bad behavior only if we understand them.[43]

III. Repetition and Empire

Throughout its history the Roman empire was marked by an almost obsessive urge to reproduce itself, either by expansion (repeating the *urbs* in the *orbis*) or by assimilation (repeating the *orbis* in the *urbs*). Verbal and exemplary repetition are the rhetorical reflection of this longing for *auxesis* and *copia*.[44] Because history-as-text ratifies history-as-event (and vice versa), the structure and form of the rhetorical *urbs* become critically important. What Quint says about the *Aeneid* is applicable, *mutatis mutandis*, to Livy's annals: "Epic takes particularly literally the axiom that history belongs to the winners. Imperial conquest of geopolitical space—the imposition of a single, identical order upon different regions and peoples—becomes a process of history making ... Power, moreover, is defined by its capacity to maintain itself across time, and it therefore requires narrative in order to represent itself ... [C]onversely, the ability to construct narratives that join beginnings purposefully to ends is already the sign and dispensation of power."[45] But beginnings cannot be joined to ends unless there is progress between them. "Good" repetition, in which an original provides a model for one or more copies, facilitates both continuity and change.[46] These pro-

---

[43] Livy would certainly have agreed with the aphorism that those who do not understand the past are doomed to repeat it. On history as medicine see *Praef.* 10 with the references collected by A.J. Woodman, *Rhetoric in Classical Historiography* (London 1988) 133; as intellectual activity see *Praef.* 9 (*intendat animum*). Repetition can be thought of as death: Rimmon-Kennan 155.

[44] Jaeger (above, n. 9) 361; Kraus (1994b) 270, 281, 284-285; for the metaphorical equivalence of rhetorical and physical structures cf. Cicero's use of *fundamenta* and *exaedificatio* for historiographical narrative (*De orat.* 2.63).

[45] Quint 30, 45.

[46] On the importance of change see above, Section I and especially Chaplin 125-54, G.B. Miles, *Livy: Reconstructing Early Rome* (Ithaca 1995), and G. Maslakov,

ductive imitations are the currency of empire, made legal tender because the consensus of readers of Roman history accepts them both as representing the way the *mos maiorum* worked and (even more importantly) as offering a blueprint for future generations.[47] In the *Ab Vrbe Condita* Livy constructs an ever-expanding narrative of ever-expanding power in which descendants learn from ancestors, younger generals from older ones, while the mechanism of the state perpetuates similarity through difference in its series of repeating annual magistracies.

Livy's doublets challenge the very basis of this historiographical project by disrupting our sense of chronological progress, the teleology that history takes as axiomatic.[48] These repetitions form loops that send us back to an exact "original" without telling us how to read the relationship between the events. We have the sense that, in Yogi Berra's words, "it's déjà vu all over again," but we learn no lesson. Nobilior's second entrance into Rome sends us shuttling back to his first in a meaningless and potentially endless cycle; with Pleminius and Scipio, theoretically unique events of death and burial are multiplied beyond all logic. The narrative trajectory goes haywire as journeys—textual elements that ordinarily facilitate transitions between scenes and move the story ahead with the characters—are repeated with no apparent reason.[49] And in all

---

"Themes of Tradition and Innovation in Livy's History," *Ancient Society: Resources for Teachers* (Macquarie, School of Historical, Philosophical and Political studies, 1971-87) 15-22. Erasing the difference between twins results in civil war (above, n. 37); between father and son, in an endless oedipal loop.

[47] Bourdieu (above, n. 2) 35-102 on "The Economy of Linguistic Exchange." The *Ab Vrbe Condita* is always going forward, into the past: for contemporary Rome in the ancient city see the references assembled by R. von Haehling, *Zeitbezüge des T. Livius in der ersten Dekade seines Geschichtswerkes* (1989); for the rush to the present, announced already in the *Preface*, see J.L. Moles, "Livy's Preface," *PCPS* 39 (1993) 141-68 and on the Roman preoccupation with present/past see M. Bettini, *Anthropology and Roman Culture* (Baltimore 1991) 113-93.

[48] Rimmon-Kenan 158: "although repetition can only exist in time it also destroys the very notion of time."

[49] Kraus *ad* 6.2.14, adding J.J. van Baak, *The Place of Space in Narration* (Amsterdam 1983) 98 (the journey is the fundamental model of the plot), 118 "... the archetypal plot construction of the road or journey with its motifs of motion provides the spatial co-ordinates for the articulation of time into transitions and episodes. Their spatial analogues, points in space (places), and the interplay between them, i.e. the stages of a trajectory, are particularly powerful in the modelling of larger, or hierarchically more independent narrative units, like chapters, stories in cycles and complete texts."

of this our historian-guide abrogates his authority, often failing to help us out of the loop even by telling us, where there are variant versions, which is to be preferred.[50]

The narratological relationship between "good" and "bad" repetition is analogous to that which Quint traces between the conflicting narratives of epic winners and their opponents. The winners' narrative we have already discussed; the losers' plot is more at home in romance (though without the happy ending) than in epic or history. The Odyssean books of the *Aeneid* with their repeated failures to repeat Troy, first near the Cretan Mount Ida, which Anchises chooses because it has the same name as the place they have left (a mistake not unlike that which generates some of Livy's doublets[51]), then in miniature at Buthrotum, show a revisiting of the past that is a "debilitating regression ... with a narrative counterpart in episodic romance digression." In general, Quint writes, "[the] paths of resistance to epic triumphalism assume a common shape: in opposition to a linear teleology that disguises power as reason and universalizes imperial conquest as the imposition of unity upon the flow of history, the dissenting narrative becomes deliberately disconnected and aimless... [T]he critique of this epic and of the ideological agenda behind it appears to be condemned not so much to the construction of alternative literary form as to the dismantling of form itself."[52] Livy's two narratives work in a similar way. Alongside the ordered narrative of power that moves from the foundation to the present, building an empire by repeating the successes of the past, is a second, "secret," subversive one in which the past returns without purpose, whose trajectory we can describe as fragmented, even aimless, and which threatens to dismantle the orderly form of the *illustre monumentum*.[53]

---

[50] Tony Woodman suggests that one thing that has gone wrong in these cases is that Livy infringes the ancient canons of source-selectivity: that is, on top of his usual reluctance to tidy inconvenient variants away (see Kraus 158, 298), with these doublets he does not present us even with a way to sort out the multiplicity of possible stories. For the *topos* see D. Fehling, *Herodotus and his "Sources"* (Leeds 1989) 275 (Index) s.v. "Sources (iii)."

[51] Cf. 29.35.2 (quoted above, n. 21) and 6.42.5-6.

[52] Quint 64, 40-41; see also M. Serres, *Rome: The Book of Foundations* (Stanford 1991) 97, 140, 224, 236-237.

[53] On secret narratives see F. Kermode, *The Genesis of Secrecy* (Cambridge, Mass. 1979).

I would suggest, moreover, that the profound uneasiness produced in the reader by this fragmented trajectory is analogous to the uneasiness occasioned by the potential collapse of the *imperium* itself. The *noua* toward which Livy's pushy (and necessary) readers are heading, are anything but secure (*Praef.* 4, 9, 12). The empire is too big for itself, precisely because it is an empire: according to authoritative conventional wisdom, it is inherent in large projects that they should fail. What is more, as Livian readers have often noted, the problems of the present infect the past: despite his eulogy of the good old days before vice immigrated to Rome (*Praef.* 11), Livy's early Romans fight the same moral and political battles that engaged late Republican and early Augustan society. Imperial expansion and assimilation begin at the beginning, with Romulus' asylum (1.8.4-6)—but together with them comes the possibility of importing corruption. The textual reflection of that empire, which shares with it a seemingly boundless complexity and expansiveness, as well as a habit of assimilating foreign elements (that is, Livy's narrative sources) and sometimes finding them recalcitrant or even indigestible, is no different.[54] Livy was as aware as we are of the difficulties he faced in trying to write a followable history of Rome's *imperium.* The places where he has failed tell us as much about the world as they do about Livy: "the most critical inference from this distancing and splitting of authority could be that the child will learn from the parent things the parent never dreamed of teaching."[55]

---

[54] The sources comprise both foreign (i.e., Greek) and ancestral elements; on Roman assimilation see Kraus *ad* 6.40.4, 41.10; on 1.8 as a meta-narrative text see *ead.* (1994b) 270.

[55] S. Bygrave, "The Pursuit of Sophia," in Hunter (above, n. 6) 153. Wendell Clausen first taught me to read Livy as poetry, for which I am most grateful. Audiences at the Institute for Classical Studies in London and the Universities of Manchester and Wisconsin patiently listened to earlier versions of this; I am particularly grateful for the hospitality extended by the Classics Department in Madison, where I was a Brittingham Visiting Scholar in April 1997. I would like to thank Mary Jaeger for inspiration, John Rich, Patricia Rosenmeyer and Tony Woodman for saving me from many errors and inconsistencies, and David Levene, who pointed me in the direction of intertextuality and whose comments have materially improved my arguments throughout.

THE DATE AND LITERARY CONTEXT OF AUSONIUS' *MOSELLA*: AUSONIUS, SYMMACHUS, AND THE *MOSELLA*[1]

DANUTA SHANZER

The companion-piece to this article dealt largely with internal evidence for the dating of the *Mosella*.[2] I argued against an early (pre-368) date for the military activities described in the poem and questioned whether Ausonius really went on campaign with Valentinian I and Gratian in the summer of 368. The traditional identification of the unknown office-holder with Petronius Probus was reaffirmed. I would like now to examine other external evidence relevant to the poem and the events to which it alludes.

The *Mosella* is inextricably bound to the relationship between Ausonius and Symmachus. It used to be thought that Ausonius first made contact with Symmachus in 369, when Symmachus came to the imperial court in Trier. But now it would appear that the relationship was initiated by Ausonius, the older man, by letter, at some earlier date, and that in 369 Symmachus met a man with whom he had already corresponded.[3] S. Roda has argued very plausibly that Symmachus' *Ep*. 9.88 is Symmachus' response to an overture from Ausonius that must be dated before 369 and after the date at which Ausonius became Gratian's tutor.[4]

---

[1] This piece about friends, politics, speeches, rivers, boundaries, and a poem is dedicated with affection to Wendell Clausen. I had the privilege of hearing Wendell teach Vergil's *Bucolics* and lecture on the *Aeneid* at Berkeley in the Spring 1982. However much he may deplore Later Roman literature, he might still hear the distant echoes of Vergil in these plashings of more Northerly streams. T.D. Barnes, S.J.B. Barnish, A.R. Birley, and R.S.O. Tomlin all read drafts of this article. I am most grateful for their criticism and for helpful argument and comments.

[2] See "The Date and Literary Context of Ausonius' *Mosella*: Valentinian I's Alemannic Campaigns and an Unnamed Office-Holder," forthcoming in *Historia* (1998).

[3] See S. Roda, *Commento storico al libro IX dell'epistolario di Q. Aurelio Simmaco* (Pisa 1981) 220-21.

[4] "Una nuova lettera di Simmaco ad Ausonio," *RÉA* 83 (1981) 273-80. See also A. Pabst, *Quintus Aurelius Symmachus, Reden* (Darmstadt 1989) 3 n. 12, hereafter "Pabst."

Symmachus and Ausonius in Gaul

The two first met in Gaul. Symmachus, *Ep.* 1. 132 (from Ausonius) says *dum in comitatu degimus ambo aeuo dispari, ubi tu ueteris militiae praemia tiro meruisti, ego tirocinium iam ueteranus exercui*. The terminology in this passage refers to the *militia ciuilis,* and not to any military encounters out in the wilds of Germania.[5] Ausonius, as far as we know, never left Gaul.[6]

Military Activities

In *Epistle* 1.14 addressed to Ausonius, Symmachus reminisced about his acquaintance with the Moselle. He says: *Ep.* 1.14 *noui ego istum fluuium cum aeternorum principum iam pridem signa comitarer*. "I got to know that river of yours when I had been following the standards of the eternal princes for some time." This sentence requires close attention. As we have seen, the evidence of the *Mosella* would indicate that Ausonius did not go on Valentinian's and Gratian's campaign of 368.[7] Ausonius and Symmachus clearly spent some time at court together, presumably following Symmachus' arrival in 369, during 369/70.[8] We have seen that Ammianus explicitly attested Gratian's presence on the campaign of 368 alone. What did Symmachus do and when?

Symmachus had not yet arrived in Gaul and clearly could not have participated in the campaign of 368. But he was present in the summer of 369, and accompanied Valentinian to the Rhine frontier.[9] The most natural interpretation of *iam pridem signa comitarer* is "had already been accompanying the military standards for a while," as opposed to Ausonius' *in comitatu degere*. Yet Symmachus uses the plural, *aeternorum principum*. Is this merely out of respect, with hindsight, for Gratian's

---

[5] R.S.O. Tomlin, "Notitia Dignitatum Omnium, Tam Civilium Quam Militarium," in R. Goodburn and P. Bartholomew, edd., *Aspects of the Notitia Dignitatum* (= *BAR* 15) (Oxford 1976) 191-192.

[6] *Symmachi Opera*, MGH AA 6.1, ed. O. Seeck (Berlin 1883) lxxix.

[7] See Shanzer (above, n. 2).

[8] Symmachus spent the year in Gaul. See J.F. Matthews, *Western Aristocracies and the Imperial Court 364-425* (Oxford 1975) 32.

[9] For Symmachus as eye-witness of Valentinian's activities in 369, see *Or.* 2.3 *sed haec magis diligo, quae probaui* and 2.18 *interfui, Auguste uenerabilis.*

status as Augustus?[10] Since in the spirit of imperial collegiality wars were waged in the name of both Augusti, whether or not both were in the field, then the plural need not mean that Gratian was on the front.[11] If Symmachus' usage is exact, however, then it may be evidence that Gratian went to the Rhine in 369, where his presence is otherwise unattested.

An argument against this suggestion could be made on the basis of Symmachus' orations. There is nothing about Gratian in the securely dated *Or.* 2 (1 January 370) to indicate his presence at the front in 369. One might think that Symmachus could, and presumably would, have addressed Gratian too at points such as *Or.* 2.23, had Gratian been involved. But, as we shall see, there is another explanation of Gratian's absence from the military activities of *Or.* 2.[12]

The question of Gratian's presence on campaign is linked to the date of Symmachus' 3rd *Oration.* The piece must date to after Gratian's proclamation as Augustus in 367 (*Or.* 3.3-4 and 8); it reflects campaigning by Gratian (*Or.* 3.7 and 10 *contubernio iunioris*) and hence cannot have been written before 368. The precise date, however, has remained elusive.[13] Seeck dates the piece, like *Or.* 1,[14] to 369 on the basis of an apparent allusion to *aurum oblaticium*, which he assumes must have been the quinquennalian *munus* from the Senate.[15] But the reference to the *quinquennalia* is in *Or.* 1, and the reference to *aurea munuscula* is in *Or.* 2. There is no explicit statement that the gold is quinquennalian. It could well be a separate present offered on another occasion.

---

[10] *Aeternus* makes its appearance as a fixed epithet of emperors after Constantine. See *TLL* s. v. "*aeternus*" 1147.3-4 and 6-9.

[11] See T.D. Barnes, *The New Empire of Diocletian and Constantine* (Cambridge, Mass. 1982) 27. In this connection it is worth noting that *ILS* 771 (*CIL* 6.1175) lists courtesy triumphal titles Germanicus, Alamannicus, and Francicus Maximus both for Gratian (who may or may not have been involved) and for Valens (who definitely was not). Ammianus (16.12.69) deliberately ignores collegiality when he upbraids Constantius for pretending to have fought at Strasbourg.

[12] See below p. 287.

[13] See Pabst (above, n. 4) 152-53 for a good summary of the problems.

[14] *Or.* 1.16 *lustrum imperialium iam condis annorum* clearly refers to the *quinquennalia.*

[15] *Or.* 2.1 *libens aurea sume munuscula*; Seeck (above, n. 6) xlvii.

Seeck's dating, however, does not take into account clear allusions to Valentinian's fortifications of 369 (*Or.* 3.9 for *castella* on both banks of the Rhine; *Or.* 2 *passim* and Ammianus Marc. 28.2).[16] Maurer's late dating to 371 (Gratian's *quinquennalia*) runs into difficulties with Symmachus, *Ep*. 9.112 to Probus, which shows that Symmachus had left Trier by the end of 370, and was somewhere far from *Galliarum longinqua* (9.112.2).[17] So, unless Symmachus made two trips, the campaigns of 370 can be ruled out.

But there is no reason not to argue description of the events of the campaigns of 369 and a dating to early 370. *Or.* 3 would then parallel *Or.* 2: this might explain the omission of allusions to Gratian in *Or.* 2, and Symmachus' promise (*Or.* 2.31): *de te, Gratiane uenerabilis, narrabo grandia, sed spondebo maiora. Narrabo grandia* refers to the praise of *Or.* 3 in anticipation: Symmachus knew he would be delivering it; *spondebo maiora* to future glories. Such a dating would nicely accommodate the evident allusion to the Pons Gratiani on the Tiber (*Or.* 3.9).[18] In addition, Gratian had held one consulship at this point in 366—hence no allusion to the second consulship of 371 at *Or.* 3.2 and 3.

In section 7 of the speech Symmachus describes how Gratian combined education and fighting on campaign (*tropaeis et litteris occupatus otiosa cum bellicis negotia miscuisti*). He then cites exempla of generals who took teachers on campaign: Fulvius Nobilior and Accius (an error for Ennius), Scipio Africanus and Panaetius, and Alexander and the philosophers. In Gratian's tents arms and books are combined. Similar themes are found in Ausonius *Precatio* 1, but there it is a question of Gratian's poetry. This is unlikely to be pure topos and one cannot help but wonder whether this paragraph hints at Ausonius' presence with his charge on campaign. This interpretation may find confirmation in the fact that Symmachus trotted out much of the same exemplary material in

---

[16] Pabst's suggestion of 368/9 encounters the same objections, although she tries to minimize them by stating (153) that modern critics are overinfluenced by Ammianus.

[17] See Pabst (above, n. 4) 153. Also H. Maurer, "Kaiser Valentinians Aufenthalt am Rhein im Sommer des Jahres 369," *Zs. für die Geschichte des Oberrheins* 25 (1910) 22 and Seeck (above, n. 6) xxv, n. 49.

[18] The bridge's dedication is dated by Pabst (above, n. 4) 158, n. 44 to between 10 Dec. 369 and 1 January 370.

378 (*Ep.* 1. 20) to congratulate Ausonius on his relationship with Gratian and his consulship. We have already cast doubt upon Ausonius' participation in the campaign of 368. If *Or.* 3 dates to 370 and describes events of 369, then we may have evidence beyond Symmachus' *aeternorum principum signa comitarer* to suggest that Gratian may have been on the front in 369 also.

To summarize the military activities of the two men: neither Ausonius nor Symmachus went on the campaign of 368. Symmachus definitely went to the front in 369. Ausonius and Gratian were arguably also present in 369, and the former may have also gone on the campaign of 370 and/or that of 371.

Symmachus and Ausonius may thus have campaigned together. There is one objection to this, the argument *ex silentio* on the basis of Symmachus' *Ep.* 1.14 to Ausonius: had they campaigned together and seen the Moselle together in that context, Symmachus would certainly have taken the opportunity to mention their shared experience. But it is also possible that Symmachus in his mock pique was not eager to emphasize their togetherness.[19] In the preface to the *Griphus*, Ausonius states that the work was conceived *in expeditione*, says that Symmachus will know of the *licentia militaris* of such occasions, but makes it clear that Symmachus was not present at the time the piece evolved.[20] The *Griphus* may perhaps date to the campaign of 370, and also, possibly, the *Cento Nuptialis*, written at the bidding of Valentinian and certainly a product of *licentia*: was the *licentia* also "*militaris*?"[21] Since, as we have seen, Ausonius wrote to Probus from the Rhine in 370,[22] it would appear that he kept on following Valentinian after 369.

---

[19] Is there an hint of irritation in the emphatic *ego* in *Ep.* 1.14.3 *noui ego istum fluuium*?

[20] Griphus, *praef. in expeditione, quod tempus, ut scis, licentiae militaris est, super mensam meam facta est inuitatio...* Ausonius makes it clear that Symmachus is seeing the verses for the first time.

[21] Note the military metaphors at the end of the preface: *quia sub imperatore tum merui, procedere mihi inter frequentes stipendium iubebis: sin aliter, aere dirutum facies, ut cumulo carminis in fiscum suum redacto redeant uersus, unde uenerunt.*

[22] See Shanzer (above, n. 2).

After Trier

After their joint time in Gaul the two continued to correspond. The friendship is documented in Symmachus, *Epp.* 1.13-33, many of which cannot be dated precisely. Further evidence relevant to the *Mosella* is also to be found in Symmachus' fragmentary panegyrics of Valentinian. The first, delivered for Valentinian's *quinquennalia* on the 25th of February 369, deals primarily with his origins, virtues, accession, his promotion of Valens, and the revolt of Procopius. The second, and longer one dates from the first of January 370. A third panegyric, addressed to Gratian, may well also date, as I have argued above, to 370. It is the second panegyric of Valentinian (Symm. *Or.* 2) that is of particular interest, because, although it has rarely been discussed in conjunction with the *Mosella*, it shares many telling thematic similarities, that suggest a close and illuminating relationship between poem and speech.

When similarities have been mentioned, it has been assumed that Symmachus knew the *Mosella*.[23] This is impossible. At some point[24] after the "publication" of the *Mosella*, when Symmachus was no longer in Trier, he wrote to Ausonius (Symm. *Ep.* 1.14) requesting a presentation-copy. He had clearly seen someone else's at that point (he knows something of the poem's content), but is somewhat annoyed at not having received his own copy. The words of the letter makes it clear that Symmachus could not have heard Ausonius recite the poem, and simply be complaining about not having received a copy. If he had heard Ausonius' poem read, he could never have said, *Cur me istius libelli, quaeso, exortem esse uoluisti? Aut amousoteros tibi uidebar, qui iudicare non possem, aut certe malignus, qui laudare nescirem.* Symmachus first heard of the poem from others at a time when he was away from Ausonius after his departure from the *comitatus*.[25] He makes witty play

[23] See the vague observations of H. Sivan, *Ausonius of Bordeaux* (London 1993) 113: "It also seems possible to hear echoes of the *Mosella* in Symmachus' allusion to the peace and order restored by the emperor on his accession."

[24] I am deliberately agnostic. Sivan (above n. 23) 393, n. 31 wants to put in 375 for no better reason than that the "rest of the correspondence between A. and S. … seems to belong … to the reign of Gratian."

[25] See Pabst (above, n. 4) 15. It is not known how long Symmachus stayed in Trier after 1 January 370. He next appears in Africa in 373.

with poetic allusions to emphasize his deprivation. At the end of his 2nd *Oration* Symmachus had perorated: (2.31) *ibo per urbes, ibo per populos iactantior uictore laudato*. In *Mosella* 475 Ausonius said of his triumphant river: *ibis in ora hominum*. Symmachus took this promise, fused it with an allusion to a famous line of Ennius, reused by Vergil (*Epigrammata* fr. 1.4 *uolito uiuus per ora uirum; Georg*. 3.9 *uictorque uirum uolitare per ora*), to task his friend: *uolitat tuus Mosella per manus sinusque multorum, sed tantum nostra ora praelabitur*. These facts are incompatible with an assumption that Symmachus' *Orations* of 369 and 370 worked from the *Mosella*, when, as we have seen, the *Mosella* reflects events of 368 and late 370/71.[26] Instead, if there is indeed a relationship between the panegyric and the *Moselle*, it must have been Ausonius who worked from Symmachus.

Symmachus, *Or*. 2 and the *Mosella*

Although the date of the second *Oratio* is known, the location of its delivery is debated.[27] The panegyric would have been formally addressed to Gratian as well as to Valentinian,[28] and, since the pupil was present, there is a good chance that the master, a friend of Symmachus' after all, also went to hear his friend perform and his charge praised.[29] But, even if Ausonius did not hear Symmachus' speech live, various features of the *Mosella* suggest that he had nonetheless read it and reacted to it.

[26] See (Shanzer, above, n. 2).

[27] Most assume that the speech was delivered in Trier. Pabst (above, n. 4) 15 suggests that it may have marked the dedication of the new fortifications on the Rhine. I would draw attention to Symm. *Or*. 2.4 *testis est haec ipsa ripa barbariae* which sounds as if Symmachus may have been at Altrip. A. Nagl, "Valentinianus I," *PWRE* 14A (1948) 2177 says that it was delivered in Alzey, but the basis for her suggestion is completely unclear. I would assume that she misinterpreted *Or*. 2.4 as a reference to Alzey, not Altrip.

[28] *Or*. 2.31 addresses Gratian.

[29] So rightly Eduard Norden, *Altgermanien* (Leipzig 1934) 33 n. 1.

A Challenge to a Poet?

At the end of his speech Symmachus mentions other writers.[30] He self-consciously refrains from breaking into poetry (2.26), and at 2.31 suggests that others should sing what remains: *canant alii quae supersunt*. While these may be purely conventional *formulae*, given the coincidences I will discuss, one wonders whether this might not have been an invitation or challenge to Ausonius, the most notable poet present to sing the deeds of Valentinian on the Rhine and Neckar.

This interpretation is reinforced by a later exchange between Symmachus and Ausonius. In the peroration of his speech, Symmachus mentioned various notable rivers in literature. The ancients had barely heard of the Rhine or the Neckar, the latter of which Symmachus characterized as *parem maximis* (*Or.* 2.24).[31] Only after Valentinian's victories, he tells us, had a barbarian *fluuius* been publicized. It is interesting to note that in his epistle to Ausonius (1.14), a wry, mock-hurt, and teasing effort, Symmachus seems deliberately to echo his own phrase (*parem maximis*) in his belittling comment on the *Moselle: noui istum fluuium*, "that river of yours," *parem multis imparem maximis*. The latter element is the significant one. Ausonius had exaggerated in making the Moselle colder than the Don and more famous than the Tiber. It was not up to the standards of the Rhine and Neckar. Had Symmachus wished Ausonius to take up his challenge and sing the Neckar? In echoing his own phrase from his second *Oratio*, is he insinuating that Ausonius should have heard it, and should have paid attention?[32]

Lines 398-99 of the *Mosella* may also have something to contribute: *dabitur nostris quoque purpura fusis.* | *quis mihi tum non dictus erit*? Ausonius is clearly hinting that he will be given a chance to deliver an

---

[30] *Or.* 2.29 *inter tot scriptores rerum tuarum* and 2.31 *canant alii quae supersunt.*

[31] Ammianus (28.2.2) alludes to it as *praeterlabente Nicro nomine fluuio*, a reference suggesting that the name is far from familiar. Ammianus similarly glosses the Main at 17.1.6 *Menum nomine fluuium.*

[32] C. Hosius, *Die Moselgedichte des Decimus Magnus Ausonius und des Venantius Fortunatus*, ed. 3 (Marburg 1926) 18 argued that Symmachus did not know of the *Mosella* in 370, because he could speak of the Neckar as unsung. The allusion to the Neckar in the *Mosella*, however, is hardly in such high relief as to qualify for panegyric.

imperial panegyric.[33] What, however, is the force of *quoque*? The word is most naturally construed as emphasizing *nostris*: "Purple will be given to my spindles also." Symmachus, *Or*. 3.9 had been unable to resist the Golden Age and the spindles of the Fates. *Quoque* could be a gentle allusion to his friend Symmachus' imperial panegyrics.[34] Ausonius too anticipates the same honor.

"Fluminal propaganda"

At the beginning of my discusion of its historical background, I hinted at the complicated generic nature of the *Mosella*.[35] I would now like to return to the inanimate protagonist of the poem, the Moselle itself, and to some of its possible significances. The river as geographical feature has many resonances. It can join people on opposite banks (*Mos*. 293-97); it can also divide, protect, or define a border; or be the vehicle of trade or imperialism. We have seen how Ausonius exploited the Danube as a military messenger between East and West.[36] Personified, the rivers of a country could even act as ancestral tutelary gods. When Alaric was withdrawing from the Italian peninsula, Claudian evoked a vivid picture of territorial jubilation among the rivers of Northern Italy.[37] Most commonly the river symbolizes a pacified nation. The following series of parallels, mostly from Claudian, illustrates this type of language in panegyric. The river can be bridged and be subjected to *calcatio colli*,[38] pacified,[39] conquered,[40] it can be unwarlike,[41] it can serve.[42]

---

[33] See above p. 291.

[34] For Symmachus' pride see *Or*. 2.31 *ibo per urbes, ibo per populos iactantior uictore laudato.*

[35] See Shanzer (above, n. 2).

[36] See Shanzer (above, n. 2).

[37] *6 Cons. Hon.* 146-192 (the awakening of the Po) and 193-200 the summoning of the other rivers, Ticinus, Addua, Athesis, Mincius, and Timavus.

[38] *Cons. Stil.* 3.13 *Rheni calcator et Histri*; 6 *Cons. Hon.* 648 *colla triumphati proculcat Honorius Histri.*

[39] 4 *Cons. Hon.* 440 *et Rhenum pacare iubes*; *Cons. Stil.* 2. 246 *usque adeone leuis pacati gloria Rheni*? Amm. Marc. 16.1.5 *pacatisque rigentis Rheni meatibus.*

[40] 3 *Cons. Hon.* 457 *Quis uictum meminit sola formidine Rhenum*?

The river, however, can also be personified as an active human conqueror in his progress from source to sea.[43] A fine (though unfortunately lacunose) example can be found in Ammianus Marcellinus' description of the Rhine at 15.4.2-6:

> Inter montium celsorum amfractus inmani pulsu Rhenus ... scopulos extenditur ... *amnes adoptans*, ut per cataractas inclinatione praecipiti funditur Nilus. et nauigari ab ortu poterat primigenio copiis exuberans propriis, ni *ruenti* curreret similis potius quam fluenti ... iamque *absolutus* altaque diuortia riparum[44] adradens lacum *inuadit* rotundum et uastum, quem Brigantiam accola Rhaetus appellat, perque quadringenta et sexaginta stadia longum parique paene spatio late diffusum horrore siluarum squalentium inaccessum,—nisi qua uetus illa Romana uirtus et sobria iter conposuit latum—barbaris et natura locorum et caeli inclementia refragante. hanc ergo paludem spumosis strependo uerticibus amnis irrumpens et undarum quietem permeans pigram, mediam uelut funali intersecat libramento et tamquam elementum perenni discordia separatum, *nec aucto, nec imminuto agmine quod intulit, uocabulo et uiribus absoluitur integris* nec contagia deinde ulla perpetiens oceani gurgitibus intimatur.

and of the Rhone at 15.11.16-18:

> et quoniam ad has partes opere contexto peruenimus, silere super Rhodano maximi nominis flumine incongruum est et absurdum. a Poeninis Alpibus effusiore copia fontium Rhodanus fluens et *procliui impetu* ad planiora degrediens proprio *agmine* ripas occultat et paludi sese ingurgitat nomine Lemanno eamque intermeans nusquam *aquis miscetur externis*, sed altrinsecus summitates undae praeterlabens segnioris quaeritans

---

[41] *Cons. Stil.* 1.20 *tota quod imbellem miratur Gallia Rhenum*

[42] *Prob. et Ol.* 160-61 *Annue, sic nobis Scythicus famuletur Araxes, | sic Rhenus per utrumque latus*; *Epith. Hon. et Mar.* 278 *iam Rhenus et Albis | seruiet*

[43] For a miniature example of such a *cursus* see *Mos.* 470-73: *fons, meatus, ostium.*

[44] See *Mosella* 432 *alueus extendet geminis diuortia ripis*. The *iunctura diuortia* and *ripa* seems to occur only in Ammianus and Ausonius.

> exitus uiam sine impetu ueloci molitur. *sine iactura rerum* per Sapaudiam fertur ... *Ararim... suum in nomen adsciscit...* hinc Rhodanus *aquis aduenis locupletior* uehit grandissimas naues ... finitisque interuallis, quae ei natura praescripsit, spumeus Gallico mari concorporatur per patulum sinum, quem uocant Ad gradus, ab Arelate octauo decimo ferme lapide disparatum.

Words with military connotations are used such as *agmen,*[45] *impetus, uires, ruens*. Legal, political, and financial language is also to be found (*in nomen adsciscere, adoptans, absolutus, iactura rerum*) as well as words that suggest other sorts of political personification (*aquis aduenis, aquis externis*). One might call these differing literary uses of rivers "fluminal propaganda," in light of Symmachus' quasi-epigrammatic phrase in *Or*. 2.24: *cui amnes incogniti mancipantur, longinqua terra coniungitur.*

Although much of the *Mosella* is devoted to the pastoral praises of the Moselle itself, at the end of the poem the focus shifts. It would be an error to get distracted by the painterly shimmer of the river in the evening and miss its other meanings. Ausonius shaped his poem deliberately. The reader, no longer the lazy observer of the view from the banks, or of what lies on, or below the surface of the water, is swept along the course of the river towards its confluence with a far greater stream. The Moselle seeks its consummation in coupling with the Rhine, the politically important brother-river. The climax of the *Mosella* is reached when the Moselle joins the Rhine that will emerge in its northward-bound push as the true *limes*.[46] The Rhine too gains from the coupling—it acquires a *geminum nomen* and becomes triumphantly *bicornis*.[47]

> Accedet tanto geminum tibi nomen ab amni
> Cumque uno de fonte fluas, dicere bicornis.

---

[45] A clear poeticism, see *TLL* s. v. *agmen* 1340.34-56.

[46] *Mos*. 434 *accedent uires, quas Francia quasque Chamaues | Germanique tremant: tunc uerus habebere limes.*

[47] *Mos*. 436-37

This passage has caused difficulties, eliciting interpretations as different as those of Green and Ternes.[48] The former (p. 509) translates "a double name," but does not explain what the "double name" is. The alleged parallel between the double names of the Danube, Hister and Danuvius, is not instructive. Instead Green's second suggestion, "a name with a double meaning" is better: the traditional theriomorphic river-deity is a two-horned bull,[49] but the river is also "two-horned" because it has the Moselle as a tributary.

The passage makes the most sense if compared to Symmachus' second *Oratio*, where he had spoken of the personified Rhine over-familiarly and jocularly as *bicornis iste uicinus* (2.14), "that two-horned neighbor of yours."[50] Naturally he alluded to a lofty epic precedent, *Aen.* 8.727 *Rhenusque bicornis*. Ausonius, however, reinterpreted the epithet. *Bicornis* is a *geminum nomen*, both because it contains the prefix *bi-*, and because to the traditional iconography of the river-god is added the new topical element, the major tributary that can make a river with one source "two-horned."[51]

The 3rd *Oration* (9) also contains a passage that is very similar to material in the *Mosella*. Symmachus discusses the bridging of the Rhine: *ille libera hucusque ceruice repagulis pontium captiuus urgetur. en noster bicornis caue aequalem isto arbitrere Tiberino, quod ambo principum monumenta gestatis: ille redimitus est, tu subactus.* At *Mosella* 374-379 Ausonius claims that if the Moselle had had a Homer or a Vergil, both the Simois and the Tiber would have ceded their primacy. He apologizes to Rome, and makes an apotropaic appeal to Nemesis. So here again a synkrisis between the Tiber and a Northern rival.

---

48 *Ad loc.* 95 "après avoir acueilli en lui-même la Moselle, le Rhin aura deux noms." (i.e. Rhine and Moselle) I know of no parallel for one river adopting the name of its tributary; the larger river imposes its name on the tributary, see Claudian, below p. 296.

49 Compare Claudian, *Cons. Stil.* 1.220 *Rhenumque minacem | cornibus infractis adeo mitescere cogis.*

50 He employed the same allusion in *Or.* 3.9 *en noster bicornis.*

51 Servius *in Aen.* 8.727 suggests that *bicornis* refers to the two *mouths* of the Rhine: *bicornis commune est omnibus fluuiis aut proprie de Rheno quia per duos alueos fluit: per unum qua Romanum imperium est, per alterum qua interluit barbaros, ubi iam Vahal dicitur et facit insulam Batauorum.*

In evaluating the message of the *Mosella*, we must address the climax of the poem, where Ausonius turns to the Rhine (vv. 418-37). Here Ausonius comes very close to writing panegyric. The Moselle will augment the Rhine with news of Valentinian's and Gratian's recent triumph and those to come (426 *hinc alias aliasque feret*). The message is clear: the rivers combined will be more powerful and the Franks and Chamavi will fear Roman might. There is a very close parallel to this panegyrical topos in Claudian (*Bellum Geticum* 329 ff.) where Rhaetia offers two rivers, the Rhine and the Danube to the Romans:

> . . . sublimis in Arcton
> prominet Hercyniae confinis Rhaetia siluae,
> quae se Danubii iactat Rhenique parentem,
> utraque Romuleo praetendens flumina regno:
> primo fonte breues, mox alto gurgite regnant
> et fluuios cogunt unda coeunte minores
> in nomen transire suum.

Here too the rivers are personified with imperial power (*regnant*),[52] absorb the names of their tributaries,[53] and, presumably, transmit Roman rule along their courses. This section of the *Mosella*, however much Ausonius may retreat in his eirenic conclusion to his native Garonne, clearly calls for the maintenance and fortification of the Rhine *limes*. The personified river is an active participant in this effort.

Fortification and Architecture vs. War

All of this, however, represents not reality, but wishful thinking, a pastoral Moselle linked to a militaristic Rhine. Instead the reality seems to have consisted of extensive building- and fortification-programs: *haec tamen caute gesta*, as Ammianus said,[54] who (28.2) details Valentinian's fortification of the Gallic bank of the Rhine from Rhaetia to the sea, and occasional fortification on the Neckar, for example at Mons Piri (Heidelberg at 28.2.5). 28.2.2 tells of the attempted fortifications on the

[52] See Amm. Marc. 15.4.4 for the *agmen* of the Rhine: *nec aucto nec imminuto agmine quod intulit*, also 15.11.16 for the *agmen* of the Rhone.

[53] See Amm. Marc. 15.4.2 for the Rhine *amnes adoptans*, also 15.11.18 for the Rhone *aquis aduenis locupletior*.

[54] 30.7.8

Neckar. In his summary of Valentinian's achievement, he specifically mentions his building-program on both sides of the Rhine.[55]

Necessary though they might be, the building of fortifications was not an achievement that regularly brought glory to the general or pleasure to the soldiers concerned. Julian's auxiliaries despised such tasks,[56] and we may be reasonably certain that this was generally true. Even the *HA* shows Probus' soldiers rebelling after having been ordered to drain a marsh near Sirmium.[57] Ammianus makes a special point of stating that Valentinian was stubborn and that his soldiers were *morigeri* ("obliging") in working in water up to their necks (28.2.5).

Bridges and Hostages

Just as the Moselle dominates Ausonius' poem, so the newly pacified Rhine lies at the center of Symmachus' speech. So many themes are shared by both works that it is difficult to escape the conclusion that the authors are in dialogue. In particular it is worth noting that the Symmachean Rhine is a river overpowered by man's skill: he must learn to give in to ships (2.4), if he is not to be bridged. He submits like a conquered barbarian to bridges (3.9) and can dare to rival the Tiber because he sports a bridge. The Emperor's power over the landscape, such as the bridge over the waters, makes sailing a *ludus* (2.4). (To this we might compare the peaceful aquatic *ludi* described in such detail at *Mosella* 200-221.) Ausonius, for his part, spoke disparagingly of Xerxes' bridging of the Hellespont (*Mos.* 289-291), and emphasized instead the possibility of peaceful contact across the Moselle *without bridging: licet hic commercia linguae iungere et alterno sermonem texere pulsu. blanda salutiferas permiscent litora uoces, et uoces et paene manus.* Symmachus states that the Rhine, as his payment for the *pax Romana*, offered his child, the Neckar, as a hostage (*quasi quoddam pignus* 2.23). Ausonius significantly reverses the situation, apologising initially for the "invasion" of the Rhine by the Moselle—the Rhine will gain not only water, but honor: the Moselle, after all, had been eyewitness to Valen-

---

[55] 30.7.6

[56] Amm. Marc. 18.2.6 *auxiliarii milites semper munia spernentes huiusmodi.*

[57] *HA* Probus 21.2-3

tinian I's and Gratian's victory-celebrations in Trier. Rhine and Neckar are conquered foe and hostage, Rhine and Moselle are generous host and gift-bringing guest.

War vs. Architecture

The two works also share a common interest in architecture. From the remains of Symm. *Or.* 2, it is clear that the campaigns of 369 did not primarily involve ruthless military operations or successes in battle.[58] Symmachus instead does everything to emphasize the achievements of Valentinian's building program. The final paradox: where are those who for so long took German *ciuitates* by siege (2.13)? The Germans are now besieged by *ciuitates*; Germany herself, formerly an unglamorous creature of daub and wattle, will finally achieve the mural crown of a *prouincia* (2.14), and Valentinian a new sort of laurel, the *triumphus pacis* (2.30). Ammianus (28.2.1) provides a more sober statement of these facts: *nonnumquam etiam ultra flumen aedificiis positis subradens fines barbaros.*

Lopodunum

Architecture is prominent in Symmachus' *laudatio*. He devotes more time to it than he does to military force. In 2.1 and 2.28 he discusses the *ornatus* of the Rhine: *riparum margines operum corona praetexit.* In 2.2. and 2.13 new towns displace the barbarians. Two paragraphs of Symmachus' *laudatio* are devoted to Valentinian's activities at Lopodunum (modern Ladenburg).[59] The town had been incautiously sited,[60] and the Alamanni had burned down the *colonia* there. All that was left were either inscriptions commemorating the now broken *foedus* between Roman and German, or else the grave-markers of the inhabitants (2.16 *tituli sceleris proditores*). After driving the enemy

---

[58] See *Or.* 2.6 not enough troops; 2.12 the living Alamann (cf. Ammianus 28.5.15); 2.13 not troops, but towns in pursuit.

[59] See R.S.O. Tomlin, *The Emperor Valentinian I* (unpublished D.Phil. thesis, Oxford 1973) 165.

[60] *Or.* 2.16 *incaute positam ciuitatem probarat amissio.*

away and recapturing the town, Valentinian seems to have transferred the building materials of Lopodunum to Altrip.[61]

Architects

At *Or.* 2.18 Symmachus, who had been present (2.18 *interfui, Auguste uenerabilis*), expatiates on the architectural achievements of Valentinian, *artifex purpuratus*: he outlined foundations and used the plumb-line. He makes a favorable comparison of Valentinian to three famous architects of the past: Archimedes, clever though he was, could not in the end protect Syracuse. Epius all by himself brought ruin to Troy and Daedalus alone fenced in the menacing Minotaur, but what can Germany expect, since a whole army can capture and enclose her? Architecture and the skills of peace-time architects likewise furnish one of the major excursus in the *Mosella* (vv. 298-348), a passage that seems to be based on one of Varro's *Hebdomades*,[62] but could very well have been prompted by Symmachus' efforts.[63]

Military architecture

At *Or.* 2.20 Symmachus describes a fortified landing-place at Mannheim-Neckarau[64] on the Rhine, complete with loop-holes for arrows and a smooth rank of downward sloping tiles, at the intersection of two rivers. The river is confined on both sides and forced to provide safe passage for various purposes. The place combines the virtues of nature (*naturae munus*) with those of the *artifex manus*. Despite its military function, the place is ornamented: *quasi sollicitus munita fecisti, quasi securus ornasti*.

---

[61] *Or.* 2.16 *animus uictoris apparuit receptae urbis reliquias transferentis.* Pabst p. 146 n. 68 mentions *CIL* 13.6416, found near Altrip, and supposedly from Lopodunum. On the *spolia* from Lopodunum used for Valentinian's fortification of the Neckar, see *CIL* 13.2.1 p. 231. On Ladenburg see P. Filtzinger, D. Planck, & B. Cämmerer, eds., *Die Römer in Baden-Württemberg* (Stuttgart & Aalen 1976) 341-52.

[62] See below p. 304.

[63] The *Hebdomades* were known to Symmachus. See *Ep.* 1.2.2 and 1.4.1.

[64] Seeck's restoration *conditae ciuitatis* at 2.20 is unlikely to be correct, given that it is not a *ciuitas* that is described in the following paragraph.

A clear thematic parallel with the *Mosella* emerges: architecture that reflects opposing purposes and concerns. Valentinian's landing-place is fortified, as if he were worried, but ornamented, as if he were carefree. Ausonius (*Mos*. 456-57) says that he will add to his description of the Moselle camps built as garrisons for uncertain times that are now granaries for the secure Belgae, a clear "swords into ploughshares" motif:[65]

> Addam praesidiis dubiarum condita rerum,
> sed modo securis non castra, sed horrea Belgis:

The passage may well allude to harder times in Gaul under Julian when, as part of his consolidation efforts he rebuilt burnt granaries[66] and recaptured and fortified various towns, including Bingen.[67]

* * *

Symmachus had wanted the Neckar sung. Ausonius did not respond quite properly. Instead of pure panegyric, he wrote a complicated poem with many models, many modulations, but only one military section that is clearly panegyrical. Even that, far from dominating the reader's sensibilities, fades away into the dying fall of Ausonius' unembattled natal stream, the Garonne. The *Mosella*, despite its warlike climax, fails to convey a suitably fierce or militaristic message. Once the man-power is added to make the Franks, Chamavi, and Germans tremble (vv.434-35), only then will the Rhine be a true *limes*. The implication is clear: the power is not there and the Rhine is not a true defensive *limes* at the time of writing.[68]

Ideally the Romans would have wanted to push beyond the right bank of the Rhine, a traditionally sensitive point. Roman occupation had

---

[65] See the propaganda (presumably reflected back from the 4th to the 3rd C.) in the *Historia Augusta,* Probus 21.5-6.

[66] Amm. Marc. 18.1.3 *horrea quin etiam extrueret pro incensis, ubi condi posset annona a Brittannis sueta transferri.*

[67] Amm. Marc. 18.3-4 which should be compared to *Mos*. 2 *addita miratus ueteri noua moenia Vingo.*

[68] I cannot see how Green *ad loc.* can suggest that the sentence is not in some way critical of the current situation.

largely been abandoned under Claudius.[69] There were fairly consistent reverses during the tumultuous 3rd. C. Only Probus, we are told by the highly suspect *HA*, crossed the Rhine to make inroads in Alemannic territory. In the 4th C. Julian campaigned on the right bank of the Rhine,[70] and subsequently Valentinian at Solicinium. After his death, the right bank of the Rhine continued to be an issue. This is still evident slightly later in Claudian, who anticipates times when one will not be able to tell which bank is Roman, and the Belgic woman may be able to pasture her cattle among the Chauci, when the Rhine will serve Rome, or when it will lament the fact that Rome does not rule on both banks.[71] The lie resurfaces in the mid-5th C. in Merobaudes: *addidit hiberni famulantia foedera Rhenus* | *orbis et Hesperiis flecti contentus habenis* | *gaudet ab alterna Thybrin sibi crescere ripa.*[72]

But military control of the right bank, except as somewhat deviously presented by Symmachus, was unrealistic. The reality remained one primarily of fortification. This reintroduces the theme of building that one cannot avoid in understanding the "message" of the *Mosella.* Ammianus describes fortifications and dams; so too Symmachus. Ausonius however depicts domestic archictecture: luxurious Statian villas to rival the amenities of Italy, and, finally, camps built for *dubiae res* that are now granaries for the secure Belgae (vv. 456-57). Civilian architecture and swords turned into ploughshares, a vision of pastoral security belied by historical fact.

Symmachus had depicted a conquered Rhine both in *Or.* 2 and *Or.* 3 and asked for the Neckar, beyond the Rhenan *limes.* Ausonius gave

---

[69] T. Mommsen, *The Provinces of the Roman Empire: the European Provinces*, ed. T.R.S Broughton (Chicago and London 1968) 131-32.

[70] In 357 (Amm. 17.1.2), 358 (Amm. 17.10), and 359 (Amm. 18.2).

[71] See *Cons. Stil.* 1.220 *Rhenumque minacem* | *cornibus infractis adeo mitescere cogis ... geminasque uiator/cum uideat ripas quae sit Romana requirat;* | *ut iam trans fluuium non indignante Chauco* | *pascat Belga pecus...*; *Prob. et Ol.* 161 *Sic Rhenus* (*sc. famuletur*) *per utrumque latus*; *Manl. Theod.* 52 *diuersoque tuas coluerunt gurgite uoces* | *lentus Arar Rhodanusque ferox et diues Hiberus.* | *O quotiens doluit Rhenus, qua barbarus ibat,* | *quod te non geminis frueretur iudice ripis*! *3 Cons. Hon* 652 *tempus erit, cum tu trans Rheni cornua uictor.*

[72] *Pan.* 2.5-7 Vollmer. See F. M. Clover, *Flauius Merobaudes*, = *Transactions of the American Philosophical Society* 61.1 (1971) 42 for the extravagance of the claim.

him and his emperor the Moselle, admittedly a Moselle that gained status for twenty lines linked to the politically-charged triumphant Rhine, but the peaceful Moselle nonetheless, a stream that flows to the West of the Rhine. This action, in and of itself, is a *recusatio*. Although Ausonius briefly approaches the *maior stilus* (vv. 417-37) and the possibility of a conquering Rhine, he retreats quickly to promise more on his peaceful river. In *Or*. 3.9 the Rhine is put in its place with regard to the Tiber. Ausonius leaves the Rhine out of comparison to the Tiber and puts forth the peaceful Moselle—but only if she had a worthy bard. He ends with promised respects from southern streams, the Drône, the *alpini fluuii*, the Rhône, and ends with his own recommendation to the Garonne. The northeastern *limes* is abandoned.

The *Mosella* is a richly-textured and complicated poem: like so many major monuments of Latin literature, it is both personal and political. It was written after 368 (Valentinian I's victory at Solicinium) and published during or after Petronius Probus' combined Praetorian Prefecture of Italy and Consulate in 371. A second edition (our current text) appeared after 378/9 (allusion to Ausonius' consulship). One cannot tell how often or in what form it may have been read and circulated between 368 and 378/9. It reflects an early victory of Valentinian's against the Alamanni and a triumph at Trier.

The *Mosella* does not allude directly to Valentinian I's building-program on the Rhine in 369. Nonetheless, its central themes and language show clear signs of dialogue with Symmachus' 2nd and 3rd *Orations*, both, it is argued, delivered in 370. Above all it is important to note that by writing an *itinerarium-cum*-praises of a river, Ausonius intentionally invokes the larger literary topos of "fluminal propaganda." For a few lines the Rhine receives the grand treatment (Ausonius shows what he could have done), but he mentions the Neckar only once, and quickly retreats to the pastoral and civilian Moselle and its fragile beauties. Ausonius achieves a difficult balance beween sensitive description, encyclopedic learning, quasi-Alexandrian whimsy, political panegyric, and political flattery. But what he fails to do is equally important. At least three powerful men, their aspirations, and dealings with Ausonius went into the making of the *Mosella*: Valentinian I, Petronius

Probus, and Symmachus. None of them, one might guess, would have been unreservedly pleased with Ausonius' poem.

Appendix: Symmachus, Ausonius and Pliny's Naturalis Historia

At some point after they had met at court, Symmachus sent Ausonius a copy of Pliny's *Natural History* (*Ep.* 1.24):

> Si te amor habet naturalis historiae, quam Plinius elaborauit, en tibi libellos, quorum mihi praesentanea copia fuit. in quis, ut arbitror, opulentae eruditioni tuae neglegens ueritatis librarius displicebit. sed mihi fraudi non erit emendationis incuria. malui enim tibi probari mei muneris celeritate, quam alieni operis examine.

This letter should be dated to after Symmachus' departure from court. In Trier, Ausonius, as imperial tutor, would certainly have had the resources himself to command what books were available without recourse to Symmachus—had there been a copy available. Since Symmachus apologizes for the poor quality of the text, we can also assume that Ausonius did not have a copy at all. (Otherwise it might have been possible for him to have had a poor copy, but to have requested a better one.) The clear force of this letter is that *any* text is better than none.

When Ausonius first read Pliny is a question that affects the dating of the *Mosella*. Since this letter seems to postdate Symmachus' return to Rome, if the *Mosella* can be shown to be dependent on the *Natural History,* then an early date can be ruled out. Hosius stated that Ausonius had no knowledge of the *HN* in his earlier writings, but referred to it frequently in the *Mosella* and afterwards.[73]

One of the passages in the *Mosella* that is supposedly dependent on Pliny is the excursus on architects at vv. 300-17. The architects who appear there are: Daedalus (300-3), Philo (303), Archimedes (303-4), Menecrates (307), Chersiphron (308), Ictinus (309-11), and Dinochares (312-17). Pliny (*HN* 7.125) has Archimedes, Chersiphron, Philo, Ctesibius, and Dinochares. Thus four of Pliny's architects (Archimedes,

---

[73] See Hosius (above n. 32) p. 18, esp. n. 1 which cites 303, 308, 312, 314, 363, 379 (Pliny 11.251). Green, on *Mos.* 306, states that, although Pliny was "evidently familiar to Ausonius," he was not his source here.

Philo, Chersiphron, and Dinochares) are also found in Ausonius. But there is no probative evidence that Ausonius was working from Pliny.[74] Nothing in Ausonius directly contradicts what is in Pliny, so Pliny *could* have been one of Ausonius' sources, but certainly not his only one. It is unnecessary to posit Pliny as a source. Instead, given Ausonius' citation of Varro (vv. 305-7 *forsan et insignes hominumque operumque labores | hinc habuit decimo celebrata uolumine Marcei | hebdomas*) and the fact that the architects are a significant seven in number, it seems most likely that the list came from Varro's lost *Hebdomades* or from an intermediate source that cited them directly.[75] The *Hebdomades* were clearly extant both in this period and in Ausonius' circle, for they are mentioned in Symmachus, *Ep*. 1.4.1 and 1.2.2.[76] Ausonius, however, evidently elaborated on Varro: Daedalus (300), for example, is mediated, at some stage, by Vergil (*Aen*. 6.32).

Famous architects' work is unfavorably compared to Valentinian's efforts in Symmachus, *Or*. 2.18-19: Archimedes and Daedalus feature prominently. At *Or*. 2.22 Symmachus mentions the pyramid (cf. Ausonius, *Mos*. 312-13). But again, Symmachus cannot be considered a *source*, although he may well have been the inspiration behind Ausonius' excursus. There were no doubt many available sources for learned allusions to architects.[77]

---

[74] For Philo, Ausonius has more than Pliny does. In the case of Archimedes there is some overlap, but Ausonius is fuller; Ausonius' account of Chersiphron could depend on Pliny, but his account of Dinochares is again much fuller than Pliny's. He could however have conflated Pliny 7.125 with Pliny 34.148 for further details on *Timochares*. See Green on *Mos*. 312.

[75] See F. Ritschl, "Ueber des Marcus Terentius Varro Imaginum sive Hebdomadum libri," in *Opuscula Philologica* 3 (Leipzig 1877) 508-522.

[76] See Ritschl (above, n. 75) 513-514 and R. Reeh, *De Varrone et Suetonio quaestiones Ausonianae* (Halle 1916) 20.

[77] Note for example, that some of the material in the *Mosella* has loose parallels in Cass. *Var*. 1.45. At 314-315 Ausonius mentioned Dinochares who was ordered to suspend an image of Arsinoe, the token of an incestuous love, in mid-air in the her temple. The feat was achieved through the use of magnets. Cassiodorus, *Var*. 1.45.10 speaks of an iron Cupid that was suspended in Diana's temple without any attachment, but the *ars* in Cassiodorus that makes all these miracles possible seems to be the *ars mechanica* (*Var*. 1.45.4, 10 *mechanisma*, and 11).

Inuidia et Latiae Nemesis non cognita linguae (Mosella 379 before a lacuna) provides somewhat better evidence for use of Pliny's *HN*. He seems to be the only known Latin source for this particular tidbit.[78]

[78] Pliny 11.251 *est post aurem aeque dexteram Nemeseos (quae dea Latinum nomen ne in Capitolio quidem inuenit)* and 28.22 *alii Graecam Nemesin inuocantes, cuius ob id Romae simulacrum in capitolio est, quamuis Latinum nomen non sit.*

AUGUSTUS AND THE POETS IN MUSSOLINI'S ROME

CLIVE FOSS

Italiae fines promovit bellica virtus
Et novus in nostra funditur urbe decor

This couplet on a building in the very centre of renaissance Rome expresses the spirit of a recent age of intense classical revival. During its two decades in power, the government of Mussolini consciously exploited ancient Rome as a model and precedent for its own policies.[1] It conducted excavation and restoration, as well as lavish celebrations of the anniversaries of Virgil, Horace and Augustus. After the conquest of Ethiopia and proclamation of the restored empire in 1936, the Duce actively pursued the parallel with the first emperor, whose memory he glorified with a mass of propaganda.

I offer these pages to Wendell Clausen because of his interest in all periods of Rome's existence. His commentary on "dum Capitolium scandet cum tacita virgine pontifex" by evoking the image of the priest and nun ascending the steps of the Aracoeli showed this breadth of vision and made a deep impression on his first freshman class at Harvard. I hope therefore that these remarks on an age that permanently transformed Rome will merit his interest.

The belligerent tone of the verses above, inscribed in 1937, reflects the new empire, their context the simplified classicism of the fascist style that spread through the city, and their Latin the identification with a glorious past. They adorn the small palazzo of the National Insurance Association opposite the church of S. Andrea della Valle, part of a typical grandiose project.[2] A new street, the Corso del Rinascimento, lined with modern buildings, cut straight through the renaissance district. It was intended to continue south to the Tiber, where extensive demolition

---

[1] J.P. Bondarella, *The Eternal City: Roman Images in the Modern World* (Chapel Hill 1987) 172-206 is a good general introduction to this subject.

[2] See A. Foschini (architect of the project), "Il Corso del Renascimento," *Capitolium* (see below, n. 14) 12 (1937) 73-89.

would free and display the Theatre of Pompey, but this part was never completed.

The project was one of many that created the modern image of ancient Rome. When the fascists came to power in 1922, the urban fabric of central Rome was still intact. Although much of the Forum and Palatine had been cleared and excavated, a dense network of houses, many of them poor and crowded, stretched from the Colosseum to the monument of Victor Emmanuel, covered the Capitoline, and extended down to the Tiber and Velabro. Beginning in 1923 with the temple of Fortuna Virilis (favored by the Duce, who prided himself on his good luck and virility), demolition, excavation and restoration cleared a vast area of ancient Rome. Mussolini wanted to modernize the city by destroying unhealthy slums and opening up urban spaces in the Roman style, in order to ease the flow of traffic and reveal the glories of ancient Rome untrammelled by the organic growth of centuries. As he proclaimed in 1925: "I monumenti millenari della nostra storia devono giganteggiare nella necessaria solitudine." He was particularly concerned with the city of Augustus: "tra cinque anni [from 1925] Roma deve apparire meravigliosa a tutte le gente del mondo; vasta, ordinata, potente, come fu ai tempi del primo impero di Augusto."[3] As a result, the imperial fora were revealed, the Capitoline was cleared and landscaped, the theatre of Marcellus restored and the temples of the Velabrum and of the Largo Argentina as well as the Circus Maximus were exposed.[4]

The present image of ancient Rome is largely the product of this fascist ambition that included far too many projects to discuss here. Instead, I will concentrate on paper rather than stones, and consider the

---

[3] Benito Mussolini, *Opera omnia* (Florence 1951-1962) 22.47f. Both quotes are from a programmatic speech, "La nuova Roma", celebrating the inauguration of the first appointed governor of Rome.

[4] The bibliography of these and related projects is extensive. In addition to the articles that constantly appeared in *Capitolium*, see Antonio Muñoz (who supervised and designed much of this work), *Roma di Mussolini* (Rome 1935) 119-224; Gaston Bardet, *La Rome de Mussolini* (Paris 1937) 103-96; Antonio Cederna, *Mussolini urbanista* (Rome 1979), a dim view, with full references, and two works of Spiro Kostof, *The Third Rome* (Berkeley 1973) and "The Emperor and the Duce: the Planning of the Piazzale Augusto Imperatore in Rome," in H. Millon and L. Nochlin, edd., *Art and Architecture in the Service of Politics* (Cambridge, Mass. 1978) 270-325.

equally extravagant treatment the regime lavished on Virgil, Horace and Augustus. Their anniversary celebrations reveal the immense mechanism of publicity that a totalitarian government commands and the uses to which ancient writers and rulers can be put to support modern ideals and policies. The classics were very much alive in Mussolini's Italy, but often in a form barely recognizable to serious scholars.

The fascists took great interest in agriculture and in the countryside. Already in 1925, Mussolini announced a campaign to achieve self-sufficiency in wheat. The Duce himself was photographed shirtless harvesting grain, leading the charge in a widely-publicized effort that was actually successful by 1929. At the end of the campaign, he handed out awards to the "vincitori" of the "Battle of Grain" in the Teatro Argentina.[5] Concluding his speech, he announced: "Il 1930, o camerati agricoltori, sarà l' anno di Virgilio, il poeta dell' Impero e dei campi. Noi lo celebreremo fascisticamente, al lavoro e col lavoro." A new campaign had begun.

Actually, the celebration of the bimillenary of Virgil's birth had been in planning since the beginning of the regime.[6] In 1923, the Virgilian Academy of Mantua announced a program of publications and exhibitions, and in 1927 dedicated a monument to the poet in the presence of representatives of the government and of Italian and foreign universities.[7] The project gained real momentum, though, after the personal intervention of the Duce who entrusted the celebrations to the Italian Academy, itself a fascist creation.

Virgil received universal attention in 1930. He was celebrated at all levels in formal orations, commemorative poems, learned conferences, editions, books, scholarly studies, magazine articles, celebratory medals and a series of postage stamps. By the end of the year, no one in Italy could have escaped his praise. Although many of these works are sober

---

[5] Mussolini (above, n. 3) 24.178.

[6] The best introductions to the bimillennium are the article of Luciano Canfora, "fascismo" in *Enciclopedia Virgiliana* 2.469-472, with full references, and T. Ziolkowski, *Virgil and the Moderns* (Princeton 1993) 17-19. Despite its title, R. Faber, "Présence de Virgile: Seine pro-faschistische Rezeption," *Quaderni di Storia* 18 (1983) 233-72 essentially deals with Nazi Germany.

[7] For the plans, see V. Ussani, "La celebrazione bimillenaria di Virgilio," *Nuova antologia* 348 (1930) 263-69.

and objective, a large proportion are laced with propaganda, holding Virgil up as a model for the policies or aspirations of the regime. Whatever their attitude, the outpouring was enormous.[8]

The King, the Duce, and the leaders of the corporate state celebrated Virgil's birthday at the Campidoglio on 15 October 1930. Ettore Romagnoli of the Royal Academy harangued them with a speech that set the tone of the festivities by pointing out that Virgil, the poet of peace, justice and the countryside, held fast to the idea of Empire; he glorified the land, the spirit of Roman civilization and the conquests which brought civilization and peace.[9] These themes will reappear.

A congress of Virgilian studies sponsored by the Istituto di Studi Romani (founded by the regime in 1926 to propagate the ideals of Rome), convened for the anniversary. Its volumes of published essays betray their official origin and the blessing of the Party.[10] Luigi Federzoni, president of the Senate, wrote the introduction, and Giuseppe Bottai, Minister of Corporations, and Pietro Fedele, ex-Minister of Education, contributed essays. Federzoni rather candidly noted that such celebrations stirred up "copiose fioriture rettoriche dell' ignoranza improvvisatrice" which naturally led to fatigue and oblivion. These essays, though, which issued from pure founts of rigorous and experienced intellects, would keep the Virgilian poetry of the Land and the Empire alive in the sensitivity of the new Italy. He concluded that fascist Italy naturally turned to Virgil as its poet, because Fascism represented the desire for a healthy life and legitimate power—a return to the fields and arms.[11] Despite this tone, the volume contained some serious studies.

The Institute of Roman Studies also held congresses of its own, in 1928 and 1930, on a broad range of Roman topics ancient and

[8] Note the abundant entries in *l'Année philologique* 1931: 156-169, 1932:154-167 and 1933: 127-134. A striking proportion of the 4000 items in the bibliography G. Mambelli, *Gli studi virgiliani nel XX secolo*, Guide bibliografiche dell'Istituto nazionale di cultura fascista 7 (Florence 1940) relate to these years.

[9] E. Romagnoli, *Virgilio, discorso pel bimillenario pronunciato in Campidoglio il 15 ottobre 1930-VIII* (published by the Reale Accademia, Rome in 1931).

[10] *Studi virgiliani*, published by the Istituto di Studi Romani (Rome 1931).

[11] *Ibid.*, prefazione (unnumbered pages).

modern.[12] Here, too, studies of Virgil appear, with the same mixture of the scholarly and the profane. The bimillenary conference held in Mantua, though, was almost entirely serious.[13] These congresses, like the books on Virgil that abounded in these years, generally reached a specialized and informed audience. Much more important for spreading information (or propaganda) were articles in magazines of general interest (*Convivium*, *Nuova antologia*, *Capitolium*) or of wide circulation (*l'Illustrazione italiana*).[14] Whatever the medium, the message was remarkably consistent, stressing the parallels with modern Italy, and the themes of nationalism, the land, the Empire and occasionally, society. Virgil became a source the Fascists could exploit directly or subtly.[15]

For many, Virgil was Italy's first national poet and the *Aeneid* her first national poem.[16] The poet's relevance to the present was clear, since Italy was returning to him, the singer of sacrifice, labor, piety, power, love and glory—the virtues dearest to the renewed nation.[17] He interpreted the supreme needs of the people: cultivation of the land and the willingness to die with arms in hand to defend and aggrandize Italy.[18] He embodied the idea of *Romanità* (a constant fascist theme), the

---

[12] Published as *Atti del 1. (2.) congresso nazionale di Studi Romani* (Rome 1929, 1931). Some of these studies will be considered below.

[13] *Studi virgiliani* (Mantua 1930).

[14] *Convivium* was a magazine of literature, philosophy and history, *Nuova antologia* dealt with letters, science and art, while *Capitolium* was the informative and generously illustrated journal of the Government of Rome. All are still published. *L'illustrazione italiana* produced a supplement for Christmas and New Year 1930/31, *Virgilio*, edited by V. Ussani and L. Suttina.

[15] That this discussion is confined to works with a clear political message should not obscure the fact that there were many serious studies which eschew politics altogether or insert the modern message subtly or incidentally. See, for example, Canfora's comments (above, n. 6) on the books of Terzaghi and Funaioli and on an article of Rostagni. Likewise, P. Fabbri's *Virgilio, poeta sociale e politico* (Rome 1929) is almost entirely a straightforward study, which only pays lip service to the ideals of the regime.

[16] V. Ussani (above, n. 7); poem: V. Ragazzini, "Il sentimento della patria in Virgilio," *Convivium* 1 (1929) 75-85: "l'Eneido ... è certo il nostro primo e perfetto poema nazionale."

[17] The concluding words of Fabbri (above, n. 15) 225.

[18] F.P. Mulè, "Virgilio," *Capitolium* 6 (1930) 468-81, conclusion.

civilizing mission of Italy, the profound sentiment of the family and the devoted attachment to the land.[19] Therefore, there was something in him that responded exactly to a basic program of fascism.[20]

I much regret not being able to locate a copy of the work which carries these ideas to an extreme. The heroic poem (in 56 sonnets) of Maria Ricciardi Bosi, *Sulle grandi orme*, apparently makes all possible parallels.[21] Some are obvious enough: Virgil and the Duce, the Aeneid and fascism. Others display a vivid imagination: the refugees from Troy and the veterans returned from the trenches; the Harpies and the politicians of Versailles; the games of Book V and the competitions of the new youth now being trained; and the arrival of Aeneas in the promised land and the march on Rome. Virgil, it seems, had foreseen everything. The poem won a prize in a competition whose theme was "Virgil's heroic poem and its Latin and Mediterranean feeling in relation to the feelings and ideals of the new Italy of Benito Mussolini".

The most recurrent themes are the Land and the Empire. The fascist government had been carrying out a massive program of land reclamation (draining the Pontine marshes were its most signal accomplishment) and establishment of new agricultural settlements, trying to keep people on the farms and out of the cities. Its apologists easily found inspiration in the Georgics. For Pietro Fedele, Virgil's voice, which called the Italians back to work in the fields, was being heard as never before since he affirmed that the farmer who worked his own land sustained the family and the nation. Fedele saw this in the context of Augustus' policies of restoring Italian agriculture and repopulating the countryside. Consequently, the most effective and worthy celebration of Virgil was the recently passed law on agricultural reclamation.[22] Yet, in another article,

---

[19] For the concept of *romanità*, see L. Perelli, "Sul culto fascista della romanità," *Quaderni di Storia* 3 (1977) 197-224, with many further examples of interpretations of Virgil, Horace and Augustus in works on the history and literature of Rome.

[20] P.E. Santangelo, "Virgilio e lo spirito della nuova Italia," *Atti e memorie della reale accademia Virgiliana di Mantova* 22 (1931) 19-49. This is probably the most comprehensive statement of the connection between Virgil and Fascism.

[21] Published in Milan in 1932; summary in Mambelli no. 299.

[22] "Il ritorno alla terra nell'insegnamento di Virgilio," *Studi virgiliani* (above, n. 10) 57-75

he found himself in an awkward position, since he started with a quote from Mussolini about the peace of Augustus being a brilliant facade.[23] He therefore had to admit that, though the emperor's agricultural policies were not a total success, Virgil nevertheless could make the affirmation just noted. The Duce would in fact soon change his opinion of the Emperor.

Giuseppe Bottai, a leader of the party's left wing, found a stranger and more specific inspiration.[24] Virgil, he claimed, favored the labor of free individuals over slavery, just as the fascists opposed American standardization, socialism, and the excesses of Taylorism.[25] The Italy of today, when capitalism was moving to its inexorable crisis, could turn to Virgil for ideas worth applying to give labor back its place in the moral and spiritual hierarchy.

In 1930, Italy was not yet the aggressor of the Ethiopian war, or partner in the Axis. Yet it already had an African empire that should make it a great power and serve as a basis for colonization. Virgil, who glorified the Roman empire, provided an excellent justification for imperialism. He showed how the empire was the natural result of perfecting Roman institutions, so that Italy became the invincible instrument of imperial power.[26] It was inevitable (to some) that the fascist movement, the restorer of traditional values, should return to the "official poet of Latinity, the celebrator of the Empire". Yet this did not entail a threat to European peace, for Italian imperialism was an element of orderly and peaceful development, not an attack on the independence of neighboring countries.[27]

In particular, Virgil was the poet of Latin Africa.[28] His episode of Dido reflected Augustus' refoundation of Carthage, and innumerable

---

[23] "Virgilio e la terra" in Ussani and Suttina (above, n. 14) 3-5.

[24] "L'esaltazione del lavoro nell'opera di Virgilio," *Studi virgiliani* (above, n. 10) 19-34, reprinted in Bottai's *Incontri* (Rome 1931) 7-40.

[25] An American method for increasing industrial efficiency and mass production, which had some influence on Lenin's Russia.

[26] E. Bodrero, "Virgilio e l'impero," in Ussani and Suttina (above, n. 14) 6-8.

[27] Santangelo (above, n. 20) 41f.

[28] V. Ussani, "Il culto di Virgilio nell'Africa latina," in Ussani and Suttina (above, n. 14) 37-42.

remains in the country showed the strength of the Roman presence there.[29] One of the most vivid of all these discussions recorded that with the bimillennium, Virgil's light shone more brightly on the African lands of Italy. As in the time of Augustus, the ferocious Garamant was ready to submit quietly to Roman rule, and Ceres and Minerva, as then, were making the fertile fields of Libya bloom with harvests and olives: "may the song of Virgil resound sweetly and softly in the ear of the soldier who watches in arms on the now firm frontier of the desert, may it rise to the soul of the colonist who sows the seeds of good hope in the fertile furrow."[30] A final piece of imperialistic Virgilian propaganda was more subtle when the Italian excavations of Buthrotum were described. This place, beside being mentioned in the Aeneid, was in Albania, where the Italians had already established a virtual protectorate and were soon to make greater claims.[31]

One subject lent itself far less successfully to this kind of propaganda. Mussolini had long advocated a policy of strong families and population increase (he himself had five children) so that Italy could assume her rightfully powerful role. The childless bachelor poet proved difficult to identify as an advocate of family values. A brave attempt by one of the regime's leading archaeologists began by denouncing the unnamed author of a scandalous work, *Virgilio anormale*, but could only lamely note that Virgil celebrated the love of people and animals, especially of children (Iulus and the mysterious child of the fourth eclogue), and so could be considered as having a feeling for the family.[32]

ITALUS: Natalis tibi bis millesimus, ecce, Menalca
Hoc Itali certant te celebrare die
MENALCAS: At cur me celebrent? malim mea carmina versent
Italiam ut discant quo sua fata vocent.

---

[29] Dido: P. Romanelli, "Reflessi virgiliani dei rapporti tra Roma e l'Africa," *Studi Virgiliani* (above n. 10) 199-218; remains: V. Ussani, "Virgilio e l'Africa latina," *Atti del 2. congresso nazionale di studi romani* (Rome, 1931) III.161-71.

[30] The peroration of Romanelli (previous note), 218.

[31] L. Ugolini, "Enea a Butrinto e gli scavi archeologici italiani," in Ussani and Suttina (above, n. 14) 9-14. Note that these excavations were given a remarkably prominent place in the most widely diffused of these publications.

[32] R. Paribeni, "Virgilio e la sensa della famiglia," *Studi virgiliani* (above n. 10) 151-64.

Et postquam ipsa omnis e finibus expulit hostis
Iam renovare parent pascua rura duces.

These verses, whose tone will seem familiar, introduce one of the great scholarly productions of the bimillennium, the magisterial edition of Virgil's works by R. Sabbadini.[33] His volumes inaugurated an impressive series of editions issued "Iussu Beniti Mussolini".[34] Even here, in a work of pure scholarship, the aggressive didactic tone of fascism intruded. Despite it, reviewers were universally enthusiastic, though one did point out that the watermark of the fasces tended to break through the paper.[35]

Another work of permanent value was the grandiose reproduction of the Medici-Laurentian codex of Virgil. A copy was presented to the Duce on 27 October 1930. After "admiring the magnificent spectacle of the enthusiasm of the strong and healthy workers of the State Polygraphic Institute," the dictator addressed the assembled dignitaries. He claimed (perhaps implausibly) that he had often wondered what was the oldest manuscript of the Aeneid, and promised that the volume would find an honored place in his library. He also noted that Virgil was born in a furrow, an evident symbol of this poem of the earth and the empire.[36]

The celebration of Virgil took many forms beside the printed word. Most substantial was a work of permanent benefit, the Parco Virgiliano, in the northeastern part of the city off the Via Salaria. With its small lake, Roman ruins and a playground, it was dedicated in time for the bimillennium.[37] On a different scale, the poet's image appears on a medallion struck by the government. The obverse shows him enthroned,

---

[33] Rome, 1930; 2nd edition, 1937.

[34] The series eventually included editions of the *Res Gestae* (Barini), Florus (Malcovati), Livy 41-45 (Giarratano), Tacitus (Giarratano and Lanchantin), Synesius (Terzaghi), Xenophon (Pierleoni), and the *Tabulae iguvinae* (de Voto). Many others were planned: see the notice in *l'Urbe* (a popular journal celebrating Rome and fascism edited by Muñoz) 3 (1938) 6.43.

[35] F. H. Sandbach in *CR* 46 (1932) 26.

[36] Mussolini (above, n. 3) 24.287 f.

[37] See the article of B. Braschi in *Capitolium* 7 (1931) 108-14, with illustrations. The architect was Raffaele De Vico, who also designed the Rome zoo and the monument to the war dead in the Campo Verano.

crowned by Victory with Roma standing behind, facing scenes symbolic of the countryside. It is inscribed *salve magna parens frugum Saturnia tellus, magna virum*, while on the reverse, *tu regere imperio populos, Romane, memento* runs around the scene of Anchises showing Aeneas the images of the future great Romans, with Augustus prominently in front.[38] These images stressed the two salient points about Virgil: care for the land, and the inevitable power of the empire.

The same themes recur in the most widespread form of propaganda, a set of commemorative postage stamps, which circulated in Italy and reached collectors all over the world.[39] Each stamp featured a verse of Virgil and an appropriate scene. The verses, by implication, clearly connect the poet and the regime. All have one common inscription, *antiquam exquirite matrem*, seemingly innocuous enough until the next line comes to mind : *hic domus Aeneae cunctis dominabitur oris*.

The stamps stress patriotic feelings, love of the countryside, and the empire. Patriotism is easily illustrated with *ecce tibi Ausoniae tellus, hanc eripe velis; Italiam laeto socii clamore salutant* and *salve, fatis mihi debita tellus: hic domus hic patria est*. This shades into love of Italy and its fertility: *salve magna parens virum, Saturnia tellus*, and then the happiness of rural life (in accordance with the fascist policy of discouraging emigration to the cities) *o fortunatos nimium sua si bona norint agricolas; hic focus et taedae pingues, hic plurimus ignis*. Love of the home naturally brings love of the family (and by implication, the fascist policy of encouraging large families), *interea dulces pendent circum oscula nati*.

As usual, Virgil's words could support more aggressive policies. The stamp which shows Turnus in full battle dress accompanied by warriors proclaims *et opes nobis et adhuc intacta iuventus*, which could hardly fail to bring to mind the highly militarized fascist youth organiza-

---

[38] The medal is illustrated in G. Q. Giglioli, "Roma e l'Italia in Virgilio," *Capitolium* 6 (1930) 482-92.

[39] The stamps (which may be found in Scott's *Standard Postage Stamp Catalogue* as nos. 248-56 and C23-6) are illustrated in *Enciclopedia virgiliana* 2.496f, pl. xxxv. Their types are discussed in "Vergilzitate auf italienischen Briefmarken," *Gnomon* 6 (1930) 670f. and R. Dehn, *Italian Stamps* (London 1973) 112f. For the propaganda on the stamps of the fascist period, see E. Diena, *Un secolo di francobolli italiani* (Bologna 1966) 54-66.

tions, while two other stamps leave the imperial future open: *tu regere imperio populos Romane memento* and *his ego nec metas rerum nec tempora pono*. Peaceful as the country may have been in 1930, it had vast ambitions easy to justify by the poet's works.

The fascists also approved of Horace, regarding him as one of the two great poets of their Roman past, but they never built up much enthusiasm for his bimillennium in 1935, the year of the Ethiopian war. The long poems of Virgil, especially the epic, could easily be mined for propaganda, but the short lyrics of Horace, despite their fine sentiments, were less tractable. The poet's genius for sliding from one subject to another and his constant refrain of *carpe diem* were not at all suitable to fascist doctrine. Only the *carmen saeculare* generated unambiguous praise. Nevertheless, some writers managed to find Horatian precedents for current policies amid studies which were generally serious. The volume of this literature, though, is a mere trickle compared to the flood of works on Virgil.

The inevitable Ettore Romagnoli set the tone in an oration delivered on the Capitoline.[40] This time, however, neither King nor Duce was present. The speaker managed to find many virtues in Horace that suited the present. The poet favored fertile marriages; agriculture; reclamation of land and distribution of water; the equal social distribution of goods; a disciplined youth; military preparation (for peace more than war); and especially the dignity of civil life and the Patria. Fides, Pax, Honor, Pudor, Virtus and Abundance were the stars of his firmament, now rising again on the Italian horizon.

The Istituto di Studi Romani carried the main burden of the celebrations. In first half of 1935, it offered two series of lectures, a course of Roman archaeology that traced Horace's route along the Appian Way, a radio program on the *carmen saeculare*, and a tour to Horace's villa.[41] The first lectures were published in a volume called *La figura e l'opera di Orazio*.[42] In general, the articles emphasize Horace's praise of a renewal of morals and religion, of the harmony of social life, and of Italy;

[40] *Orazio, discorso pel bimillenario pronunziato in Campidoglio il 17 novembre 1935-XIV*, published by the Royal Academy in Rome.

[41] See *Istituto di Studi Romani* (Rome 1935) 14f.

[42] Published by the Institute, Rome, 1938.

politics rarely intruded. As in the case of Virgil, his *romanità* (now reborn to new glories under the Cross of Savoy and the Fascio Littorio), which led to the ideal of dying for the country, was exalted.[43] So was his love for Rome and its Leader ("il Duce come egli lo chiama") and his conscience of the destined grandeur of Rome.[44]

Magazine articles also dwelt on Horace's suitable attitudes, particularly his patriotism and praise of Augustus. Parallels between the Horace's works and the *Res gestae* showed that poet and emperor shared the same ideals.[45] Horace could also be considered "una voce limpida e alta della grandezza di Roma imperiale" for whom the Empire was the perfect regime, in which civilization reached its peak.[46] For him, Augustus was the light of the world, and the empire the natural product of an inevitable law. The author of these sentiments returned to the attack in 1938, after Mussolini's new Roman empire had been proclaimed and the world was on the edge of war.[47] He quotes Juno's prediction in *Odes* III.3 of the universal Roman empire, cites Horace on necessary and just wars, and concludes with the favorite verse of the time, that the Sun can never see anything greater than Rome.

This occasion also produced a set of commemorative stamps with appropriate designs and quotations.[48] Some refer to the poet, without further implication: *me doctarum hederae praemia frontium* and *non omnis moriar*. The rest dwell on the now familiar themes: the land (*fertilis frugum pecorisque tellus*), the city (*stet Capitolium fulgens* all the more appropriate since the Capitoline had recently been cleared), and martial virtue (*impavidum ferient ruinae* and of course *dulce et decorum est pro patria mori*). *Odes* III.3 seemed especially appropriate for expressing Horace's healthy attitude. Only *diffugere nives redeunt iam gramina*

---

[43] C. G. Paluzzi, "La romanità di Orazio" in *La figura* (above, n. 42) 51-68.

[44] G. Q. Giglioli, "L'Italia antica nella poesia di Orazio, " in *La figura* (above, n. 42) 79-92.

[45] O. Tescari, "Orazio poeta civile," *Convivium* 8 (1936) 58-74.

[46] M. Cerini, "Ideali morali e civili di Orazio," *La rassegna italiana* 42 (1936) 250-56.

[47] M. Cirini, "L'impero di Augusto nella poesia di Virgilio e di Orazio", *La rassegna italiana* 47 (1938) 15-28. I have not been able to locate A. Malpicati, *Orazio poeta dell'impero* (Milan 1936) which no doubt expands on these sentiments.

[48] Listed in the Scott catalogue as nos. 359-66, C84-88; issued on 1 July 1936.

campis arboribusque comae seems out of place in this positive moralizing, since the rest of the poem reveals an altogether un-fascist sentiment of the transitory nature of life. At first sight, the airmail stamps of this series appear to be relatively free of propaganda, until the quotes they use are put into context. *Expertus vacuum Daedalus aera pinnis*, obviously appropriate to airmail, is part of a poem that stresses bravery and the importance of training; *duris ut ilex* reflects the invincibility of Rome; and *hic in reducta valle* offers the usual praise of rural life. *Alme Sol, possis nihil urbe Roma visere maius*, though, speaks for itself. Despite the emphasis on Latin in the fascist educational system, it seems doubtful that the intention of most of these references would have been easily recognised.

On 9 May 1936, the Duce proclaimed to a huge enthusiastic crowd the annexation of Ethiopia and the recreation of the Empire. A new imperial age had begun, bringing with it even closer identification with ancient Rome and the founder of its first Empire. Mussolini took this parallel very seriously and consciously exploited the image of Augustus. This exaltation, expressed in works of all kinds, reached its height in the lavish celebrations of the 2000th anniversary of the emperor's birth, celebrated in 1937 and 1938.

Planning began already in the time of Virgil's anniversary. The Istituto di Studi Romani, at its second congress in 1930, announced a program of celebrations. Its spokesman noted that the bimillenary of Caesar had not been celebrated twenty years earlier because Italy had not yet been regenerated by fascism. Now, though, the victories of fascist Italy would be commemorated in the anniversary of Augustus, pacifier of the ancient world and founder of wise social and political measures. Honors for the great past would prepare for the greater future.[49] The ambitious proposals included conferences and publications about Augustus; restoration of two of the emperor's greatest monuments, the Ara Pacis and his mausoleum, and excavations and celebrations all over

[49] G. Q. Giglioli, "Per il secondo millenario di Augusto," *II. Congresso nazionale di Studi Romani* (Rome 1931) I.277-80.

Italy.[50] As it turned out, Augustus provided one of the greatest shows of the entire period.

Three learned conferences discussed the Emperor, in Rome at the Accademia dei Lincei, and at Milan and Padua. These were all published, as was an overwhelming mass of books and articles.[51] This material as usual varied from the most serious to the most tendentious. Much of it made open or implicit parallels between the Emperor and the Duce, who took a direct personal interest in the proceedings. One of the most important publications was a new major edition of the *Res gestae*, by Concetta Barini, in the series "iussu Beniti Mussolini".[52] Its dedication page "in Augusti natalem bis millesimum" bore the following passage "ex Beniti Mussolini Italorum ducis oratione": "levate in alto o legionari le insegne, il ferro e il cuore a salutare, dopo quindici secoli, la riapparizione dell' impero sui colli fatali di Roma".

Giuseppe Bottai, now Minister of National Education, set the tone in a speech to the Istituto called "L' Italia di Augusto e l' Italia di oggi".[53] He dwelt on the various accomplishments of the Emperor, then drew a general parallel with the present. His themes, reflected in so much of this literature, may serve as representative of the entire body of work.[54]

In this view, Augustus carried out a peaceful revolution, preserving the forms of the republic, but modifying them profoundly to adapt to the current situation. The Senate kept its authority, though purged of hostile elements, and the popular assemblies lost their power. The administration of Italy was reorganized and centralized. Augustus did all this not from a desire for absolute power, but following a logical and necessary development. Mussolini, of course, also claimed to have led a peaceful

[50] See the comprehensive study of M. Cagnetta, "Il mito di Augusto e la 'rivoluzione' fascista," *Quaderni di Storia* 3 (1976) 139-81, part of a volume dedicated to 'Classicismo nell'età dell'imperialismo'.

[51] The extensive note of Cagnetta (above, n. 50) 168-71 gives an idea of the volume of this material.

[52] Rome 1937; an expanded version of the first edition which had been published by Mussolini's newspaper *Il popolo d'Italia* (Milan 1930). For other editions, see Cagnetta (above, n. 50) 168ff.

[53] Published in *Accademie e bibliotece d'Italia* 11 (1937) 207-22.

[54] I have made the parallels more specific than did Bottai.

revolution, which took power away from the Parliament and abolished representative government. In other words, the regimes of Emperor and Duce both represented the supremacy of one man.

Augustus was especially concerned with people of Italy. He encouraged agriculture and strengthened the population by internal colonization and by encouraging large families. Close parallels for these policies were easy to find in the legislation of Mussolini. They were all part of Augustus' restoration of moral and social values, an effort in which the fascists also took great pride.

Religion received the determined and favorable attention of Augustus, as it did from the Duce who restored close relations with the Church. Augustus created the *iuventus*, as the fascists did an all-embracing set of youth organizations. He regulated the *collegia*, for which a far-fetched parallel was found in the fascist Corporations that were supposed to organize economic activity. Finally, both emperor and Duce took great interest in the city of Rome, rebuilding the city and creating innumerable public works.

Others added further parallels. Augustus and Mussolini both headed revolutionary parties whose victory brought peace to Italy under the rule of one man. They both reformed the administration of Rome by putting it under their own control, Augustus through the *praefectus urbi*, Mussolini by an appointed governor. They both reconciled the army and civil authorities, and fought major campaigns in Spain.[55]

Emilio Balbo, in a work of slavish aduation called *Augusto e Mussolini*, went even further.[56] In addition to the parallels already noted, he pointed out that Christ was born during the first empire, just as the Concordat with the church marked the second; that the chaos after the Ides of March was like that which followed the World War; and that both Emperor and Duce had solved the unemployment problem through public works. The ancient empire, like the modern, was based on the rule of one party, the totalitarian spirit, moral renewal and a militarized nation.

---

[55] For these and related points, see Cagnetta (above, n. 50) 155-57, 162-67.

[56] Rome 1938. Kostof (above, n. 4) 302f, cites another work of this author, *Protagonisti dell'Impero di Roma, Augusto e Mussolini* (1941), which I have not seen. It seems to convey the same message.

Yet for Balbo, Augustus was inferior to Mussolini. After all, he could make use of the resources of the entire world, while the Duce had only Italy. Augustus had inherited power, while Mussolini created his; he was slow and deliberate while the fascists never hesitated. Clearly, Augustus was not a great statesman like Julius Caesar, Napoleon or Mussolini. In fact, the better parallel was between the Duce and Caesar: both had marched on Rome, showed clemency to their enemies and created an empire. Mussolini, in other words, combined the virtues of Caesar and Augustus.[57]

As far as the public were concerned, all this propaganda was secondary to the great exhibition, the *Mostra augustea della Romanità*, which the Duce himself opened with a splendid ceremony on the Emperor's birthday, 23 September 1937. The show, which focused on Augustus but embraced all aspects of Roman life, had been in planning since 1932, when the archaeologist Giulio Quirino Giglioli suggested the idea to Mussolini as another great exposition, the *Mostra della rivoluzione fascista*, celebrating the tenth anniversary of the regime, was closing.[58] The new exhibit, with copies and casts of thousands of objects from all over the ancient Empire and foreign museums, was intended to manifest the glories of ancient Rome as a precedent and parallel for those of the new Empire.[59] In the words of one admirer, "Anche il piu umile ed incolto dei legionari di Mussolini, che effetuarono al modo cesareo la conquista dell' Impero Etiopico, possiede una sua viva coscienza della romanità ed ha il senso preciso del legame storico fra l'

---

[57] The parallel between Caesar and Mussolini, though often evoked, had serious problems as official doctrine, for Caesar was not the first emperor and his assasination was hardly a favorable omen.

[58] It was no coincidence that the Duce inaugurated a second exhibition of the Fascist Revolution at the Villa Giulia on the same day as the Augustan show: see V. Talarico, "La mostra della rivoluzione fascista a Valle Giulia," *Capitolium* 12 (1937) 513-18.

[59] For the planning and aims of the exhibition, see G. Q. Giglioli, "La mostra augustea della romanità", *Capitolium* 12 (1937) 519-28 (with good illustrations), Cagnetta (above, n. 50) 147-51, and especially the official catalogue, *Mostra augustea della Romanità*, 3d edition (Rome [1937]), with introduction by Giglioli. Most of the exhibit can still be seen in the Museo della Civiltà romana at EUR outside Rome.

odierna rinascita dei valori nazionali e il multisecolare dominio, imposto da Roma al vecchio continente, e non con la sola forza delle armi."[60]

The exhibition filled the Palazzo delle Esposizioni on the Via Nazionale, where the fascist exhibition had been held. The facade was transformed into a gigantic triumphal arch, flanked by walls which bore the enormous repeated titles REX and DVX and long inscriptions in praise of the empire by Livy, Cicero, Pliny, Aelius Aristides, Tertullian and St. Augustine. This concept of *romanità* included the Christians. Over the entrance arch were the pointed words of the Duce: "Italiani, fate che le glorie del passato siano superate dalle glorie dell' avvenire". From there, a stairway led to the main floor, dedicated to the entire history of Rome. Its first major hall, the Sala dell' Impero, lined with reliefs and monumental inscriptions, formed an "antologia di opere trionfali" to impress the viewer with the might of the Empire.

The historical rooms led to a great hall dedicated to the only figure of the republic who merited special attention, Caesar, the Dictator (a favorable term in those days) whose military genius was stressed. Next to it a smaller room featured the writers of the Augustan era: Virgil (in a mosaic from Tunisia), Horace and Sallust (contorniate medallions), Maecenas (from the Ara Pacis) as well as a model of the Ara Pacis and the text of the *Res Gestae*.

The central hall, the focus of the exhibition, glorified Augustus. It featured statues, portraits, coins, a map of the empire, and a great glass cross with St. Luke's description of the birth of Christ. Typical of the fascist interpretation of this age was the connection between Rome and Christianity, as true in the time of Augustus as that of the Duce. Inscriptions on the walls concluded with the words of Mussolini: "nella silenziosa coordinazione di tutte le forze, sotto gli ordini di uno solo, è il segreto perenne di ogni vittoria".

Several more rooms were devoted to Augustus, his family, the Roman army and navy, law and government, the empire and its defense, and Christianity. They led to the last of this floor, dedicated to "Immortalità dell' idea di Roma. La rinascita dell' impero nell' italia

---

[60] A. Neppi, "La mostra augustea della romanità," *Rassegna italiana* 47 (1932) 29-35.

fascista". A statue of Victory and busts of the King and Duce confronted the visitor. The walls bore photographs which stressed the parallels between the ancient and modern empires: the arch of Constantine and the fascist arches of Bolzano and Cyrenaica; Augustus' obelisk in the Circus Maximus and the obelisk from Axum; ancient and modern rural settlements; and the Foro Mussolini where the youth was to be educated in the Roman style. Monumental inscriptions from medieval and modern writers, but especially from Mussolini, stretched above the exhibits. Central among them was the Duce's proclamation of the Empire.

The lower floor dealt with cities, architecture and engineering, while the upper featured private life. The show was intended to give a complete image of the Roman empire, focusing on Augustus, and his relevance to the present. This was propaganda at its most dramatic and effective. No one was to leave the exhibit without a sense of pride in the achievements of ancient Rome and the confidence that they would be equalled or surpassed by the modern regime.

Ancient Rome, of course, had left many important traces behind, not the least of them associated with Augustus who had transformed the face of the city. As part of his identification with the founder of the "first Empire", Mussolini devoted special attention to the excavation and restoration of the greatest surviving monuments of his predecessor, the Forum, the mausoleum and the Ara Pacis.

Excavation of the Forum of Augustus began in 1924, soon after the fascists came to power. The surrounding buildings, which had left only the tops of a few columns and part of the retaining wall visible, were demolished and the forum cleared and excavated by 1930, in time for the bimillenary of Virgil. Ironically, most of its remains were to disappear beneath the broad Via dell' Impero, the route of fascist parades.

The huge tomb of Augustus by the Tiber was almost completely surrounded by houses. It had been in continuous use, most recently as a concert hall where two congresses of the Fascist Party had been held.[61] Giglioli, who excavated the building between 1926 and 1930, expressed the hope that it would be exposed in time for the bimillennium of

[61] For what follows, see the detailed and entertaining study of Spiro Kostof, "The Emperor and the Duce" (above, n. 4). For the "decayed tooth" and further references, see Cederna (above, n. 4) 209-15.

Augustus. The project advanced with exemplary speed after the Duce himself intervened in 1934, wielding a pick to great publicity, to begin the vast demolitions. In three years, 120 houses—27000 square metres of structures—had been cleared and the tomb stood isolated in its grandeur, freed from its modern accretions and standing as a real Roman ruin. Its stark appearance led unkind critics to call it the "decayed tooth". The work was completed during the bimillennium.

The mausoleum stood in the middle of a new square designed in the fascist imperial style by Vittorio Morpurgo, with buildings decorated with Roman-style mosaics and reliefs and a monumental Latin inscription in raised letters honoring the emperor and "Mussolini dux". It begins with a curiously Lucretian touch, "hunc locum ubi Augusti manes volitant per auras". The open side of the square toward the river provided a suitable location for the *disiecta membra* of Augustus' Ara Pacis, reassembled on the site in 1938 by order of the Duce and adorned with a copy of the text of the *Res Gestae*. With that, this part of Rome had an impressive collection of Augustus' work, the suitable culmination of one of the largest restoration projects of the regime.

The celebration for Augustus produced another series of postage stamps, featuring images were derived from real or restored Roman monuments, with exceptionally apposite inscriptions from the *Res gestae.*[62] In keeping with the times, several were bellicose: *mare pacavi; bella terra et mari toto in orbe terrarum saepe gessi victorque omnibus; classis mea per Oceanum* and, most appropriate of all, *meo iussu et auspicio ducti sunt exercitus in Aethiopiam.* The issue for the colonies bore the simple *imperium sine fine dedi.* These needed no comment. Less obvious is *laurum de fascibus deposui in Capitolio votis solutis.* Mussolini, in self-conscious imitation, had actually deposited the laurel of the fasces in the Capitoline after the victory in Ethiopia.[63]

Some deal with the restorations of antiquity: *templa deum in urbe refeci; coepta profligataque opera a patre meo perfeci* (a reference to Mussolini's restoration of the forum of Caesar), and *aram pacis augustae senatus pro reditu meo consacrandam censuit.* The rest reflect poli-

---

[62] Scott 377-86, C95-9, issued 23 September 1937.

[63] See Cagnetta (above, n. 50) 143 n.15.

tics: *censum populi egi* (the regime was anxious to count and increase the population) and *iuravit in mea verba tota Italia et me ducem depoposcit* (the Dux, like many of his kind, was very form of plebiscites).

The air mail stamps return to the poets, especially Horace, using their words, as usual, to glorify Rome, Italy with its agriculture and people, the empire, and Augustus: *tua Caesar aetas fruges et agris rettulit uberes; Romulae genti date remque prolemque et decus omne; alme sol possis nihil urbe Roma visere maius* (a fascist favorite), *qui mare qui terras omni ditione tenerent* and *tutela praesens Italiae dominaeque Romae*.

With these stamps, the emperor and his two greatest poets come together in what was to be a final glorification of the regime and its empire. A few months after the end of the bimillennium, Italy entered into the Pact of Steel which soon dragged it into the war. Two years later, the regime was at an end. With it went the ephemeral dream of the new Roman empire, so grandly celebrated.

During his twenty years in power, Mussolini promoted an ambitious classical revival, which brought Virgil, Horace and Augustus to the attention of millions. Much of the effort, though, was mired in the demands of a propaganda that sought to exploit ancient Rome as a way to glorify and justify its fascist successor. Some of the works it produced are of lasting value, though most of them have earned merited oblivion. The edition of Sabbadini and some of its fellow volumes, as well as the serious studies of the two poets and of Augustus, will retain their place as distinguished scholarship. The excavation and restoration of the Augustan monuments, however dubious their motives and methods, have made ancient Rome accessible in a way that was not possible before. The copies of Roman monuments amassed for Augustus' birthday can still be seen and studied. Yet, on the whole, the vast effort can only be seen as a distortion of scholarship. It shows more the powers of a totalitarian government to bend the past as well as the present to its will, by extracting real or imagined present realities from the writers of Antiquity, and by marshalling ancient monuments to glorify a modern regime.